MANAGING HIGH TECHNOLOGY:
An Interdisciplinary Perspective

MANAGING HIGH TECHNOLOGY: An Interdisciplinary Perspective

Based on Papers from the Third International Conference on Interdisciplinary Research
Seattle, Washington, U.S.A., 1-3 August, 1984

edited by

Brian W. MAR
Civil Engineering
University of Washington
Seattle, Washington, U.S.A.

William T. NEWELL
Management and Organization
University of Washington
Seattle, Washington, U.S.A.

Borje O. SAXBERG
Management and Organization
University of Washington
Seattle, Washington, U.S.A.

N·H
P C

1985

NORTH-HOLLAND
AMSTERDAM · NEW YORK · OXFORD

ISBN: 0 444 87762 2

Publishers:
ELSEVIER SCIENCE PUBLISHERS B.V.
P.O. Box 1991
1000 BZ Amsterdam
The Netherlands

Sole distributors for the U.S.A. and Canada:
ELSEVIER SCIENCE PUBLISHING COMPANY, INC.
52 Vanderbilt Avenue
New York, N.Y. 10017
U.S.A.

Library of Congress Cataloging in Publication Data

Main entry under title:

Managing high technology.

Includes indexes.
1. High technology industries--Management--Congresses.
I. Mar, Brian W. (Brian Wayne), 1933- . II. Newell, William T. III. Saxberg, Borje O. (Borje Osvald), 1928- . IV. International Conference on the Management of Interdisciplinary Research (3rd : 1984 : Seattle, Wash.)
HD62.37.M36 1985 621.381'7'68 85-10191
ISBN 0-444-87762-2

PRINTED IN THE NETHERLANDS

PREFACE

INTERSTUDY, an international association to foster the study of interdisciplinary problem solving groups, was founded after the First International Conference on the Management of Interdisciplinary Research held in Schloss Reisenberg, the Federal Republic of Germany in April 1979. INTERSTUDY's goals are to disseminate information and maintain a network of individuals studying interdisciplinary problem solving via the INTERSTUDY Bulletin (a news-letter) and to hold conferences every two or three years to examine the advancement of knowledge. The Second International Conference was held at the Manchester Business School, Manchester, England in July 1981. Proceedings of the first conference were edited by Richard Barth and Rudy Steck and published in a book entitled *Interdisciplinary Research Groups: The Management and Organization.* For the second conference S.R. Epton, R.L. Payne, and A.W. Pearson edited and summarized some of the papers and presented other papers in their entirety in their book *Managing Interdisciplinary Research.* The present book represents the presentations and discussions from the third of this series of international conferences on interdisciplinary processes.

The title of these conference books as well as their contents reflect the evolution of the understanding of interdisciplinary problem solving efforts. The first conference was devoted to the development of a nomenclature to describe the phenomena of interdisciplinary problem solving and to observe these phenomena in university settings. Research prior to the second conference was able to resolve most of the nomenclature issues, and study of phenomena in other settings increased. The period between the second and third conferences was characterized by the growing interest of several professional organizations in interdisciplinary processes, by the heterogeneity of the nature of the interdisciplinary issues encountered in the spectrum of case studies being developed, and by a growing question of whether interdisciplinary activities are a cause or a symptom of complex problem solving in changing environments.

The theme of the Third International Conference "Managing High Technology", was selected to focus the discussion on activities where (1) the rate of change of ideas, management styles, technologies, etc. was high, (2) the environment for the activities was also in a state of flux, (3) the systems being developed are complex, and (4) there is a scarcity of individuals with the ability to contribute to the solution of these problems. In this book the insights presented by practitioners of high technology suggest that interdisciplinary issues reported in early conferences are not the problem of greatest concern. There are many other interfaces between individuals, groups, organizations, and nations that may present greater problems. Many of the problems may be related to the ability to cope with change rather than disciplines. Members of three other professional organizations were invited to join the INTERSTUDY delegates at this conference to present their views on the conference theme. From COLRAD, the College of Research and Development, of the Institute of Management Sciences, individuals who were practitioners of high technology management as well as researchers with backgrounds in management or technology joined the conference. Society for Social Study of Science, 4S, was represented by researchers with backgrounds in social sciences and humanities who observe high technology processes, and from AIS, the Association for Integrative Studies, educators came to develop interdisciplinary courses on subjects including high technology processes. This broad spectrum of views on MANAGING HIGH TECHNOLOGY provides a unique insight and clarification on the real and imagined interdisciplinary issues. More importantly, it provides guidance and shares lessons on the management of high technology.

The papers are not presented in the exact order they appeared at the conference, but have been regrouped to emphasize specific issues and solutions. The editors have not altered the substance of any of the papers without the consent of the authors, but do accept responsibility for the final organization and flow of the information presented.

This manuscript could not have been completed without the efforts of Jane Lybecker, Lori Nagle and Theresa Sauer in word processing and layout and Diana Leytham's assistance in proofing and editing the contributions.

We are most grateful to the National Science Foundation, Office of Interdisciplinary Research, and especially to Richard Goulet, the Program Manager of the grant providing the conference funding, for their support. Travel for conference participants was supported under NSF Grant No. OIR-8403975.

Any opinions, findings, and conclusions or recommendations expressed in this publication are those of the authors and editors and do not necessarily reflect the views of the National Science Foundation.

Brian W. Mar
William T. Newell
Borje O. Saxberg

CONTENTS

Section I:

THE NATURE OF HIGH TECHNOLOGY

MANAGING HIGH TECHNOLOGY
An Interdisciplinary Perspective
B.W. Mar, W.T. Newell, and B.O. Saxberg (Editors)

THE NATURE OF HIGH TECHNOLOGY

Editorial Perspectives

INTRODUCTION

High technology is a term applied to many activities, in this book the phrase managing high technology is used to describe the activities associated with the research, development, production, marketing, sales, and maintenance of products or services that have some or all of the following characteristics:

1. All or part of the product or service based on recent innovations or new technologies

2. Rapid changes in technology, thus shortened product life cycles

3. Strong pressures for shortened schedules with limited resources

4. Scarcity of human resources familiar with the technologies needed for the production or marketing of the product or service

5. Turbulent environment characterized by realignment of competitors, changing customer needs, changing regulations, new competing products, and others.

Products and services that fit these characteristics and that are classified as high technology are computers, software, electronics, military, aerospace, biotechnology, and energy. The list of products and services that are labeled high technology varies with the region and the group making the definition. The Western Interstate Commission for Higher Education in the United States (1983) limits the grouping to electronics, computers, energy, biotech, and aerospace. The high technology stock index includes control systems, test equipment, security systems, lasers, word processing systems, etc., which may or may not satisfy all the characteristics previously specified. All these technologies are evolving rapidly, and individuals participating in these technologies may find themselves obsolete in a few years. Research and development are major activities in high technology efforts since it is a race to provide the latest technological advances. This creates shortened product cycles and improved products are often being developed before a new product can be marketed. This race to develop new innovations, to produce and market these innovations before a competitor introduces a better product is limited by the number of people familiar with the new technology, the production process, and the marketing and maintenance processes required of these new products. The shortage of skilled or knowledgeable work force may be solved by new alignments between organizations, by creating new types of cooperation, and by new interfaces between groups, organizations, and institutions. Similarly, the cost and complexity of high technology products may force shifts in consumer operations and institutions that finance or regulate high technology products sales and purchases.

From a management perspective, these characteristics of high technology activities suggest the need to improve the ability to cope with many types of change. Changes associated with high technology require more attention to the management of researchers, the need to create new mixes of disciplines, to know when to alter research goals, and to know when to proceed with development. They require more careful attention to the development of an individual's ability to change roles, to learn new skills, and to adjust to changes in organizations and changes in the competitive and consumer environment. Management of new and changing interfaces between all levels of internal management, and between the firm and new partners must be developed. All activities within a firm may face changes in technology, organization, and operation if new high technology products are to be produced profitably.

Most high technology organizations have work forces that have a higher percentage of college educated employees than most other organizations. Also many of these college educated employees are engineers and scientists. The basic theme of prior conferences was the exploration of interdisciplinary interfaces, and the problems that are encountered at these interfaces. Since the disciplinary bias is developed during a college education, the high technology organizations, with their high proportion of college educated employees, may also be characterized by interdisciplinary interface conflicts. Nayak (1984) identifies another conflict between the technologists and the managers in high technology organizations, by noting that the technologists do not understand management and that most managers do not understand the technology they manage.

One of the common organizational forms encountered in high technology organizations is a matrix structure. Individuals in a matrix organization have two masters to serve. An individual belongs to a functional or disciplinary unit, where their technical knowledge and skills can be nurtured and developed but is assigned to one or more projects that provide resources for the individual to work. There are many variations to this basic configuration, but the interfaces faced by an individual in a matrix organization are more complex than those encountered in a vertical organization (either product or department). Since high technology organizations have rapidly changing projects and technologies, the matrix elements are continually changing, and the interfaces change as well.

It could be argued that the set of characteristics used to define high technology is not unique to high technology, but that the characteristics may be typical of any entrepreneurial effort. The most common denominator between the management of high technology and the management of entrepreneurial activities is the management of change. If change is the real management issue in high technology, then not only is the literature and knowledge associated with entrepreneurship a potential source of assistance, but information on behavior under changing conditions, studies of revolutions and many other bodies of knowledge concerned with the management of change form a potential source for guidance. The papers presented at this conference represent the insights of researchers studying not only interdisciplinary processes, but those studying the process of science, those studying management in general, and practitioners of entrepreneurial high technology activities.

PREVIOUS CONFERENCES

The papers in this book have been organized into five groups. The first group contains case studies of selected high technology firms that characterize the management issues that are of concern. To place these issues in perspective several papers are included that contrast the management problems in university organized research units and in government organizations. Next, a set of papers is presented which have examined these issues across high technology industries to generalize the management problems in high technology activities. The final three groups of papers examine the three most commonly prescribed approaches to these management issues. These approaches should not be considered to be alternative solutions, but must be integrated and molded to meet the needs of each situation. Included in this final group of three are improved planning, effective communication and improved ability to manage information.

Readers familiar with this conference series recognize that this set of papers is a departure from previous proceedings (Barth and Steck, 1980; Epton, Payne and Pearson, 1983). The earlier proceedings focused on the management of interdisciplinary research, the first proceedings emphasizing university environments and the second, both university and other research establishments. This set of papers is focused on high technology firms to determine if the management of interdisciplinary research in such efforts is similar to the management of efforts in university or other research settings, and to determine if the management of research in high technology activities is a significant issue compared to the overall management issues.

CASE STUDIES

The following set of papers are case studies to illustrate the types of management issues encountered in high technology firms, to compare the role of research management with other management issues, and to review the findings from management issues encountered in university and government research and development efforts. Electronic, computer, aerospace, and pharmaceutical firms were selected as case studies to illustrate the characteristics of high technology management problems. The reader should evaluate whether the processes encountered in high technology management are unique or whether the problems are common to the management of most activities. If the latter is true, there is a substantial set of knowledge (recognizing that this knowledge set may be fragmented and difficult to locate) that can be consulted when high technology management problems are encountered.

REFERENCES

Barth, R. and Steck, R., eds. 1980. Interdisciplinary research groups: their management and organization. Seattle: University of Washington.

Epton, S.R., Payne, R.L. and Pearson, A.W., eds. 1983. Managing interdisciplinary research. Chichester: John Wiley and Sons.

Nayak, P.R. 1984. Too many U.S. technologists are managerially illiterate. High Technology September:10.

Western Interstate Commission for Higher Education. 1983. Profiles of high technology education and manpower in the West. October, Boulder, Colorado.

MANAGING HIGH TECHNOLOGY
An Interdisciplinary Perspective
B.W. Mar, W.T. Newell, and B.O. Saxberg (Editors)

MANAGING HIGH TECHNOLOGY R&D: SOME CRITICAL STRATEGIC ISSUES

Alok K. Chakrabarti

Associate Professor & Director
Center for Research on Technology & Strategy
Drexel University
Philadelphia, PA 19104

ABSTRACT

The importance of technology in the economy in terms of creating jobs, increasing the competitiveness of the U.S. industry in the world market, increasing the quality of life as well as increasing our national security, has been well recognized by many. The campaign themes of the major candidates and the parties are laced with the big plans for nurturing high technology. The proponents of industrial policy view technology as the main panacea for the solving of many of the economic and social ills of the country.

While high technology is here to stay and further develop our economy, we must face the reality of the difficulty of managing technology. While new technology has been responsible for increasing the productivity of many industries, notably agriculture, it has also caused many socio-economic and political problems. At the microfirm level, generation of technology is not a guaranteed means of economic prosperity. In this paper, we intend to consider some strategic factors in management of high technology R&D organizations.

TECHNOLOGY AND ORGANIZATION

Bruce Merrifield (1982), the former Vice President of Technology at the Continental Group, commented that the capital intensive commodity manufacturers had failed to recognize the inroads of new technology. High inflation eroded their asset base, productivity decreased due to obsolete facilities and low R&D generated few new technologies. The high growth specialty areas, electronics, communications, engineering, plastics, energy, specialty chemicals, biogenetics pharmaceuticals, benefited from strongly patentable proprietary technology.

Ward Haas, Vice President of Corporate Research at Chesebrough Ponds (1982), identified the emphasis on short-term profitability as the main serious problem for investment in long-term research projects. George White at Xerox commented that research direction must be a function of the strategic charter of the firm, business perceptive technological plan, and the interactive budgetary process.

Lewis Branscomb (1982), Chief Scientist at IBM, as well as William Hawkins at Lockheed pointed out several key issues in the context of corporate research:

a. For successful utilization of technology, an effective interface is needed between various functional groups.

b. Research base is needed to develop the corporate technical skills in successful implementation of technology.

c. Strategic guidance must be provided to the research administration for selection and completion of business relevant projects.

d. Scale up of manufacturing involves complex interdependence between research, product development and manufacturing engineering.

e. Corporate research is a costly venture often beyond the means of small companies.

Alan Frohman reviewed nine companies who had good track records of using technology as the competitive weapon. His findings indicated the following as the factors contributing to the success of these companies:

a. Top management orientation to technology as a major competitive weapon. Generally top management having technical background have favorable attitudes toward technology as a key competitive tool. However, the management must not emphasize technology at the expense of other areas. Secondly, the management should not be overinvolved with the technical decisions.

b. Project selection criteria. The criteria used to determine the support of specific developmental projects are:

   to suport business goals
   to protect and establish technological leadership
   to solve customer's problems
   to develop a window of technology
   to pursue a technological advancement

c. Systems and structure. The organization structure and procedures provide a close connection between business and technological decisions in a consistent fashion.

Thus far we have observed that management of the technological programs at the firm level is a complex task, involving the different functional groups in a coordinated fashion. We have reviewed a number of studies to identify the causes and consequent types of conflict between functional groups (Souder and Chakrabarti, 1982). The management of the research organizations has to integrate the efforts of the various interacting groups and balance the power and authority bestowed on them.

FORMULATING TECHNOLOGY STRATEGY

Formulation of technology and the R&D strategy depends critically on the overall business strategy of the corporation. It must be remembered that technology is only the means and not the end for a corporation. The technical management group is often faced with the key issue like the following:

> Should the organization be aggressive in its technological strategy?
>
> Should one become a technological leader or be a follower?
>
> Should one develop technology internally or procure it from outside?
>
> How should one proceed with acquisition of technology from external sources in terms of licensing, joint venture, and acquisition?
>
> How should an organization be positioned in the technology life cycle?

These issues are complex. For making appropriate decision on these issues, one has to consider many factors. For example, the aggressiveness of the technical organization should be linked with other competitive strengths of the corporation. The technical posture will lead to appropriate R&D posture, such as technology avoider to innovator to inventor. The technology life cycle is intimately related to the market condition and customer's preference functions. In the early development stage of the technology, generally the market structure is relatively free from price competition; customers emphasize peformance maximization. The innovations are more radical in nature at this point. As the technology matures, the price competition intensifies; customers emphasize cost minimization. The innovations become incremental and routine in nature at this stage (Utterback and Abernathy, 1975).

Ford and Ryan (1981) have identified six stages of the technology life cycle: technology development, technology application, technology launch, application growth, technology maturity, degraded technology. The first stage is set as soon as the R&D identifies a potentially beneficial technology. A firm continues further development of the technology if (a) it has the resources; (b) if the technology is compatible with firm's production and marketing skills; and (c) the projected return is favorable in comparison with other investments. The second stage, technology application involves application of the technology in some tangible product form. Selling and licensing can be a way to get into the technology application stage. Dolby went through the licensing route. Digital Research, Microsoft Corporation, etc., followed the licensing process to apply their technology on computers manufactured by others.

The third stage in the technology life cycle is application launch which corresponds to the stage of performance maximizing stage in Utterback and Abernathy's model. At this stage the firm is likely to be developing the technology for product modification and applying the technology to wider product areas. The fourth stage, the applications growth corresponds to the sales maximizing phase in Utterback and Abernathy's model. Up to the third stage the primary concern remains the development costs. Once the firm makes enough sales to recover the development costs and a considerable profit; the emphasis shifts to maximize the sales efforts of the technology. The fifth stage, technology maturity, marks the maturity of the technology as well as the market conditions. The competitors have mastered the technology or a competing technology, further growth potential is limited. Transferring this technology to developing countries becomes an attractive proposition at this stage.

Degraded technology is the last or sixth stage. At this stage the technology sees the end as it has reached universal exploitation. New technology appears in the market place to replace it.

Ford and Ryan (1981) advocated technology sales as a part of the exploitation program along with product applications. They conclude that the following conditions are necessary for successful marketing of technology:

Development of a coherent strategy for a full portfolio of technologies

Clear early decisions on acquisition and divestment of individual technology

Awareness of the value of the developing technology primarily for direct sales without incorporation into products

Understanding of the relationship between the sale of a technology through license and the sale of products based on that technology

Recognition that a technology buyer often has a better idea of his needs and opportunities

Reliance on technology marketers charged with the responsibility of marketing technology

The key element in all these life cycle models is time. It is extremely difficult to define the interval between various stages. The lag between invention and innovation is becoming shorter. In many rapidly advancing and changing technologies, notably in the semiconductor and the microprocessor industries, the product life cycle is extremely short. The maturity stage is extremely short lived. This poses additional problems to the management. This means that the firm will not have a nice long period to harvest the profit by making the product a cash cow.

Abernathy and his colleagues at Harvard have pointed out the impact of competition from abroad, particularly from Japan and other advanced nations, on the importance of technology to increase the productivity at home (Abernathy et al., 1983). They commented:

> The time comes, inevitably, when both regular and niche creation innovations are not enough to support a more permanent competitive vitality, for neither is intended

> to break out of the existing dominant design or even to challenge core concepts. Especially in mature industries, such innovations might help keep a company in business, but only a rather limited point. When new technology develops or preferences change, firms will lose competitive position unless they are willing and able to apply substantially new design concepts to their established markets (p. 113).

CONCLUSIONS

In this paper we have pointed out some key issues which go much beyond the micro-management of a research and development laboratory. The decisions related to R&D management involved many questions which need to be examined in light of interdisciplinary knowledge. Successful running of a laboratory involves many management and behavioral principles. The investment decisions regarding funding of R&D projects need careful judgment about the corporate objectives and strategies as well as the competitive environment of the firm. The competitive environment depends on many factors, such as technical, economic, socio-political ones. Competition from manufacturers abroad is here to stay. No longer can we make strategic business decisions in isolation. A proper understanding of the global competition is needed now more than ever--if we want to succeed at home.

REFERENCES

Abernathy, W.J., Clark, K.B. and Kantrow, A.M. 1983. Industrial renaissance: producing a competitive future for America. New York: Basic Books.

Branscomb, L.M. 1982. Comments on corporate research. Harvard Business Review, May-June.

Ford, D. and Ryan, C. 1981. Taking technology to market. Harvard Business Review, March-April.

Frohman, A.L. 1982. Technology as a competitive weapon. Harvard Business Review, January-February.

Frohman, A.L. and Bitondo, D. 1981. Coordinating business strategy and technical planning. Long Range Planning 14(6):58-67.

Haas, W. 1982. Comment on corporate research. Harvard Business Review, May-June.

Lawrence, P.R. and Dyer, D. 1982. Renewing American Industry. New York: Free Press.

Magaziner, I. and Reich, R.B. 1981. Minding America's business. New York: Vintage Paperback.

Merrifield, D. B. 1982. Comments on corporate research. Harvard Business Review, May-June.

National Science Foundation. 1983. Emerging issues in science and technology 1982: a compendium of working papers. NSF publication no. 83-51.

Reich, R.B. 1983. The next American frontier. New York: Times Books.

Souder, W.E. and Chakrabarti, A.K. 1980. Managing coordination of marketing and R&D in the innovation process. In Management of Research and Innovation - TIMS Studies in the Management Sciences Vol. 15, eds. B.V. Dean and J.D. Goldhar. New York: North Holland.

Steele, L. 1983. Managers' misconceptions about technology. Harvard Business Review, November-December.

Utterback, J.U. and Abernathy, W.J. 1975. A dynamic model of process and product innovation. Omega 3, Vol. 3.

White, G. 1982. Comments on corporate research. Harvard Business Review, May-June.

MANAGING HIGH TECHNOLOGY
An Interdisciplinary Perspective
B.W. Mar, W.T. Newell, and B.O. Saxberg (Editors)

MANAGING THE HIGH TECHNOLOGY FIRM: FOUR CANADIAN ELECTRONICS CASES

Philip J. Rosson[1] and Michael J.C. Martin[2]

[1]Associate Professor of Marketing
[2]Professor of Technological Management
School of Business Administration; Dalhousie University
6152 Coburg Road; Halifax; Nova Scotia, B3H 1Z5; Canada

This paper is concerned with management and business strategy in high technology firms. These are truly difficult matters, for these firms face turbulent external environments. Among the interrelated phenomena that have to be dealt with are: rapid technological change which shortens product life cycles, the breakdown of industry boundaries which brings forward new competitors, and a shortage of qualified technical and commercial personnel which bids up costs and leads to high turnover rates (as well as the spreading of proprietary information). Managing these firms and devising strategy for them involves complex interdisciplinary processes. At the very least, there are the scientific and commercial streams of the business to draw together. Many would argue that it is more meaningful to subdivide the commercial element into, say, finance, marketing and general management. Whatever the case, managing these firms involves interdisciplinarity, for contributions from numerous sources must be melded together. As such, they merit continued research attention.

Three sections make up this paper. In the first, the experiences of four Canadian electronics firms at different stages of corporate development are described. In the second, an attempt is made to relate these experiences to those of other technology-intensive firms at similar stages of corporate development--both within Canada, and elsewhere. In the final section, a number of concluding comments are made.

FOUR CANADIAN ELECTRONICS CASES

The four Canadian electronics firms may be categorized within Cooper's (1979) typology of the evolution of high technology companies. He suggests that such companies grow through three stages:

1. The start-up stage when the firm is founded and its initial technology-market niche is identified.

2. The early-growth stage when the technology-market niche strategy is tested and the president (or founding team) maintains direct contact with all major activities.

3. The later-growth stage characterized by manufacturing diversification, the instituting of one or more levels of middle management and some delegation of decision making.

At the time of our research in 1982/83, two of the firms, Micronet and Internav, were seeking to enter the later-growth stage. Micronet failed in its attempt, while Internav is still trying. Of the two larger firms, ASDIC has experienced continued difficulties in trying to diversify its product-market base and consolidate itself in the later-growth stage, whereas CAE Electronics has been very successful in this respect. Each of the four firms is now examined in turn.

MICRONET LIMITED

Micronet was founded in 1977 when two young men saw a window of opportunity opening for them. Two trends were of significance. First, the sharply declining price of microprocessor technology, and second rapidly escalating energy costs. Given these trends, considerable potential was seen for computerized energy controllers. A condominium apartment building provided a test bed for the cofounder's first controller which they designed, built, and operated on a part-time basis. This contract was successful and provided them with useful experience. One of the founders switched jobs at this time. Since his new employer was not too demanding of his time, he was able to devote considerable effort to building Micronet, before going full-time two years later.

Although an attempt was made to concentrate on energy controllers, like many specialized, fledgling companies, Micronet involved itself in a variety of projects through this period. Sales grew well, from $10,000 in 1977/78 to $100,000 in 1979/80. During the summer of 1980, the company learned of a contract for 860 controllers to be used in a study of electrical usage patterns in Nova Scotia. The contract for the controllers was worth $600,000, and was part of a $3 million load management project. After much soul-searching and discussions with numerous bodies, Micronet decided to co-bid with the Harris Corporation, a large U.S. multinational. Micronet was to produce the controllers, and Harris the deck computer, which would read the controller-generated data and produced management information. Micronet and Harris won the contract. With the boost this gave the young company, the second founder became a full-timer, and six production workers were hired.

However, despite successfully completing the contract, things seemed to go wrong at this stage of the company's development. A sales target of some $2 million was set for 1981/82, to be achieved by sales of the energy controller in the U.S. as well as in other parts of Canada, and sales of two products spun-off from the controller. These sales were not attained. The concept of one of the new products proved to be ill-founded, whereas the other new product was not marketed sufficiently aggressively. At the same time as these products were failing, numerous new avenues were being explored by the R&D group. By the winter of 1981, Micronet was fighting for survival; despite an injection of venture capital, the company closed its doors in the summer of 1982. One of its founders has gone on to start another venture.

INTERNAV LTD.

This company was established in 1976 in Sydney, which is on Cape Breton Island in Nova Scotia. Its founder was born and raised in New York City but had strong family roots in Cape Breton. By 1976 he had some 15 years engineering and management experience with several U.S. high technology companies in aerospace navigational systems, particularly Loran C systems.

Loran C navigational systems were subsequently applied to marine navigation and (as with Micronet) the opportunity offered by the technical paradigm change was firmly grasped. In 1971, a company was founded in Massachusetts to design and manufacture low-cost Loran C receivers for use in fishing boats and other small craft. Later it was anticipated that the system would be adopted by Canada, so it was decided to set up a Canadian company. The decision to locate the plant in Sydney was partly due to commercial reasons, but the prime factor was a personal one--the founder wanted to return to his father's birthplace.

For the first years of its operations Internav was essentially a manufacturing branch plant of the U.S. company. However, an R&D endeavor was initiated in 1978, and from this point forward the U.S. and Canadian companies have operated on a world product mandate basis. R&D efforts concentrate in two areas: contract work to determine the propagation characteristics of Loran C transmitters in various geographical locations, and microprocessor development and programming as applied to shipboard Loran C systems. This work has made possible a series of incremental improvements to the Loran C technology.

By the end of 1981, sales had reached $2 million, realized from various products and services based on the generic technology. More growth opportunities were sought, however, because the Canadian marine market for these products was becoming saturated. Internav's founder anticipated a technical paradigm change in airtransport landing systems, the current generation of instrument landing systems giving way to microwave landing systems (MLS) and formed a new company called Microwave to exploit this opportunity. The worldwide market for MLS is expected to reach $3-4 billion with the Canadian market about $200 million. Microwave's goal is to achieve sales of $20-30 million per year by 1990, having achieved a significant share of these markets.

ASDIC LIMITED

ASDIC was set up in 1947, initially intended as a marketing subsidiary of its British parent which has an enviable record in the fields of radio, radar and navigational systems. Two years later, however, it began to develop its own production operations, making electronic equipment required in antisubmarine warfare.

The company's generic skills were built up over the next 15 years based upon variable depth sonars, sonarbuoys, ionospheric sounders, VHF transmitters, and related products. Its prime customers were the Canadian and U.S. defense departments, and it experienced varying fortunes reflecting changes in defense budget levels and priorities. In 1969, its U.K. parent planned to terminate operations, so ASDIC was refinanced as a wholly owned Canadian company.

Wide fluctuations in product demand continued to bedevil ASDIC in the 1970s and early 1980s, leading to frequent crises and several ownership changes. Consequently, the company has been unable to reach the latter-growth stage, with a diverse product base selling to a widely spread group of customers. Several reasons explain this situation. First, profits on defense work have been relatively slim so that the company has lacked the cash to invest in significant R&D. Second, the company has historically lacked a commitment to market research and other contacts with private sector customers. Therefore, it has failed to uncover new technology-market niches for sustained and profitable exploitation. Finally, the lack of a marketing/entrepreneurial orientation has also led to its failure to capitalize on products which are the fruits of past R&D. Thus, there appears to be no long-term business strategy or overall management of innovation or new product development in this firm.

CAE ELECTRONICS LTD.

The origins of this firm go back to 1947 when Canadian Aviation Electronics Ltd. was incorporated. In the early days it was principally engaged in the repair and overhaul of electronic and electromechanical equipment. In the 1950s it began to make flight, radar, and weapons simulators for Canadian defense requirements. The technical skills and facilities developed during the production of some 80 military flight simulators for 12 western nations led CAE into the commercial simulator field.

Over the years, CAE Electronics has developed a variety of products and systems including: simulators for military and commercial airline pilot training and nuclear power plant operator training, avionics, and air traffic control. Behind each of the product moves it has made lay CAE's skills in "real-time computer control systems." This has been the common denominator to all of their activities since the first digital flight simulator in 1965. It also guides their current R&D work.

Diversification has not only been achieved within CAE Electronics. Because management wanted to have a broader portfolio of products and customers, from 1962 onwards it mounted an acquisition and diversification program. This has taken the company into the automotive industry as well as other sectors of manufacturing. As a result, sales to the military made up 20 percent of total sales in 1982, as opposed to 90 percent in 1962. Today, CAE Electronics is but one (albeit very important) company in the CAE Industries Ltd. family.

CAE Electronics has been a very successful operation. It is at the leading edge of technology and has grown from about 400 employees in 1976 to 1700 in 1983. Some 600 of those employees are highly qualified engineers and scientists, with 300 working on computer software. Sales revenue and income for 1983 were $147 million and $18 million, respectively. These figures represented 55 percent of CAE Industries' revenues and 100 percent of its income for the year in question.

Key characteristics of these four firms are shown in Table 1. Three of these electronics case studies (all but CAE Electronics) are more fully described in Martin and Rosson (1982).

THE BROADER CONTEXT

In order to provide a better perspective on management and business strategy in high technology firms, the discussion is now broadened. We briefly review research findings on the experiences of other technology-intensive firms, both in Canada and elsewhere. Then we relate the four case study firms to this wider body of writings.

RESEARCH ON CANADIAN HIGH TECHNOLOGY FIRMS

The first point that should be made is that anything that can be done to nurture high technology firms in Canada is to be welcomed. Between 1964 and 1976, Canada's balance of trade in technology intensive products was in deficit, and its performance in electrical (including electronic) products was particularly worrying: the deficit reaching $0.9 billion by 1976 (Canada, 1978). Naturally, this has become a major concern giving rise to considerable research. Steed's (1982) study of medium-sized firms, and a major investigation of innovation and technical change in five industries (DeMelto, McMullen and Willis, 1980), provide useful and recent introductions to the subject.

In this paper, where four electronics firms at various stages of corporate development have been described, it is particularly instructive to note the work of Litvak and Maule (1971; 1980). These researchers studied 47 small technology-based Canadian companies in 1970/71 and then performed a follow-up study in 1980. Thus, the history and development of these firms was tracked for a 10 year period. Of the 47 companies, 18 were categorized as failures, 9 were marginal survivors, and the remaining 20 were successful.

The failure companies--bankrupt, in receivership, or had ceased business for several years--tended to have single products, and while some were technically successful, most lacked enough technical novelty to achieve good market penetration. In at least one-half of these ventures, the owner managers were insufficiently prepared managerially, psychologiclly, and financially to cope with the protracted and mounting problems of launching and consolidating a business.

The nine marginal survivors had failed to show any profits over 10 years, or, at best, had produced only spasmodic profits. All nine firms have narrrow product lines and have achieved only modest sales, six under $1 million, only one over $5 million. Sheer determination and endurance characterized the managers of these firms. Indeed, their survival may have been dependent on their maintaining a Dunkirk spirit organization climate.

All 20 successful firms had achieved sales in excess of $1 million, five in excess of $10 million. Compared to the failures and marginal survivors, the successful companies had diverse product lines

Table 1
Characteristics of Four Canadian Electronics Firms

Charac-teristics	Micronet	Internav	ASDIC	CAE Electronics
Established	1977	1976	1947	1947
Sales -Total -Military	 $1 M 0	 $2 M 5%	 $20 M 90%	 $147 M 20%
Employees -Total -% Science, Eng.	 15 33%	 40 25%	 300 25%	 1700 33%
Products	Energy Controllers, Contract R&D	Loran C Navigation Systems; Contract R&D	Sonar Buoys: Environmental Data Systems; Communications Systems	Training Simulators; Avionics; Supervisory Systems; Space
Mission/ Operating Mode	Develop "Smarter" Software	Develop New Navigation Products, Military R&D Microwave Landing Systems	Production of Military Equipment, Components and R&D Contracts	Becoming a "Real-Time computer Control Systems" Firm
Production	←	In House		→
Marketing	Nonexistent	Dealers in Canada, World Agent Offshore	Little Real Marketing Agents, Help Direct Sales	Reps & Agents Supplement Direct Sales
Planning	Nonexistent	Intuitive	Extrapolative	Careful, Based on Scanning

which incorporated a higher degree of technological novelty. They concentrated their technical and marketing efforts on specialized, expanding industrial markets where they enjoyed comparative advantages and avoided competition with larger companies. They were less obsessed with the technology of their products and less prone to mousetrap myopia.

These study findings seem in harmony with other Canadian studies which point out that while most technology-based firms are headed by owner-managers that are technically proficient, general management, marketing and finance are key areas of weakness which often exact a heavy toll (Charles and MacKay, 1975; Calantone and Cooper, 1977; Hecht and Siegel, 1975; Knight and Lemon, 1978; Robidoux and Garnier, 1973).

Recent research by Dermer (1984) is also critical of the entrepreneurial and marketing skills of managers in Canadian technology-based firms. These are worrying deficiencies when one considers that most Canadian firms, lacking low cost advantages, serve fairly narrow market niches, where a leadership position is only maintained through being sensitive to the need for change, and good timing. In this respect, Dermer is also critical of Canadian government policy which is focused on the supply side, e.g., R&D assistance, acquisition of capital equipment, and testing laboratories when he sees serious market-side problems.

RESEARCH ON HIGH TECHNOLOGY FIRMS ELSEWHERE

There is a wide measure of agreement in the literature concerning the sorts of problems that firms experience as they develop over time. Useful discussions are provided by Steinmetz (1969) and Thain (1969). As far as technology based firms and concerned, management (in the broadest sense) is the most acute initial problem area. This is reflected in the tendencies listed below:

1. Technological obsessions. Frequently too much emphasis is put on technical matters to the detriment of commercial considerations (sometimes referred to as the toy-train syndrome).

2. Ill-considered diversification. The pressure to generate income often leads new firms into unforeseen fields where its skills may not truly lie.

3. Poor pricing. New firms often fail to set the right price. Too high a price results from over-engineering the product, and too low a price from the view that customers are price sensitive.

4. Sales equated with order predicted. New firms find it hard not to treat customers' estimates of orders as fact, and gear up their operations accordingly.

5. Overreliance on part-time staff. Sometimes the company becomes too dependent upon the outsiders that it uses to make good the necessary expertise it lacks internally.

6. Financial and cost measurement and control. Most new venture teams do not include a professional accountant, relying instead on part-time assistance in this area. This arrangement seldom satisfies the firm's acute need for financial control (Martin, 1984).

Beyond the initial stages what are the problems most often faced? Getting to Cooper's (1979) later-growth stage requires that the firm continue to match its skills and resources to emerging opportunities. There is, of course, the continual risk that success will be followed by failure. Assuming that, by now, the management team has acquired the commercial maturity and judgment to identify suitable growth opportunitites and to avoid illogical diversification, financing often becomes a difficulty at this stage. Firms frequently have to balance the risks of being underfinanced during a rapid growth phase against being overfinanced if, and when, sales level out. Moreover, until it has achieved a sufficient size and fiscal performance to attract the attention of the financial community, it is unlikely to be able to attract the required capital by going public. The growth must be sustained either from ploughing back profits, or from secondary financing by the initial investors.

Expansion, of course, means that more staff have to be hired, possibly with more managment and bureaucratic orientation, to monitor management standards and control procedures. This can have a disruptive effect on the climate of the organization, and unless managed carefully, stifles innovation (Martin, 1984).

COMPARING THE EXPERIENCES

How similar are the experiences of the four Canadian electronics companies to those of other firms? Broadly speaking, we would have to say very similar. Consider first the two firms attempting to move from the early to later-growth stage.

A number of factors contributed to Micronet's failure (not only to make the transition but also as a business). Principal among these was a lack of realism about the kind of company it was, and could be in the next few years. Its initial success in gaining the large contract for energy controllers led to grandiose plans for the future. Too many R&D projects were undertaken, almost no marketing was being done, and too much time was wasted on nonessentials, like designing a logo, and drawing up an employee handbook when the workforce only totalled 15. A typical and classic case of a firm that lacked good management. Interestingly, the second venture of one of Micronet's founders appears headed in the same direction.

Internav seems to have more focus than Micronet. Without doubt, this reflects the owner's background. Not only is he a competent scientist, but he also had several years of managerial experience prior to establishing this company. As a result, there is more evidence of planning and marketing activities than at Micronet. This shows up in at least two areas: (1) concentrating R&D and marketing activities so as to exploit the Loran C technology as long as is possible, and (2) developing a new slant to the business, microwave landing systems, before existing markets become saturated. Internav has had its share of troubles, however. Profits have not been very satisfactory, the business is very seasonal, and some difficulties have been experienced in attracting senior personnel to its off-center location. Also, considerable government financing has been used by Internav, both to get the company started and to enable it to move in its new direction.

Turning to the larger firms, ASDIC has been trying to consolidate its position as a later-growth stage company for many years, but success still eludes it in 1984. A number of factors which explain this state of affairs were mentioned earlier. The biggest problem the company faces is its high dependence on sonarbuoy production for the military. Diversification attempts have been unsuccessful, either because acquisitions were poorly chosen or insufficient commitment was made to internally-generated diversification projects. Business strategy and marketing are more absent than present in this company. As a result it swings from (relative) boom to bust, shedding skilled employees along the way, and waiting for money or contracts to bail it out.

By contrast, CAE Electronics has very definitely made it to the later-growth stage. Through a series of well planned moves, it has been able to reduce its dependence upon military contracts, in a way that ASDIC has not. Active scanning of the technological and market environments take place on a continuing basis, and this has enabled CAE to cleverly apply its R&D skills and experience to emerging technology market niches. Thus the transfer of interests from military flight simulators to commercial flight simulators, and on to nuclear power station simulators, and so on. Although technical personnel dominate the senior management positions in the firm, this is a company that <u>is</u> managed: the diverse elements are very well orchestrated.

CONCLUDING COMMENTS

We began with the observation that, for a number of reasons, management and business strategy are inherently difficult matters in the high technology firm. The case studies of four Canadian electronics firms bore this point out, and the literature review showed that the problems faced by these particular companies were fairly typical of those faced by technology-based firms at similar stages of development.

It seems, therefore, that the problems involved in managing the high technology firm are fairly well documented. Ideas about how these problems might be solved are also to be found, although the scattered and scholarly nature of much of the literature may not facilitate as much access or use by managers as is desirable. It could be argued that a need exists for a compendium text on these problems and solutions, based on solid research, yet written in a straightforward manner (see Cohn and Lindberg, 1972, pp. 25-27, for example, on small companies and innovation).

Turning to the future management of technology-based firms, the environment will not likely be any less turbulent given the technological leveling among nations and companies, the increasing international flows and sharing of technology, the rising R&D costs, times, and accelerating rate of technological substitution, and the slow down in economic growth (Grandstrand, 1982). Even more than in the past then, the future will be marked by the struggle of firms to find and maintain a viable market position.

Space limits preclude much discussion of developments that will help firms achieve such viable positions. Noteworthy (at least to us), is the growing acceptance and practice of explicitly incorporating technology as a vital element in business strategy (see recent papers by Crawford, 1980; Horwitch and Sakakibara, 1984; Liberatore and Titus, 1983; Pappas, 1984; and Wilkinson, 1983). Continued research into new product development processes (Cooper, 1979) will also add to our understanding of innovation, <u>the</u> essential function of high technology firms.

ACKNOWLEDGEMENTS

The authors gratefully acknowledge the financial support of the Federal Department of Industry, Trade and Commerce, Technological Strategy Branch in the development of the case studies upon which this paper is based.

REFERENCES

Calantone, Roger J. and Cooper, R.G. 1977. A typology of new product failures. In Educator's conference proceedings, eds., B.A. Greenberg and D.N. Bellenger, pp. 492-497. Chicago: American Marketing Association.

Canada. 1978. Canadian trade in technology-intensive manufactures, 1964-1976. Ministry of State, Science and Technology, Background paper 5. Ottawa: Supply and Services Canada.

Charles, M.E. and MacKay, D. 1975. Case studies of industrial innovation in Canada. Ottawa: Technological Innovation Studies Program, Department of Industry, Trade and Commerce.

Cohn, T. and Lindberg, R.A. 1972. How management is different in small companies. New York: American Management Association.

Cooper, A.C. 1979. Strategic management: new ventures and small business. In Strategic management, eds. D.E. Schendl and C.W. Hofer, pp. 316-327. Boston: Little, Brown and Co.

Cooper, R.G. 1979. The dimensions of industrial new product success and failure. Journal of Marketing 43(Summer):93-103.

Cooper, R.G. 1983. The new product process: an empirically-based classification scheme. R&D Managment 13(January):1-13.

Crawford, C.M. 1980. Defining the charter for product innovation. Sloan Management Review 22(Fall):3-12.

Dermer, J. 1984. Growing Canada's threshold technology-producing firms. The Business Quarterly 49(Spring):37-45.

De Melto, D.P., McMullen, K.E. and Willis, R.M. 1980. Preliminary report: Innovation and technical change in five Canadian industries, Discussion Paper 176. Ottawa: Economic Council of Canada.

Granstrand, O. 1982. Technology, management, and markets. New York: St. Martin's Press.

Feldman, L.P. and Page, A.L. 1984. Principles vs. practice in new product planning. Journal of Product Innovation Management 1(January):43-45.

Hecht, M.R. and Siegel, J.P. 1973. A study of manufacturing firms with special emphasis on small and medium-sized firms. Ottawa: Technological Innovation Studies Program, Department of Industry, Trade and Commerce.

Horwitch, M. and Sakakibara, K. 1984. The changing strategy-technology relationship in technology-based industries: a comparison of the United States and Japan. Unpublished working paper No. 1533-84, Massachusetts Institute of Technology.

Knight, R.M. and Lemon, J.C. 1978. A study of small and medium-sized Canadian technology-based companies. Ottawa: Technological Innovation Studies Program, Department of Industry, Trade and Commerce.

Liberatore, M.J. and Titus, G.J. 1983. Synthesizing R&D planning and business strategy: some preliminary findings. R&D Management 13(October):207-218.

Litvak, I.A. and Maule, C.J. 1971. Canadian entrepreneurship: a study of small newly established firms. Ottawa: Technological Innovation Studies Program, Department of Industry, Trade and Commerce.

Litvak, I.A. and Maule, C.J. 1980. Entrepreneurial success or failure: ten years later. Ottawa: Technological Innovation Studies Program, Department of Industry, Trade and Commerce.

Martin, M.J.C. 1984. Managing technological innovation and entrepreneurship. Reston, VA: Reston Publishing.

Martin, M.J.C. and Rosson, P.J. 1982. Four cases on the management of technological innovation and entrepreneurship. Ottawa: Technological Innovation Studies Program, Department of Industry, Trade and Commerce.

Pappas, C. 1984. Strategic management of technology. Journal of Product Innovation Management 1(January):30-35.

Robidoux, J. and Garnier, G. 1973. Factors of success and weakness affecting small and medium-sized manufacturing business in Quebec, particularly those using advanced production techniques. Ottawa: Technological Innovation Studies Program, Department of Industry, Trade and Commerce.

Steed, G.P.F. 1982. Threshold firms: backing Canada's winners, Background Study 48, Science Council of Canada. Ottawa: Supply and Services Canada.

Steinmetz, L.L. 1969. Critical stages of small business growth. Business Horizons 12(February):29-36.

Thain, D.H. 1969. Stages of corporate development. The Business Quarterly 34(Winter):33-45.

Wilkinson, A. 1983 Technology--an increasingly dominant factor in corporate strategy. R&D Management 13(October):245-259.

MANAGING HIGH TECHNOLOGY
An Interdisciplinary Perspective
B.W. Mar, W.T. Newell, and B.O. Saxberg (Editors)

EVALUATING PHARMACEUTICAL R&D

Nicolas V. Danila

Professor and Consultant
Institute of Public Management
Centre de Recherche et d'Etudes Doctorales
46, rue Barrault
75634 Paris Cedex 13
France

ABSTRACT

The results of an empirical study on the evaluation of R&D activities in a French pharmaceutical company are presented. A sample of more than a dozen managers expected a priori to be involved in the evaluation of research was drawn and personal interviews were conducted. Extensive demograhic data on the respondent and on the organization of his department were collected and related to the criteria and to the methods used to evaluate and to select research projects. Two specific checklists are presented. The great diversity of the criteria used, the multiplicity of their utilizations, and the continuity of the decisional path explain the necessity of establishing several committees or informal working groups. These committees make the synthesis of all the information, internal and external, and they play a decision making role. More than twenty more or less formal committees were identified. Some financial methods are used for development projects, ROI (Return on Investment), NPV (Net Present Value) and IRR (Internal Rate of Return). A kind of PERT (Program Evaluation and Review Technique) was adopted for the scheduling and the control of projects, but no formal mathematical programming models for R&D evaluation or resource allocation model are used. As a result, the evaluation of R&D activities is considered to be one of the most important decisions in the company and is discussed at the highest level. Nevertheless, the need of a model integrating all of the criteria used was explicitly expressed by the deciders. A visual evaluation is proposed, the cubic evaluation and a new evaluation and selection method ALADIN (AnaLyse About Development and INovation) is suggested.

INTRODUCTION

Managing R&D in general and high technology R&D in particular is a very complex process. It involves the search of the environment for opportunities, the identification of projects or alternatives, the evaluation and selection of projects, and finally their implementation.

One of the most important problems in this process is that the terms in which managers think about R&D are not yet broadly agreed upon. Last year, Merten and Ryu (1983) tried to define an industrial research system utilizing the IRI (Industrial Research Institute) definitions. The answer for the communication problem as Merten and Ryu point out is a common language between management and technical staff. The role of knowledge in R&D efficiency and communication process was also underlined in the literature by Nelson (1982).

Research organizations have a strong tendency to propose problems for investigation at a faster rate than the resources can be supplied to support the work. This poses a fundamental issue for the managers of research because there are always more ideas and problems to be investigated than the available resources to investigate them. For this reason the greatest difficulty in the R&D management is the evaluation and the selection of worthwhile projects. The managers often lack a strong research background and the R&D staff lacks the skills to communicate comprehensibly and credibly with managers. The answer is to try to convert the uncertainty, which is sure to be costly, to risk, which measures the chance of success and permits clear choices at every stage of the R&D activities (Roussel, 1983).

This paper will focus on the evaluation and selection of R&D activities in the field of pharmaceutics. Generally speaking, there are three purposes for a program evaluation. First is to learn about the operations and effects of the program. Second is to control the behavior of those responsible for the program implementation. Last is to influence the responses of outsiders in the program's political environment (Goldenberg, 1983).

There are numerous papers which contain descriptions of R&D evaluation and selection problems. Most are speculative, but this portion of the literature nevertheless represents a substantial body of knowledge. During the last quarter of a century, about two hundred of these models have been presented. Literature surveys and reviews abound.

In 1964, Baker and Pound (1964) could already cite over 80 papers dealing directly with normative models for R&D project selection and resources allocation. The same year, Kiefer (1964) presented the models used in the industrial chemical research. Cetron, Martino and Roepcke (1967) compared thirty methods to a standard set of characteristics relating to ease of use and to scientific or technological area of applicability. A description of evaluating, selecting, and controlling R&D projects was presented by Dean (1968).

Gear, Lockett, and Pearson (1971) analyzed, some portfolio selection models used in the R&D field. A scoring methodology, using five major criteria (realism, flexibility, capability, ease of use, and cost) for evaluating the suitability of models for R&D project evaluation and selection was proposed in 1972 by Souder. The same year, Pearson (1972) analyzed the use of ranking formulae in R&D projects in the United Kingdom. One year later, Augood (1973) proposed his own survey and suggested that a good evaluation and selection method might be gained by combining impact and risk analysis with considerations of R&D costs and probabilities of success. Baker (1974) made a new assessment and wrote that the R&D

project selection decision can be characterized as having multiple criteria with no common underlying measure.

Souder (1978) presented eight characteristics types of project evaluation models: index, portfolio, decision theory, risk analysis, frontier, scoring, profile, and checklists models. Danila (1983) briefly analyzed about seventy models; giving their respective formulae, strong and weak points, along with a bibliography. Danila (1984) also presented twelve types of R&D project evaluation and selection methods: ratio, index, programming, portfolio, matrix, systemic, checklists, relevance trees, tables, multi-criteria, consensus, and graphic.

Recently, some mathematical R&D models, such as DPM (Decision Process Model) try to be useful for policy analysis both at the level of the company and of the federal government (Winkofsky et al., 1981). Few papers have appeared in the literature concerning the usage of the R&D evaluation and selection methods (Baker and Pound, 1964; Dean, 1968; Liberatore and Titus, 1983). Empirical studies have rarely been undertaken to determine the extent of use of these models in the industrial setting. Mansfield (1981) wrote that the precise use of these models in the United States was unknown. However, Liberatore and Titus (1983) published a paper on the practice of management science in R&D project management for a sample of 29 American firms. In France, in 1970, the head of the Battelle's Office in Paris, Mahieux, conducted an empirical study about the current usage of R&D methods in French industry (Mahieux, 1970). Since then, no empirical studies have been reported in France, but work on one such project began in 1983 (Danila, 1983).

The literature in the evaluation of R&D pharmaceutical projects is not very large. Given the uncertainties and obvious cost pressures in pharmaceutical R&D, some managers of Sandoz, a Swiss pharmaceutical company, proposed in 1978 a measure of evaluating the technical success of individual projects as a way of quantifying the unquantifiable (Balthasar et al., 1978). Four years before, Faust and Ackerman (1974) described the management approach in another Swiss pharmaceutical company. Upjohn's goal oriented organization (coordination matrix) of eleven years ago was compared by Stucki (1980) to the current organization consisting of a leadership matrix and a simultaneous coordination matrix. In France, the style of leadership and success in a public pharmaceutical laboratory was analyzed by Levy-Leboyer and Pineau (1981). No empirical study in the pharmaceutical field about the evaluation of R&D in France has been reported. That is precisely the aim of this paper.

METHODOLOGY

First, a French pharmecutical company must be chosen. The company must have excellent performance in R&D activities, outstanding financial results, and a good reputation. All the classical functions of a company must also be represented: research, manufacturing, marketing, distribution, and so on. Some companies responding to these criteria were contacted by telephone and by a letter confirming the purpose of the study. Two companies agreed to participate in the study. The first, which boasts the best financial results in France, was, after four interviews, dropped because a part of its research was done by the parent establishment, and further, the pharmaceutical subsidiary did not have its own distribution network. The second company seemed to be a good site for exploratory research.

An appointment was scheduled for a personal interview with the General Manager. A list with the major functions supposed to be involved in the evaluation of the R&D activities was presented to him:

Research and Development
Marketing and Distribution
Financing and Planning
Regulatory and Legal Factors
Manufacturing

He was asked to give the name of these managers and other deciders that he thought it would be necessary to meet. He accepted and expressed interest in obtaining the final report. He also wrote an internal paper to announce this collaboration of all the participants in the study. In this manner, the climate of the study was very favorable. Fifteen interviews were conducted with the following actors: Chairman, General Manager, Financial Director, Therapeutic Strategies Director, Scientific Director, Marketing Director, Personnel Director, Production Director, Exports Manager, Patent Manager, Plant manger, Chemical Manger, Biochemical Manager, Pharmacokinetic Manger, Researcher. The interviews were scheduled during a two month period, September and October 1983, and averaged about two hours. They were conducted with the aid of a semistructured questionnaire. This questionnaire sought the following data: demographic information on the respondent; the organization of his department; the place of the respondent in the organizational chart; the information environment (internal and external); the relationship in the reporting of information up and down; the life cycle of a research project; the different typologies used; the current R&D Department budget; the criteria used to evaluate and select the R&D projects, the strategic planning, the committes used to evaluate and select the R&D strategic planning; the committees and working groups involved; the respondent's major problems (whether or not research evaluation related) and, finally, other comments.

A fully structured questionnaire was not used. Rather, major information to be sought was noted on a single page, and other questions remained committed to memory. This alleviated respondents' unease at a long series of structured questions and allowed the researcher to adapt particular question wordings to avoid inducing certain answers. This type of interview procedure, which may be called a flexible structured interview requires a very good knowledge of the substance and memory in order to insure that no important points are overlooked even while limiting one's own questions to a minimum to let the respondent express himself fully. Between ten and twenty pages of notes were obtained for each interview in this manner. The interpretation is more difficult than with a classical structured questionnaire, but the information collected is incomparably richer.

Globally, four hypotheses have been chosen to be tested in the study:

1. Are pharmaceutical R&D evaluation activities interdisciplinary?

2. Which are the mst important criteria used in this evaluation?

3. Are any formal methods used for the evaluation and, if not, how is the aggregation done?

4. Is it possible to imagine an interdisciplinary method for the evaluation of research?

RESULTS AND DISCUSSION

The report, covering only part of the information collected in the interviews, is a monographic study on the pharmaceutical research system. Four points were developed:

1. the description of the global research system
2. the comparison between the formal organizational charts and the informal but actual decision patterns in the research field
3. the identification of the criteria used by the different managers
4. the decisional path on the R&D activities

For lack of space, all of these subjects cannot be considered here; only some of them will be pointed out in this paper.

Organizational Structures

The present structure of the company is organized by function: finance, therapeutic strategies, marketing and production, legal advice, medical service, and scientific direction (see Figure 1). One result is that the real research decision system has a tetraedic form with the chairman and general manager at the apex (CGM) at the center of the therapeutic strategies director (TS) and at the three other points the medical and scientific director (MS), the production director (P), and the commercial director (C) (see Figure 1).

Checklists

Further analysis makes it possible to establish checklists of criteria more or less explicitly referred to by those involved. Some criteria are common to two functions. Examples of checklists are given in Tables 1 and 2.

Research Decisional Path

The decisional path from the idea to create a new product to its production and commercialization has been described, and the criteria used at each level for the evaluation were mentioned. Finally, there is a continuity of the criteria through the chemistry, biochemistry, pharmacology, toxicology, galenic medicine, production, and commercialization. Figure 2 shows the most important entities which occur in the research process. The

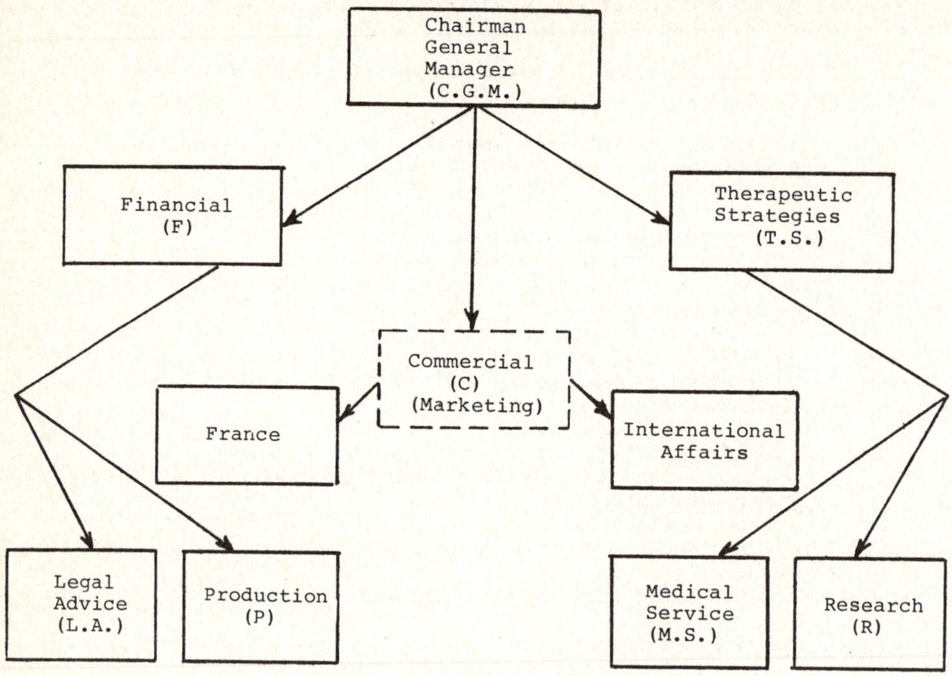

FUNCTIONAL ORGANIZATIONAL CHART

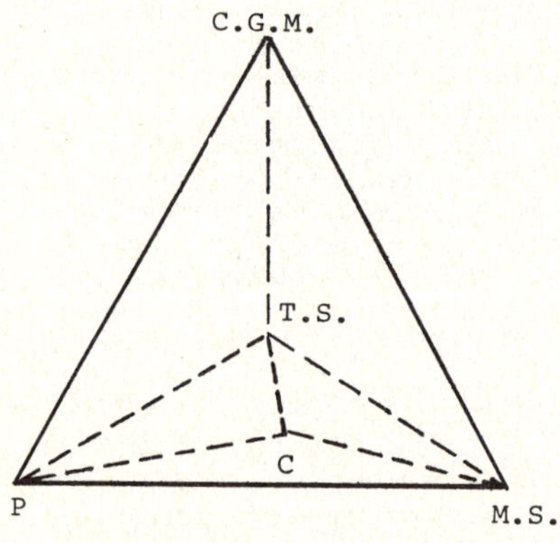

RESEARCH DECISIONAL SYSTEM

Figure 1

Table 1
Checklist No. 1
Scientific Research Criteria

A.	<u>Therapeutic Interest of the new drug</u> Necessity to answer to a unsatisfied therapeutic need Activity and the efficiency of the new drug Innoculty Adverse reactions Galenics Originality of the methobolism action
B.	<u>Physico-chemical characteristics</u> Stability of the new drug Ease of the chemical synthesis and the possibility to industrialize the chemical reaction Originality of the molecule Newness of the chemical mechanisms of the reaction
C.	<u>Influence of the environment</u> R&D activities of the competition Patent status Scientific community's interest
D.	<u>Common criteria</u> With the production: Theraputic's indications of the new drug Technical feasibility With the commercialization: Quality of the scientific arguments in favor of the new drug and possibility to show objectively drug's characteristics Pharmaco-vigilance (clinical supplies after the launch) Possibility to change the drug and its dosage Importance of the training for the medical commercial agents, Biodisposibility of the new drug. With the finance: Price of the patent With the strategy: Research activity level of competition Opinion of the external research teams (secrecy agreements) and of academic researchers Difficulty to obtain national or international registration

Table 2
Checklist No. 2
Production Criteria

Importance of the investment required (production and packaging tools) and possibility to use present installations

Difficulty to package the new drug

Industrial price

Common criteria

With the commercialization:

Size of potential market
Market trend and growth
Market share possible with the drug characteristics
Selling price

With the finance:

ROI (Return On Investment) for the investment to be done
ROI for the research project
Previsional account for the new drug
Pay back
Break-even of the profitability

With the strategy:

Dimensions of the equipment required
Familiarity of the company with the industrial process,
Compatibility with the present equipments.

With the research:

Therapeutic instructions of the new drug,
Technical feasibility.

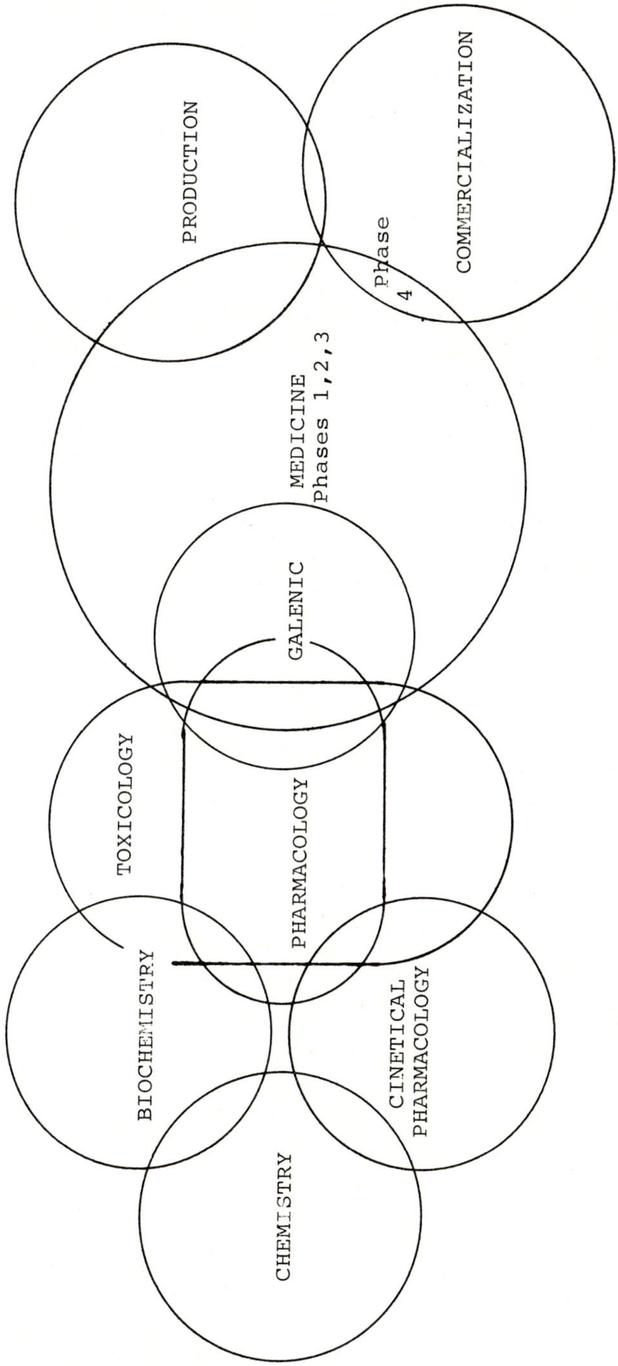

Figure 2
Entities of Research Decision

research decisional path is not linear, and it is difficult to overemphasize the complexity, the interactive nature of the process, and the feedback loops which are characteristics of pharmaceutical research.

Committees

More than twenty formal or informal committees were identified covering all levels of the hierarchy and the different aspects of the process (management committee, coordination commmittee, research committee, reflexion group, planning and information committee, new opportunities committee, medical committee, medical-promotion committee, galenic committee, social affairs meeting, patent meeting, Friday or research information meeting, and so on). All of these diverse committees, meetings or working groups are not interdisciplinary, of course.

A purely interdisciplinary research team may possess, from the standpoint of the literature definitons eight characteristics (Birnbaum, 1981). All of these attributes can be found in the research committee of the French laboratory analyzed in this study. Different bodies of knowledge including organic chemistry, analytical chemistry, physical chemistry, clinical chemistry, galenical chemistry, biochemistry, chemical engineering, bioengineering, computer engineering, biostatistics, molecular biology, microbiology, immunology, pharmacology, clinical pharmacology, cinetical pharmacology, medicine, are represented in this committee.

The different group members use different problem solving approaches in attempting to obtain approval for marketing of new drugs by the relevant government regulatory agencies. For instance, the chemists try to obtain new molecules by synthesis, biologists try to understand the mechanisms of the chemical product in the human body, and so on.

The members of the research committee perform different roles in solving problems. Physicians conduct the application of drug therapy in illness through clinical studies to demonstrate the safety and the efficacy of the new drug. They perform the role of evaluator from the point of view of the patient. The chemists perform the role of builders of molecule for the new drug and the pharmacologists and toxicologists, the role of screener for the innocuity and the adverse reactions.

All the members of this committee work on a common problem, on a common whole which is to obtain a good drug supporting the successful drug regulatory agency approval. Pharmaceutical companies express that it is more and more difficult to satisfy escalating drug regulatory agency demands for data.

The research committee has a group responsibility for the final product, the new drug and its approval. All the researchers belonging to the committee are concerned with the result of their work, in concert with their colleagues from other disciplines.

Common facilities are shared by all the members of the research committee. The HMR or IR equipment may be used by chemists, biochemists, physical chemists and other researchers concerned with this type of analysis.

The selection of the researchers asked to participate in the research committee depends on the nature of the problem to be solved. For a drug preventing the heart attack, the committee must have a cardiologist and not a dermatologist.

The members of the committee are influenced in their work by how the other members perform ther tasks. This implies that the toxicology tests can change the type of molecule sought by the chemists, for example.

Figure 3 shows the research decisional path, the most important criteria and committees, and the moment of their intervention.

Methods

Financial methods are used for the development projects (ROI, NPV, IRR), but there is no formal method for the research projects. In any case, the ROI is not used as some authors have indicated (Goldenberg, 1983) as restraint or innovation but as one financial criterion among others. The same assumption may be made for the other financial methods NPV and IRR. Mathematical programming models, portfolio, matrix models or some other formalized methods are not used in this French pharmaceutical company. There is no multicriteria method to aggregate all the criteria from the checklists. For the scheduling and the control of the projects a PERT is used.

Cubic Evaluation and the Aladin Method

As can be seen in Figure 4, it is possible to imagine a cube for the research evaluation with six aspects: research, production, economics, finance, marketing, distribution, and the seventh, strategy, as a vector going out of the cube. The direction of this vector determines the weight of each dimension. For example, fundamental research will be evaluated more with research and economic criteria, a development research more with financial, marketing, and distributional criteria.

The main dimensions of research evaluation are presented in Figure 4, the evaluation cube. This graphic presentation is easy to understand and makes possible a visual aggregation of all the criteria used by this pharmaceutical company.

A new, interdisciplinary model was built in 1983 (Danila, 1983) and was called ALADIN (AnaLysis About Development and INovation). This model would be useful in the pharmaceutical R&D evaluation. ALADIN is a multidimensional model which works on checklists, matrix, multicriteria, consensus, and graphic aspects. The central point in this model in the constitution of an interdisciplinary working group called GES (Group of Evaluation and Selection). This group must be interdisciplinary because the evaluation and the selection of research prospects need an interdisciplinary approach to be complete and accepted by all other actors. The whole of the GES is to require the information to negotiate this information, to sort it

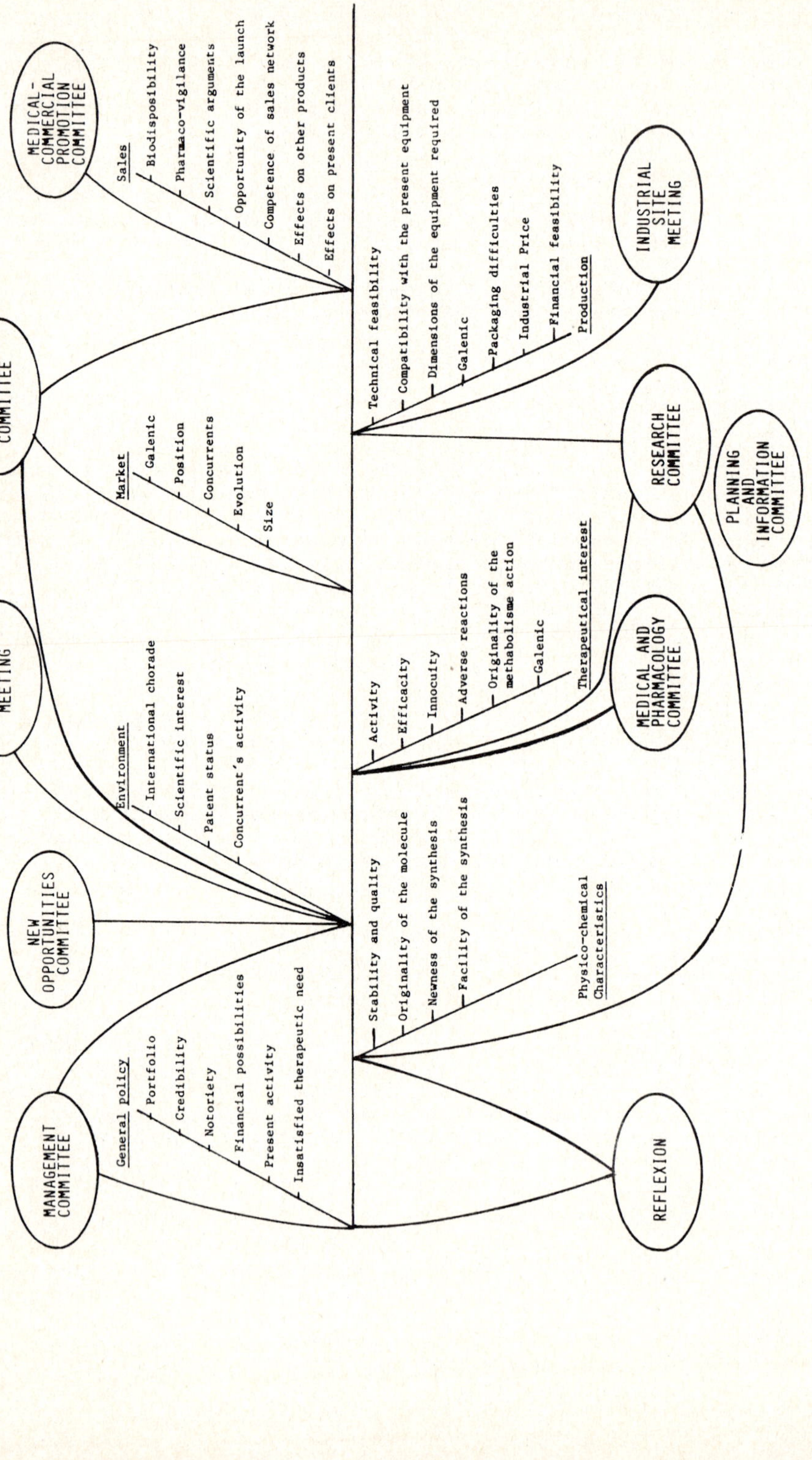

FIGURE 3
RESEARCH DECISIONAL PATH

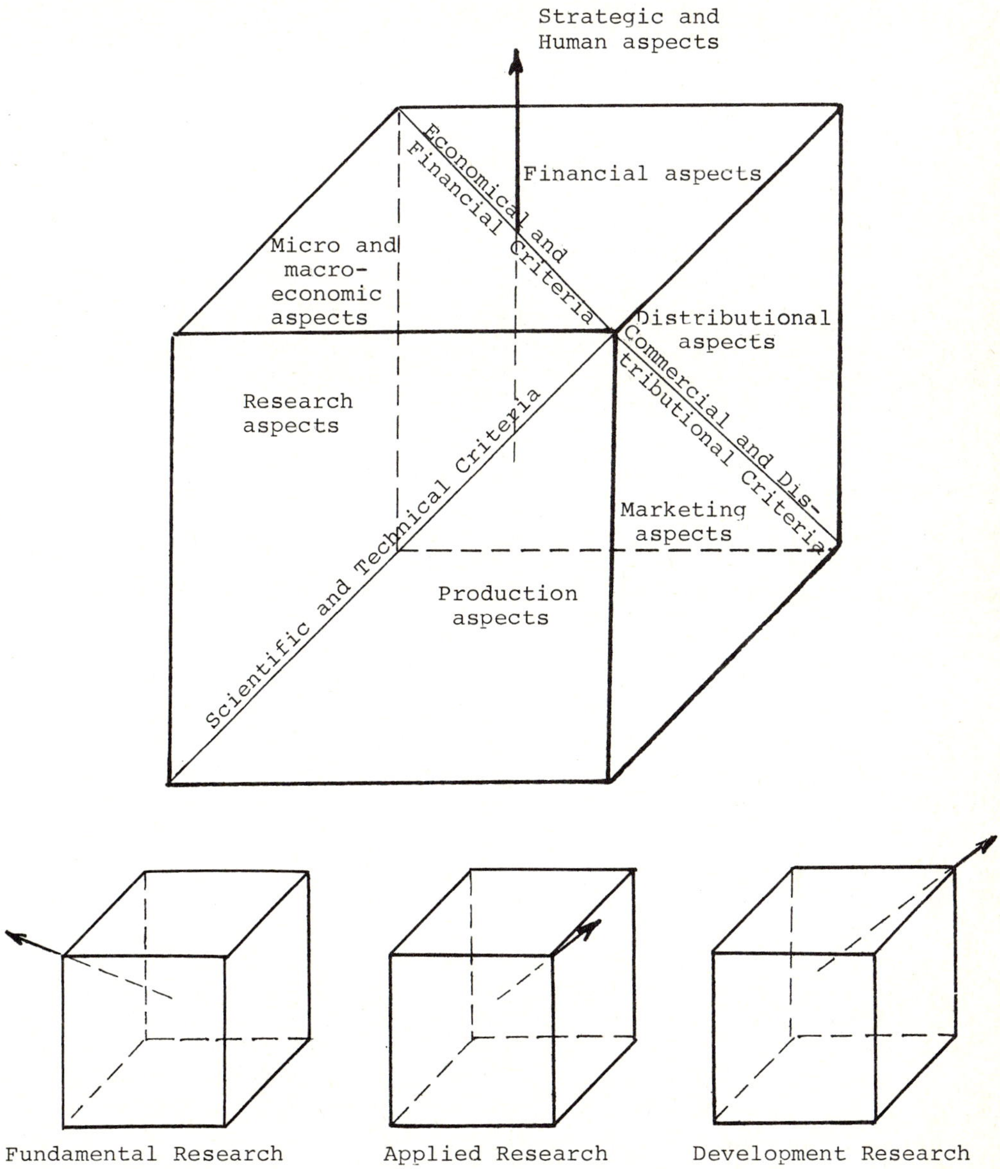

Cubic Evaluation Method

Figure 4

out, to complete it if it is necessary and to feed the model.

CONCLUSION

The empirical study has shown the complexity and the interdisciplinary character of the R&D process in the field of pharmaceutics. It is possible to assert that on one hand the pharmaceutical R&D is interdisciplinary and on the other hand the evaluation of research must be equally interdisciplinary. In this manner, the pharmaceutical R&D evaluation projects which occur by the intersection of these two fields is a good area for investigation in interdisciplinary research. The criteria used are very heterogeneous and the entrance in the decisional path is specific. There are no aggregation models to consider all these criteria in a single formal approach. An interdisciplinary method can be very useful for the evaluation of pharmaceutical research, in particular as well as evaluation research in general.

ACKNOWLEDGMENT

Portions of the work reported here were supported by a grant from the Center of Prospective and Evaluation (no. CPE 83.293.044) from the French Ministry of Industry and Research.

REFERENCES

Augood, D.R. 1973. A review of R&D evaluation methods. IEEE Transactions on Engineering Management EM-20:4:114-120.

Balthasar, H.U., Boschi R.A., and Menke M.M. 1978. Calling the shots in R&D. Harvard Business Review 56:3:151-160.

Baker, N.R. 1974. R&D project selection models: an assessment. IEEE Transactions on Engineering Mangement EM-21:4:165-171.

Baker, N.R. and Pound, W.N. 1964. R&D project selection: where we stand. IEEE Transactions on Engineering Management EM-11(4):124-134.

Birnbaum, P.H. 1981. Contingencies for interdisciplinary research: matching research questions with research organizations. Management Science 27:11:1279-1293.

Cetron, M.J., Martino, J., and Roepcke L. 1967. The selection of R&D program content - survey of quantitative methods. IEEE Transactions on Engineering Management EM-14:1:4-13.

Danila, N.V. 1983. Strategies technologiques. Paris, France: FNEGE/IDMP.

Danila, N.V. 1984. Methodes d'evaluation et de selection des projets de recherche. Revue Francaise de Gestion Janvier-Fevrier:46-52.

Dean, B.V. 1968. Evaluating, selecting and controlling R&D projects. AMA Research Study 89, American Management Association.

Faust, R.E. and Ackerman, G.L. 1974. Program/project management at Hoffman - La Rouche. Research Management 18:1:5-9.

Gear, A.E., Lockett, A.G., and Pearson, A.W. 1971. Analysis of some portfolio selection models for R&D. IEEE Transactions on Engineering Management EM-18:2:66-76.

Goldenberg, E.N. 1983. The three faces of evaluation. Journal of Policy Analysis and Management 2:4:515-525.

Kiefer, D.M. 1964. Winds of change in industrial chemical research. Chemical Engineering 77:4:88-109.

Levy-Leboyer C. and Pineau, C. 1981. Organizational characteristics, style of leadership and success. In Bio-Medical Reseach. Second International Conference on Interdisciplinary Research, July 19th-23rd 1981, at Manchester, England.

Liberatore, M.J. and Titus, G.J. 1983. The practice of management science in R&D project management. Management Science 29(8)962-974.

Mahieux, F. 1972. Le Calcul de la rentabilite' de la recherche selection des projects. Paris, France: Editions Eyrolles.

Mansfield, E. 1981. How economists see R&D. Harvard Business Review 59:6:98-106.

Merten, U. and Ryu, S.M. 1983. What does the R&D function actually accomplish? Harvard Business Review 61:4:24-28 (Ideas for Action).

Mechlin, G.F. and Berg, D. 1980. Evaluating research - ROI is not enough. Harvard Business Review 58:5:93-99.

Nelson, R.R. 1982. The role of knowledge in R&D efficiency. Quarterly Journal of Economics 98:3:453-470.

Pearson A.W. 1972. The use of ranking formulae in R&D projects, R&D Management 2:2:69-73.

Roussel, P.A. 1983. Cutting down the guesswork in R&D. Harvard Business Review 61(5):154-160.

Souder, W.E. 1972. A scoring methodology for assessing the suitability of managment science models. Management Science 18:10.

Souder, W.E. 1978. A system for using R&D project evaluation methods. Research Management 21:5:29-37.

Stucki, J.C. 1980. A goal-oriented pharmaceutical research and development organization: an eleven-year experience. R&D Management 10:3:97-105.

Winkofsky, E.P., Baker, N.R., and Sweeney, D.J. 1981. A decision process model of R&D resource allocation in hierarchial organizations. Management Science 27:3:268-283.

MANAGING HIGH TECHNOLOGY
An Interdisciplinary Perspective
B.W. Mar, W.T. Newell, and B.O. Saxberg (Editors)

MANAGING THE R&D/MARKETING INTERFACE: A PROBLEM OF INTERDISCIPLINARITY

A.W. Pearson,[1] D.F. Ball[2] and W. Dench[3]

[1]R&D Research Unit; Manchester Business School; Manchester, U.K.

[2]School of Economics & Accounting; Leicester Polytechnic; Leicester, U.K.

[3]The Welcome Foundation Ltd.; Langley Court; Beckenham, Kent; U.K.

ABSTRACT

In most research and development establishments matching technical capability to market need is a necessity with timing being a key factor in successful innovations.

It is therefore important that activity at the R&D/Marketing interface should promote rather than hinder progress towards project objectives. This necessitates that as much attention be paid to communication and coordination at the interface as in the separate technical and marketing functions.

In this paper the findings of a recent study of six R&D projects in the pharmaceutical industry are presented. Some potential areas of concern are identified and compared with those reported in other research studies and with the literature on managing interdisciplinary and interfunctional groups.

Recommendations for management are made as well as some suggestions for further research.

INTRODUCTION

The importance of coordination between various organizational groups within a firm has long been recognized and for R&D and Marketing has been extensively reviewed by Souder and Chakrabarti (1978). It is not the intention to repeat this in any detail here but to extend the discussion by incorporating some of the conclusions arising from a recent study with which the authors were involved (Phillips, Pearson and Ball, in press).

In such discussions the term interface has often been used to refer to such joint action as that between R&D and Marketing. The dictionary describes interface as being between two faces. However there is nothing between R&D and Marketing in the way in which, for example, an arbiter is between union and management in a dispute. In the physical sciences interface represents a boundary between two components and interest focuses on the position of the boundary and the movement of material across it. The R&D and Marketing interfaces might therefore be associated with actions and decisions influencing the two parties' unilateral decisions, some of which may profoundly affect the way in which the other is stimulated or constrained to work. This does raise the question of the role of top management. Their first implied role is to determine the organizational relationship within the company and within that the form of top management responsibility. There are a number of alternative modes. The first is where top management is responsible for directing projects and assumes all the responsibilities. It acts both as coordinator and as arbiter at the interface. Some other decisions made by either R&D or Marketing can be classified as specialist whereas others can be described as interactive. A specialist decision is one taken by R&D or Marketing which affects the project but does not significantly affect the other party, e.g., a decision by Marketing to open up Malaya rather than Burma is unlikely to affect R&D. An interactive decision is one which affects the other party as well as the project, e.g., if R&D decide to concentrate work on a substance of limited solubility which can only be formulated as a suspension. When the project is directed by top management it needs to pay special attention to the interactive decisions. There are two further modes which can be described as customer-contractor: in one R&D is directing and in the other Marketing. In addition there is a fourth mode which can be descibed as collaborative.

Which of the four modes is selected will depend upon the organizational relationships which exist within the firm; and this is likely to depend upon the nature of its business, its traditions, and the size of the individual functions. For example, in small high technology companies R&D and Marketing are not always separate. For any given firm it is worth considering which of the four modes described is most appropriate and whether or not there should be any changes as the project develops. Whether or not the best choice of mode is being used should be the responsibility of top management as it monitors progress.

Such monitoring should recognize that the various modes of progressing the project give rise to a differing set of problems of interpersonal relationships. These derive not only from the differing organization forms but also differences between R&D and Marketing. They have differing subcultures expressed in terms of breadth of vision and time horizons and they also have differing rates of staff turnover which may preclude long standing relationships. Attention to detail is high in R&D and lower in Marketing. R&D personnel often enter via the PhD route whereas Marketing people come in from sales. Scientists are tolerant of uncertainty, whereas marketing people are tolerant of catastrophe. Marketing will have their tails up for a big launch for which they have a current but not long standing commitment. If it fails they are able to forget and remain enthusiastic for tomorrow's launch, i.e., they have the capacity to forget. With R&D the opposite is the case: they have had a long and detailed involvement in the development and have to build up a good memory.

PROCEDURE

Against this complex background the study (Phillips et al., in press) set out to identify the extent to which the characteristics of the involvement of R&D and Marketing and its interaction with top management affected the eventual outcome, the rate of progress, and hence the rate of return. It did this by looking at six major and long-term development projects in three companies in the pharmaceutical industry. Each industry has its own particular characteristics which will, no doubt, moderate the findings. For the pharmaceutical industry, discovery and development of a new product involves a substantial element of chance with the screening of many new chemical entities for both efficacy and side effects. The market is special in being divided between products sold directly to the consumer and those where the product is prescribed by the medical profession. In addition there are formidable regulatory procedures which govern the use of new and existing drugs. The model of the new product and development process described by Cooper and Little (1974) was modified for use in this pharmaceutical study and is shown in Figure 1.

Marketing and technical tasks are independently identified with the coincidental management decisions interfaced between them. Five distinct stages (A to E) can be distinguished and used to assess the effectiveness of the R&D/Marketing interface at particular intervals in the development process. For a new drug it is not possible to limit the tasks to those identified by Cooper and Little. Idea generation and the screening of new chemical entities (NCE's) does not involve Marketing to a significant degree. Only when a technical assessment has been made on the NCE will it be assessed jointly by the two functions. When a likely candidate for development has emerged the subsequent stages of development must follow guidelines set by regulatory requirements. The technical and market assessments provide the preliminary business assessment. At this stage marketing should be focusing on identifying the target market, determining the nature of the product needs in the target market and estimating the market potential for the product (Cooper and Little, 1974). If development is to be continued, the next phase is the preclinical work required for a clinical trial certificate (CTC) or clinical trial exemption (CTX). In adapting the Cooper model to new drug development there is a need to make way for preclinical work. At about this time marketing research needs to define the market information more closely with a more rigorous approach to the acquisition of information. Once the CTC or CTX has been obtained clinical trials are undertaken in accordance with the product license application envisaged. The Cooper model proceeds via "prototype endurance" but for drug development this designation is not included since clinical trials undertake a comparable role.

Six case studies involving the development of a new product from three companies were documented and information presented within the framework described. The information concerning the development of each new product, the manner in which decisions were taken, and the extent of communication within the project group and company were collected. The main source of information was company documents supplemented by interview. Structured questionnaires were not used since circumstances varied considerably. Involvement of individuals in the different stages of product development varied as the project progressed. The interviews sought to quantify this involvement to produce an accurate timetable of the different inputs during the product development period.

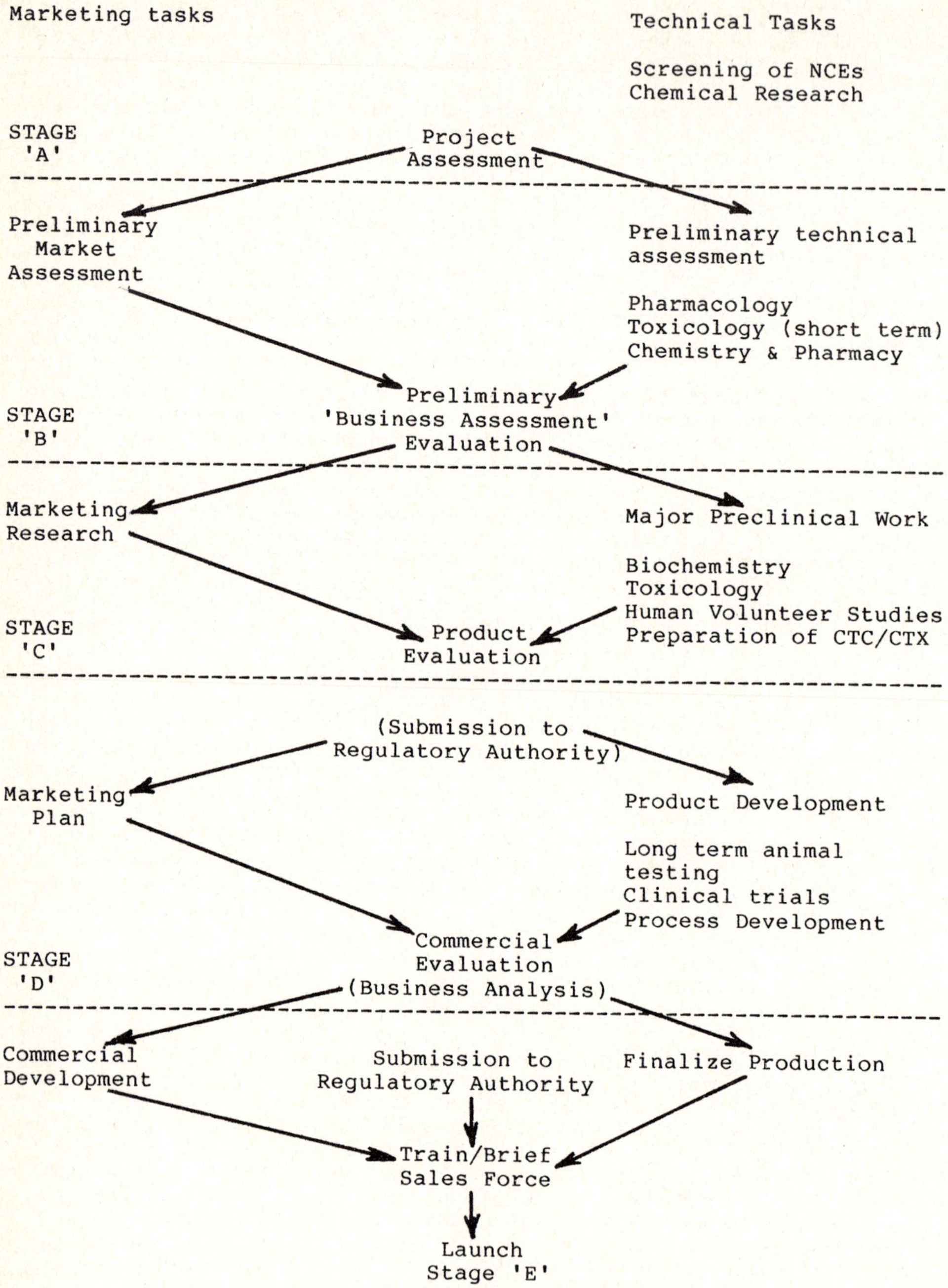

Figure 1
Model of New Product Development

For each case study the inputs by R&D and Marketing were recorded and an evaluation of the interface undertaken at the main points of the development. Each stage was assessed independently and the interaction between R&D and Marketing noted.

ANALYSIS

The detailed information collected for each project was summarized in terms of the model. A representation of the degree of interaction through the use of a time-based analysis is illustrated in Figure 2. Hatched areas indicate where a significant activity is taking place. This shows in one specific case the inputs from both Marketing and the technical functions (which include R&D), and management evaluation as a function of time.

Four types of R&D/Marketing interface problems have been specifically emphasized by Souder (1978). These were modified slightly for this study given the nature of the projects and the time scales involved as follows:

a. Influences of STATUS: the extent to which relative position in the organization hierarchy or management structure was perceived to be satisfactory by each of the disciplines

b. Effectiveness of COMMUNICATION

c. APPRECIATION of each other's issues and

d. TRUST, mutual and interactive, in this case between R&D and Marketing

These are the characteristics which would certainly appear to present a summary of the areas of difficulty as perceived by R&D and Marketing in this study. Accordingly an attempt was made to use them as guidelines to assess and to compare the individual cases. This was done by the use of a five point scale ranging from 1 for excellent to 5 for inadequate.

DISCUSSION

The work described confirmed the potential value of examining the R&D/Marketing interface using the semistructured approach. Despite all the cases being associated with pharmaceutical and related product areas, the companies and projects differed considerably. Nevertheless there are certain basic problems which were identified and which exerted an influence on the overall outcome. In the pharmaceutical industry, it is essential to recognize the dominant role exercised by the technical activities, particularly in defining the nature of the products based on new chemical entities and in satisfying the demands of the regulatory bodies. Those operating at the R&D/Marketing interface must have regard to this position, but not allow Marketing to assume a subservient role when technical considerations emerge. Marketing tasks, in those areas where R&D has the major role, need to be precisely specified in order to establish clearly their relevance and importance within the overall new

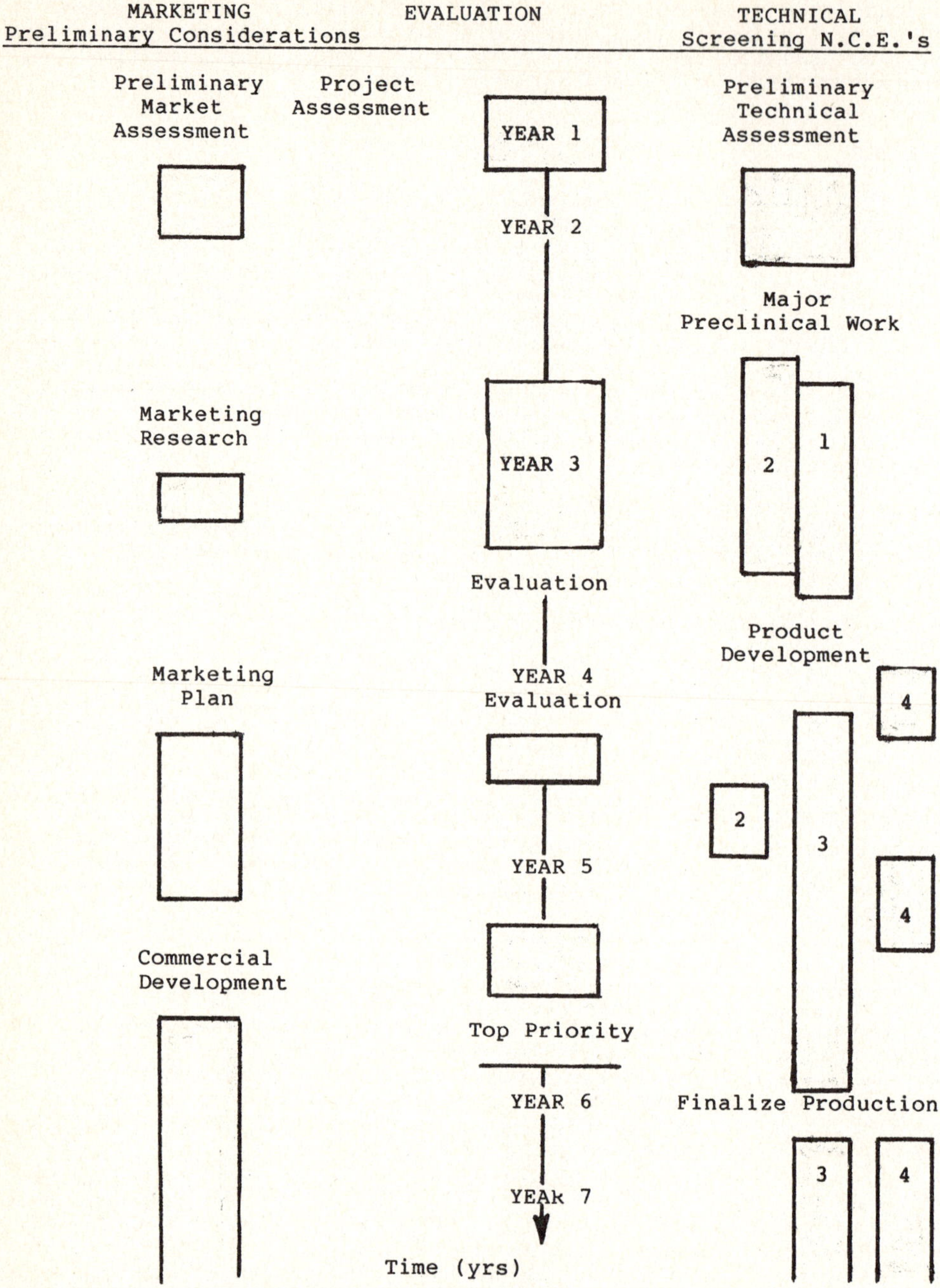

Key to Technical Activities: 1. Biochemistry; 2. Toxicology; 3. Process Development; 4. Clinical Trials

Figure 2
Correlation of Marketing and Technical Tasks

development process. Whereas certain tasks are capable of independent action by both groups, it is this sensitive, yet vital point of interaction which is an important focus. Certain problems can be observed which are common to all the projects, despite their inherent difference.

Market Research. A basic tool of marketing is the techniques which have been developed for assessing the viability of new areas of involvement for the company. It is an activity which needs to be undertaken at a very early stage in the life of a new project. Lack of agreement of conflict at this stage can influence the interface throughout the project. In several of the projects studied there was a definite lack of accord about the potential for the product. The scope for disagreement is most when there is a paucity of information about the disease in question and the current treatments are deficient or nonexistent. Marketing interpreted the absence of information as evidence of lack of need, whereas R&D regarded such a position as signifying dissatisfaction with current therapy and thereby offering an opportunity to fulfil that need. These approaches must be reconciled if a policy credible to both parties is to be pursued. There were indications that R&D was ready to criticize Marketing for this reluctance to explore new areas. Marketing needs to recognize that conventional market research is inadequate in some instances. In one project, initial market research provided little information about the potential market whereas comparison with related areas and products indicated what possibilities might exist.

Technical research and development. The technical R&D activity is fraught with setbacks and difficulties, so much so that these are accepted as almost inevitable. There is belief that such problems can eventually be overcome. Marketing needs to recognize that similar difficulties will inevitably arise in its activities and as such require the same type of persistence and determination in identifying potential market segments. This approach can provide a common bond, as opposed to becoming the root of disagreement as occurred in some of the projects in this study.

Clinical trials. Joint involvement of R&D and Marketing can greatly assist the planning and conduct of clinical trials, despite the fact that at first sight these may be regarded as primarily a technical activity. When well conducted, such trials do a great deal more than demonstrate the safety and efficacy of a drug. A properly planned and executed Marketing input can help increase their usefulness. The study described identified a definite and direct relationship between the degree of Marketing involvement in clinical trials and the project's success. In one of the projects complete responsibility for planning and structuring the clinical trials rested with R&D. The eventual launch was comparatively unsuccessful to the point that a new strategy had to be drawn up and further trials organized in support. Marketing provided suggestions to be included in a further clinical trial. Little notice was taken by R&D. As a result when the product was finally cleared by the regulatory bodies, authorization did not cover the claims Marketing wished to make. The project subsequently suffered in terms of sales and public acceptance. Due to the restricted claims which can be made for a new drug it is vital that R&D and Marketing fully coordinate their input in systematically planning clinical trials--otherwise resources are wasted.

Basic attitudes. Marketing and R&D within some companies, regrettably, often have a poor opinion of each other. Such opinions stem to a large extent from their different backgrounds and training. Inevitably, it dictates the manner in which these groups deal with each other. In one project the project leader was satisfied with the input from Marketing simply because he expected nothing from them. He did not recognize that they had skills and expertise which would contribute to the success of his project. When one group is heavily loaded with staff with first and post graduate degrees while the other is mainly experience trained such an attitude might be expected and needs to be maintained throughout the whole project with the appropriate weight given to recommendations from whatever source they derive.

Organizational structure. In a large organization the structure will be a dominant influence in establishing the R&D/Marketing interface. If the Marketing group operates as an independent unit from the product development team, inevitably lack of communication results. Lack of daily routine contact sets up misconceptions and reduces cooperation. Within one company Marketing had a supportive specialist role with R&D being the dominant influence. As a result Marketing had to strive continually to establish its credibility. The Marketing specialist was invited to attend all project meetings. However, being in the minority, the effectiveness of the Marketing input was dependent entirely on the personality and attitudes of the individual representative. In another company there were no clear formal guidelines set out for new product development. Thus individuals could make decisions in isolation from an organizational framework. For the two projects investigated in this particular company, the eventual progress was dependent on such an individual input, contributing in one instance to success and in the other to its comparative failure. There is clearly a need for a structure within a company which recognizes and controls potential negative effects while capitalizing on the positive ones. Such a structure needs to be sufficiently flexible to allow operational groups to change in response to the needs of the project.

Management. The structure of management has a prime role in monitoring and controlling the entire process of new product development. Once a collaborative structure has been set up and is operating to a fixed timetable, management can and should provide evaluation of the process as well as monitoring scientific and technical progress. One project demonstrated how excellent communication ensured that all views were taken account of at each stage. Another project, in contrast, continued without any apparent regular control by management of the technical and commercial efforts. Consequently the resources were not well utilized, with tests having to be repeated, clinical trials reorganized, in order to obtain more information from them. Such control of overall resources is a major function of management, which in turn will depend on synchronization of the technical and commercial requirements.

Management philosophy and attitudes. These eventually reflect upon all the activities within the company. Certain observations from the case studies illustrate this. For example, in a company where the approach was market oriented, R&D's awareness and understanding of commercial issues also increased. Moreover when senior management in another company was dissatisfied with Marketing's contribution, this was echoed by the R&D staff. It is vital, therefore, that senior management itself is clear about its expectations and policy towards the R&D/Marketing interface. A clear definition of their separate roles and interrelationships is necessary. When the importance of one or the other is emphasized by management, dissatisfaction inevitably results, with corresponding loss of effectiveness. For development to proceed smoothly and creatively, agreement and harmony throughout the organization is essential.

Balancing the input. Even in the most technically oriented areas of new drug development, R&D and Marketing have important roles to play and a balanced input from both disciplines is more likely to lead to success. When this does not occur, problems follow.

One project was unbalanced in terms of input: R&D and Marketing went in different directions. R&D dominated and took the product to the point of marketing. Another project similarly went on to its later development stages with no discernible marketing input. In each instance it was shown that the projects suffered. A third project on the other hand was an excellent balanced partnership, yielding a product which fulfilled a particular market segment need with excellent technical characteristics.

CONCLUSION

This initial study of the interface between R&D and Marketing in the pharmaceutical industry confirmed that benefit can be derived from paying more attention to management at the interface not at the expense of that within the individual functional areas but as a high level and complementary activity. Satisfactory interfacial performance is likely to lead to a speedier outcome. If the outcome is successful, then earlier completion increases the profitability; and if the outcome is unsuccessful, then a speedier outcome reduces the losses.

In one project which was terminated the quality of the interface was assessed as above average at all stages but the project failed. This could be attributed to the technical or marketing inputs failing to achieve expectations. It demonstrates that even if interaction at the interface is adequate, the inputs from R&D and Marketing must meet the target goals. This leads to the question of how top management recognizes the quality of individual inputs to each project. Since an essential characteristic of R&D is uncertainty, the judgment by top management of individual projects is a poor guide to overall capability. It is top management's job to monitor functions and not their input to individual projects.

In the cases which have been examined in this and much previous work the cases were analyzed near to or after completion and in isolation from one another. It is proposed to extend this work by taking a more dynamic approach in which projects will be analyzed at different stages of development within the same company to determine the extent to which relationships at the R&D/Marketing interface in one project interact with others. In addition to this, account needs to be taken of the size and timing of workloads within both R&D and Marketing. Under conditions where either function has a very high workload it is the job of top management to monitor the choice of priorities since those selected by the functions themselves may not represent the best interests of the organization.

Although not included in this study the interviews which formed part of the case studies did draw attention to the importance of other interfaces such as those with process control. It may well be that top management needs to pay as much attention to those interfaces as to the one focused on here.

REFERENCES

Cooper, R.G. and Little, B. 1974. Reducing the risk of industrial new product development. Canadian Marketer Fall:7-12.

Phillips, E.A., Pearson, A.W. and Ball, D.F. In press. The R&D/ Marketing interface in the pharmaceutical industry.

Souder, W.E. 1978. Effectiveness of product development methods, Industrial Marketing Management 7(5):299-307.

Souder, W.E. and Chakrabarti, A.C. 1978. The R&D/Marketing interface: results from an empirical study of innovation projects. IEEE Transactions on Engineering Management EM25(4):88-93.

MANAGING HIGH TECHNOLOGY
An Interdisciplinary Perspective
B.W. Mar, W.T. Newell, and B.O. Saxberg (Editors)

JOINT VENTURES

F.P. Vince

Control Data Corporation
8100 54th Ave. S./ HQWIZY
Minneapolis, Minnesota 55400

Good afternoon. I've been asked to talk this afternoon about Control Data's experiences in joint venture activities. Let me start by first providing a little background information on Control Data for those of you who are not familiar with the company.

Control Data was founded by Bill Norris and a number of others in 1958 for the specific purpose of developing and marketing large scale scientific computers. Over the past 26 years that objective has been revised and expanded somewhat, so that major business areas now include computer systems, computer peripherals, computer services, and financial services. Control Data is today one of the largest companies in the data processing industry, with 1983 revenues of over $4.5 billion. It ranks in the top 100 in the Fortune magazine list of the largest industrial companies in the United States, and it employs over 55,000 people in 47 countries.

In order to remain competitive in this fast moving, high technology industry, Control Data spends hundreds of millions of dollars each year on research and development projects. These projects range from small relatively simple ones to very large, complex programs. To give you an idea of the complexity of some of these larger programs, let me cite a couple of examples.

Several months ago Control Data announced a new family of computers--the CYBER 180 series. Included were six compatible computer mainframes, a new operating system, and various systems and applications software packages. Development of this product line cost well over $500 million and took a number of years. It involved thousands of people in over a dozen divisions located in various areas of the United States and Canada. The number of technical disciplines involved was extensive. Included were specialists in the areas of system architecture, logic design, circuit design, power system design, cooling, packaging, reliability, safety, maintenance software, software design, systems programming, applications programming, system test, technical writing, and a number of others. Managing a program of this complexity is obviously a challenge. Because Control Data's divisions are typically organized along functional lines, a form of matrix management was used for this project as well as for most other large projects.

Another example--this time a single product from our Peripheral Products area--is a large high performance disk subsystem that recently was placed in volume production. Development and manufacturing startup costs were in the neighborhood of $100 million, and hundreds of people in various organizations (including several joint venture organizations) were involved. The list of technical specialties involved was again extensive, and included chemical engineering, mechanical engineering, logic design, analog circuit design, power system design, cooling design, maintenance software, systems software, and many others. I think that these examples amply illustrate that Control Data has substantial experience in managing complex research and development projects involving a multitude of technical disciplines.

Now let's talk about cooperative or joint venture activities in a high technology industry--and more specifically, Control Data's experience in this area. Control Data has long viewed cooperation and joint venture activities as a sensible and responsible approach to many types of research and development and business situations. The rationale for pursuing a cooperative approach involves many interrelated factors--let me briefly mention several of them.

First of all, every organization (whether in business, government, or education) sooner of later faces a resource constraint. In a high tech industry, these constraints are typically availability of skilled technical personnel and availability of funds for the latest (most advanced) capital equipment and facilities. It simply makes good sense to share these resources and avoid duplication of effort whenever possible. By so doing, we make the best use of available resources, and (possibly) set resources free to pursue other high priority projects.

A second good reason for supporting cooperative efforts is risk sharing. High technology research and development projects almost by definition include a significant risk factor. In many cases a single organization cannot afford to pursue a high risk project simply because the consequences of failure are too costly or too severe. However, if a number of organizations cooperate on such a project, the risk to each is reduced to a more reasonable level and the consequences of failure are much more acceptable.

A third reason for mutual cooperation is to achieve some kind of critical mass in order to increase productivity. A small scale operation is, in many instances, ineffective and/or inefficient, simply because equipment is under utilized and key technical skills are either under utilized or not available within the organization. Combining several small operations into one larger operation can in such cases result in significant economies of scale and improved efficiency.

Yet another reason for mutual cooperation is that it creates the possibility of pursuing projects that are beyond the ability of any one organization to pursue. In many instances, the potential benefits of these large projects are enormous, but no single organization has the resource to tackle the project. Therefore, the only way such projects can be pursued is via cooperative ventures or government support.

I have pointed out just a few of the potential benefits of cooperative ventures, and there are a number of others that I have not discussed because of time limitations. To be fair, however, I must admit that there are a number of obstacles that frequently must be overcome in order to achieve these benefits. The obstacles include:

Fear of unknowingly giving a competitor some kind of advantage

Fear of antitrust action (at least in the United States)

The "not invented here" (NIH) syndrome

Added complexity of projects due to different goals and cultures of the cooperating parties

The effort involved in transferring the output (technology or product) of the joint venture back to the partners

Despite these obstacles, I can assure you that the benefits of cooperative ventures are almost always worth the effort involved, provided that the partners approach the venture with the appropriate attitude and that the venture is properly planned. Based on our experience with hundreds of cooperative ventures, let me mention a few factors that are key to the success of any joint venture.

First, the venture partners must have an appropriate attitude toward, and motivation for, participating in a cooperative venture. If a prospective partner views cooperation as a sign or weakness, as a risk to his independence, or, worse yet, as an opportunity to exploit other partners, the venture is bound to encounter difficulties and will probably fail. The proper attitude for a cooperative venture is, naturally, one of cooperation for the mutual benefits of all the partners.

Second, a cooperative venture should have well documented and well understood objectives and goals that are agreeable to all the partners. Furthermore, accomplishing these objectives and goals should satisfy some real need of each partner in the venture. In other words, all concerned parties must have their own selfish motivation to cooperate.

Third, there must be mechanisms established as part of the cooperative venture arrangement that provide for (a) review of and agreement on operational plans, (b) implementing changes in objectives, goals, and program plans as necessary, and (c) resolving conflicts that may arise among the partners.

Fourth, for any cooperative venture to be successful, there must be an equitable sharing of benefits derived from the venture. Obviously, a venture in which benefits are distributed in a one sided or inequitable manner cannot survive.

And finally, for ventures involving technology development, another key factor is the ability to transfer the resulting technology back to the parent organization, while at the same time preventing dissemination of the technology to outside organizations (not participating in the venture). Such technology transfer is often difficult and requires considerable planning and effort by both the joint venture organization and the parent organizations.

Having covered all that, just what has Control Data done in the area of cooperative ventures? Well let's start by talking about several of our oldest and most successful joint ventures. I indicated earlier that Control Data started out developing high performance scientific computers. Within a few years the company also started to develop some of the peripherals (disks, tapes, and so on) needed to support these computers. Although these peripheral products were well designed and compared very favorably with similar products available from other sources, manufacturing volumes needed to support Control Data computers were relatively low. As a result, development costs had to be amortized over a relatively small number of units, and manufacturing costs were relatively high.

To resolve these problems, Control Data approached a number of other computer manufacturers and suggested a cooperative venture in the peripherals area. Two joint venture companies resulted; Magnetic Peripherals Incorporated (MPI) was formed to produce mass storage equipment, and Computer Peripherals Incorporated (CPI) was formed to produce magnetic tape, card, and printer equipment. These joint ventures have proven to be extremely successful, and today are the basis of a Peripheral Products organization that generates yearly revenues of about $1.5 billion. It offers one of the industry's broadest lines of peripheral equipment, and sells peripheral products to almost every major computer manufacturer in the world.

Semiconductor Research Corporation (SRC) is a much more recent example of a cooperative venture that Control Data supported and participated in establishing. Founded in 1982 and headquartered in Research Triangle Park, North Carolina, SRC has 28 members and a 1984 budget of about $13 million. Membership includes companies that produce or use semiconductor devices.

SRC was form to plan, promote, coordinate, conduct, and sponsor basic research in microelectronics. Specifically, SRC's goals are to produce (1) new knowledge of semiconductor materials and phenomena, (2) new and more efficient design and manufacturing technologies for semiconductor devices, and (3) and increase in the supply of trained scientists and engineers for the semiconductor industry.

To achieve these goals, SRC awards research contracts to universities and individual academic researchers. However, the work of SRC does not end with research. Results must be communicated and new technologies must be transferred to SRC members. In addition to the time honored academic method of publishing papers, technology transfers are being facilitated by industry/academic visits, workshops, and conferences.

A third cooperative venture that Control Data was instrumental in forming is the Microelectronics and Information Sciences Center (MEIS) at the University of Minnesota's Institute of Technology. MEIS was started in 1980 with funding of $6 million from Control Data, Honeywell, 3M and Sperry. MEIS was established to promote mutually beneficial collaboration between the Institute of Technology and Minnesota industry and to advance microelectronic and information sciences. The primary means of accomplishing these goals are (1) sponsoring and conducting advanced research, (2) improving course offerings in these areas and thus the quality of graduates available to industry, and (3) facilitating technology transfer between researchers and product developers.

Another very interesting venture is Optical Storage International, a joint venture of Control Data and N.V. Phillips, of The Netherlands. This joint venture was established to develop a high performance optical storage system that uses a laser beam to read and write data. Such a system is capable of storing a huge amount of data in a very small space. This venture has been successful, and an optical disk product was recently announced; the product is now being marketed to various computer manufacturers for incorporation in their systems.

The last cooperative venture that I want to discuss today is the Microelectronics and Computer Technology Corporation (MCC). Several years ago Bill Norris persuaded 19 computer and electronics companies to meet in Orlando, Florida, to discuss the possibility of a major cooperative effort to counter the fifth generation development going on in Japan. In early 1983, a dozen companies agreed to proceed with such a venture, and MCC was formed. The basic premise of MCC is to do those things that the shareholder companies cannot do individually because of high cost, high risk, or skills shortages. Members have agreed to pursue technology programs in seven areas: software technology, packaging/interconnection, CAD technology, artificial intelligence, data base management, human factors, and parallel processing.

MCC is now operating in Austin, Texas, and currently has a staff of well over a hundred people. It is attracting top notch, experienced, and highly educated personnel to pursue research in the seven areas I mentioned previously. Spend rate at the end of 1984 will be about $40 million, and that is expected to rise to approximately $80 million annually in the future. A total of 16 companies are now members and more are expected to join.

The five cooperative ventures that I have described in the past few minutes are by no means the only ones in which Control Data is involved. Control Data's corporate philosophy supports cooperative ventures as a normal way of doing business, and the company is currently involved in over 100 cooperative ventures of one kind or another. Of course, many of these are much smaller ventures than those I described to you.

And now, let me conclude by quickly summarizing a few key points. First, there are significant benefits to be gained by cooperation between business, government, and educational organizations. Second, for a cooperative venture to be successful, participants must be appropriately motivated, and the venture must be carefully planned and organized. And finally, Control Data's experience with cooperative ventures indicates that the benefits of such ventures are well worth the effort required to overcome the obstacles that sometimes arise.

MANAGING HIGH TECHNOLOGY
An Interdisciplinary Perspective
B.W. Mar, W.T. Newell, and B.O. Saxberg (Editors)

SPACECRAFT DESIGN AND DEVELOPMENT AT TRW - MATRIX MANAGEMENT APPLIED

E.I. Reeves

TRW Space & Technology Group
One Space Park
Redondo Beach, CA 90278

ABSTRACT

TRW is a principal supplier of unmanned spacecraft and spacecraft components. These spacecraft have ranged from the lunar probes launched in the late 50's, which weighed less than 100 lbs, to the NASA Tracking and Data Relay Spacecraft (TDRS) launched in 1983 which weighs over 2-1/2 tons. The engineering and construction of these spacecraft involves applying the talent and capability of literally hundreds of people on a project team. These people span most of the engineering disciplines, management functions, and applied sciences. Since the end product is extremely complex and very valuable, the people must apply state-of-the-art techniques and hardware to maximize the effectivity of the end product. Furthermore, the spacecraft development business is highly competitive thus putting a premium on efficient resource utilization. To meet these challenges, most of the aerospace companies and TRW in particular, are organized in matrix form. By grouping people along separate functional lines, responsibility for maintenance of state-of-the-art technical skill can be effected and a cadre of trained and skilled people maintained. In juxtaposition to the functional organization, i.e. organization by technical discipline, TRW also has an organizational arm responsible for project conduct. These project offices draw their personnel from the functional organizations and meld the resources necessary for conduct of their projects. TRW has developed a number of techniques for building and managing such project teams. Particularly important are methods for facilitating communication and resolving problems. In the final analysis, the process seems to work quite well. TRW has launched 64 spacecraft over the last 22 years and accumulated over 350 years of operation in space. Equipment malfunctions averages less than one per spacecraft per year and catastrophic design error is essentially nonexistent.

INTRODUCTION

The design and development of a modern spacecraft requires the efforts of literally hundreds of people over an extended period of time. These people represent a variety of technical disciplines including most of the engineering fields, a number of areas of pure and applied science, a variety of management fields and various industrial capabilities. This paper describes the way in which TRW Space & Technology Group applies the principles of matrix management to meld together the various disciplines required to conduct spacecraft development programs.

This paper first presents an idea of the size, complexity, and capability of typical spacecraft. Next is a description of the spacecraft development program. In particular, the various phases of the program are enumerated and the tasks and manpower phasing are described. The TRW Space and Technology Group Organization is then presented and implementation of the matrix organization is described. Since a number of the management techniques used are peculiar to the industry and in some cases to TRW in particular, these formalized techniques of project management are described in moderate detail. Organizational development and team building are next discussed and finally some comments are presented on current trends and projections for the industry.

TYPICAL SPACECRAFT

Pioneer 10 was designed to measure the radiation belts around the planet Jupiter. It was launched in the spring of 1973 and accomplished its mission of investigating the Jovian radiation belts. It was one of the first spacecraft to cross the asteroid belt, and the spacecraft made further history about a year ago when it became the first manmade object to pass beyond the orbit of the outermost planet. In size the spacecraft weighed only about 550 pounds; roughly half as much as a Volkswagen Bug. Power was supplied by a radioisotope power supply located on outriggers. The spacecraft is about 9 feet in diameter and spin stablized to point the axis of its parabolic antenna toward the earth.

The TDRS spacecraft was designed to relay communications and data signals from low altitude earth orbiting spacecraft to a ground station at White Sands. The spacecraft system is designed to replace the extensive ground based network currently used for communicating with low altitude spacecraft. This includes the shuttle as well as a variety of unmanned spacecraft. The spacecraft is located in geosynchrous orbit and from this vantage point, some 19,000 nautical miles above the earth's equator, it can provide communication coverage over a large portion of each orbit of a low altitude orbiter. Thus a single TDRS can provide communication capability equivalent to a large system of earth stations. In size the spacecraft is about 60 feet long. The large parabolic dishes are 16 feet in diameter and the central equipment compartment is about 8 feet in diameter. This spacecraft weighs about 2 tons or roughly the equivalent to a long wheel based Dodge Van. The spacecraft also carries about 1000 pounds of propellant which is the equivalent of loading the Van with Boy Scouts and camping equipment. In complexity the TDRS spacecraft has about 60,000 electronic parts and produces about 2 kilowatts of electrical power.

In the design of a spacecraft, such as TDRS, weight is an extremely important parameter. Cost tradeoffs of spacecraft weight are constantly made during the development program and typically tradeoff factors of between $10,000 and $100,000 per pound are employed.

That is, a man year or more of design effort can gainfully be expended to reduce weight by a single pound.

With regard to complexity, a spacecraft such as TDRS at 60,000 electronic parts is roughly equivalent to 200 television sets. That is, a typical television set has about 300 electronic parts. If one were to operate 200 television sets continuously; that is, run them seven days a week, 24 hours a day, how soon would one of the sets fail? Yet a spacecraft such as TDRS is expected to perform without equipment failure for a design life of 10 years. The bottom line, then, regarding our current spacecraft is that they are large, typically one to two tons, and very complex, typically 50,000 to 100,000 electronic parts. They are designed to operate without malfunction for extended periods of time. Lifetimes of 10 years are the current norm.

THE SPACECRAFT DEVELOPMENT PROGRAM

Figure 1 summarizes the phases of a typical spacecraft development program. The project population during the various phases of the program is shown at the bottom of this figure. During conceptual design the project population is small and restricted to people who are working with the spacecraft sponsor to define system requirements. This may go on for several years and results in the formulation of a program concept for which bids are solicited. After contract award, the conceptual design which has been developed by a small system design team needs to be expanded into allocations to subsystems, and the personnel responsible for engineering of these subsystems need to be assigned to the program. Typically the program builds from 50 to 1000 or more during this preliminary design phase of the program. Critical areas of component construction are also examined during this program phase. At the completion of the preliminary design phase when the project team is all in place, a formal review of the design is conducted. Providing that the system design is sound and that the allocations are all correct, the program proceeds into the final design phase. During final design the first preprototype components are completed. Once the design is complete, it is reviewed at a formal design review, generally termed the critical design review, and the drawings are released to manufacturing. As described in the next figure, the first manufactured units which are called prototype or qualification units are tested to environmental conditions which are worse than expected in end use. These tests demonstrate that the design is capable of performing in the expected environment. In the aerospace business, control of the engineering data, i.e. drawings and specifications, is a vital step in assuring the adequacy of the flight equipment. That is, equipment built for flight must be constructed to the same drawings and specifications which were used for construction of the qualification units. By carefully controlling the engineering data and making sure that flight equipment is built in exact conformity to these data, one is assured that the flight equipment is suitable for the expected environment. Once the qualification phase is complete, the flight spacecraft is constructed and launched. During these latter phases of the program, design personnel are phased off and the program population reduced to roughly 200 people. These people carry the program through the launch and orbital operations phases.

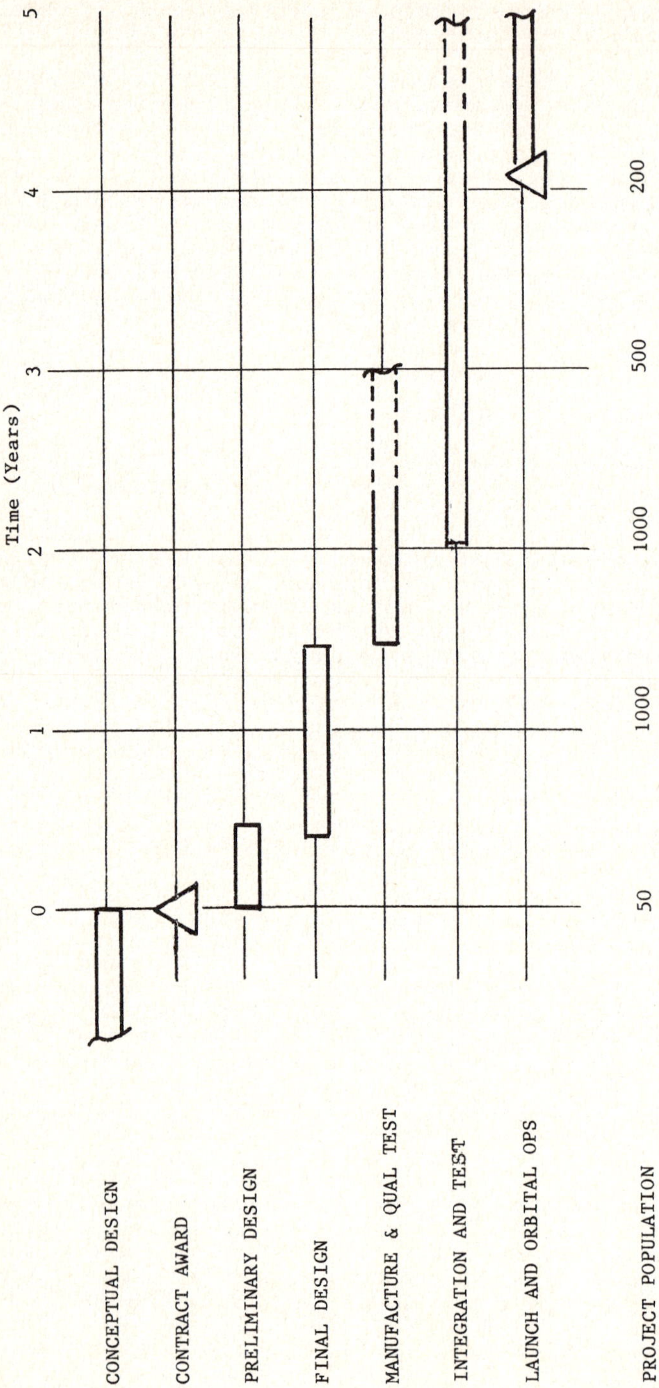

Figure 1
Spacecraft Engineering Phases

CONCEPTUAL DESIGN — TYPICALLY 2 YEARS

- Mission formulation
- Development of spacecraft concepts
- Determination of design drivers
- Preliminary performance allocation
- Typically conducted by a small team of systems engineers

PRELIMINARY DESIGN — TYPICALLY 6 MONTHS

- Performance allocation to subsystems
- Subsystem design to component requirement level
- Breadboard construction

Conducted by key engineering personnel:

Subsystem managers, lead engineers

- Engineering teams formed of key engineering personnel

Figure 2
Spacecraft Engineering Phases

FINAL DESIGN TYPICALLY 1 YEAR

- Detail design, drawing preparation, worst case analysis
- Construction and test of engineering models
- Preparation of orbital operation plans and procedures
- Conducted by full engineering team
- Support by manufacturing and test personnel

QUALIFICATION TYPICALLY 6 MONTHS

- Construction of production units
- Test to worse than expected environmental levels
- Supervised by lead engineers, conducted by manufacturing and test personnel

SPACECRAFT INTEGRATION AND TEST TYPICALLY 1 YEAR

- Assemble and test the flight article
- Training and rehearsal of orbital operations

conducted by test personnel support by design engineering

LAUNCH AND ORBITAL OPERATIONS

Figure 3
Spacecraft Engineering Phases

The program population figures presented here are typical. Some programs may be considerably smaller and some may employ considerably more. However, most programs are within a factor of two of these numbers. These population numbers are computed from the number of manhours expended on the program. That is, a person is 40 hours of work per week. In actual fact the number of people working on the program is considerably larger than this population would indicate. A number of people are employed on a part time basis because their particular talents are not required full time. Matrix managment techniques allow for this sort of personnel multiplexing.

ORGANIZATION ASPECTS

To discuss the organizational aspects of a spacecraft development program, I would like to use the organizational charts from the TRW Space and Technology Group, directly as they come from the organization book. These charts (Figures 4 and 5) show both arms of the matrix organization, namely the functional organization and the project organization. The first organization chart is an organization chart of the entire group. It shows the principal divisions which are six in number. On the left side of the chart are the three functional organizations: namely, the Systems Test Division, Applied Technology Divison and Spacecraft Engineering Division. On the right side of the chart are the three project organizations, namely, the Federal Systems Division, Defense Projects Division and the Military Space Systems Division. Carrying the organization a couple of steps lower, the next chart is the Spacecraft Engineering Division which consists of three Operations: Mechanical Engineering, Electrical Systems and Control and Data Handling. Continuing down the organization, the Mechanical Engineering Operation consists of two Laboratories: one concerned with mechanical design, and the other with engineering mechanics. The next chart (Figure 6) is the Mechanical Design Laboratory. This Laboratory is responsible for spacecraft mechanical design and is about 150 people in size. On the right is a list of staff personnel. Please note the project designation GRO. Art Goldberg is the subproject manager on a spacecraft program called the Gamma Ray Observatory (abbreviated GRO).

A similar trace can be conducted of the project organization from Division level to the GRO structures subproject. This project is in the Federal Systems Division which is shown on the next organization chart (Figure 7). The GRO project reports directly to the Division Manager. The next chart is a partial organization chart of the GRO Project which shows spacecraft design and development in the center. Under spacecraft design and development is the structures subproject and again the name Art Goldberg. From these organization charts, it is evident that the Group is organized by both project and function. The subproject represents the intersection of these two axes of the matrix.

In the TRW organization, Systems Engineering is considered a separate function which is a portion of the project organization. The people who become systems engineers are generally senior people of an analytic bent who have command of a number of technical disciplines. Furthermore, they are selected for their abiltiy to communicate and resolve requirements between the various portions of the spacecraft program. Systems Engineering has three different

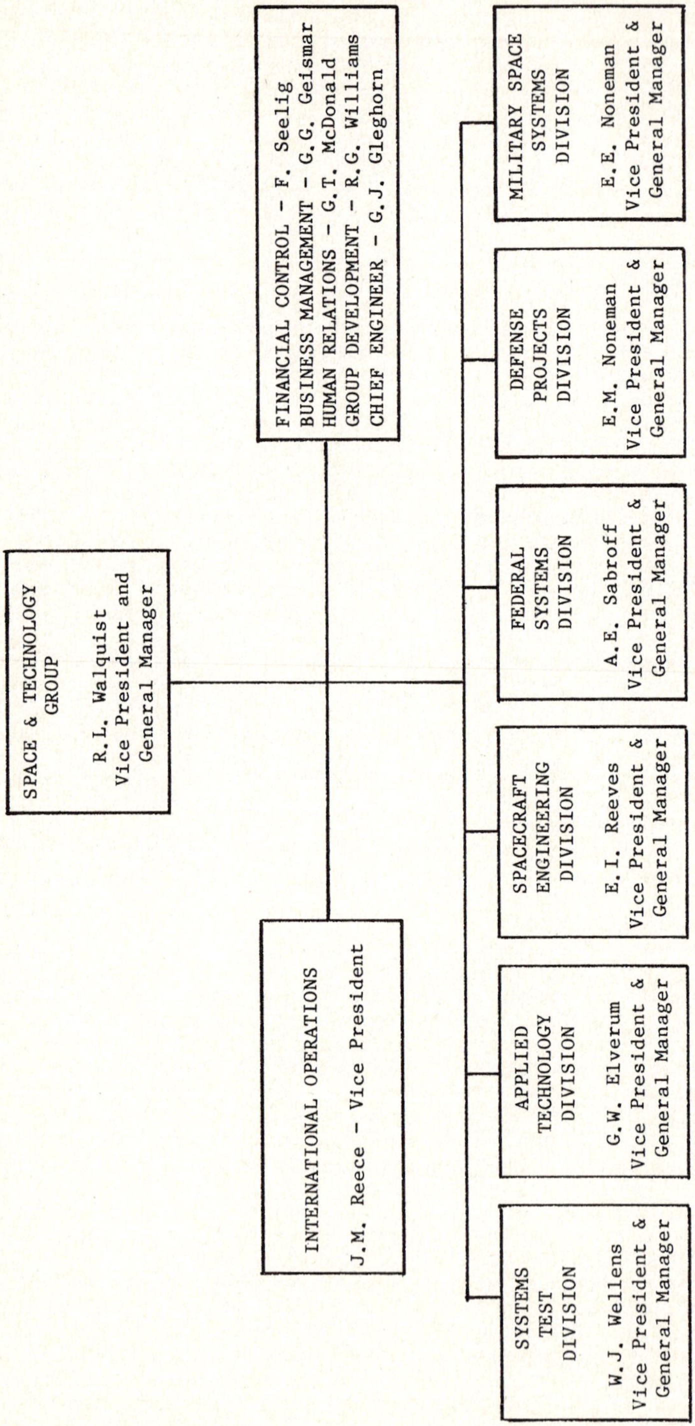
SPACE & TECHNOLOGY GROUP
R.L. Walquist
Vice President and General Manager
INTERNATIONAL OPERATIONS
J.M. Reece - Vice President
FINANCIAL CONTROL - F. Seelig
BUSINESS MANAGEMENT - G.G. Geismar
HUMAN RELATIONS - G.T. McDonald
GROUP DEVELOPMENT - R.G. Williams
CHIEF ENGINEER - G.J. Gleghorn
SYSTEMS TEST DIVISION
W.J. Wellens
Vice President & General Manager
APPLIED TECHNOLOGY DIVISION
G.W. Elverum
Vice President & General Manager
SPACECRAFT ENGINEERING DIVISION
E.I. Reeves
Vice President & General Manager
FEDERAL SYSTEMS DIVISION
A.E. Sabroff
Vice President & General Manager
DEFENSE PROJECTS DIVISION
E.M. Noneman
Vice President & General Manager
MILITARY SPACE SYSTEMS DIVISION
E.E. Noneman
Vice President & General Manager

Figure 4

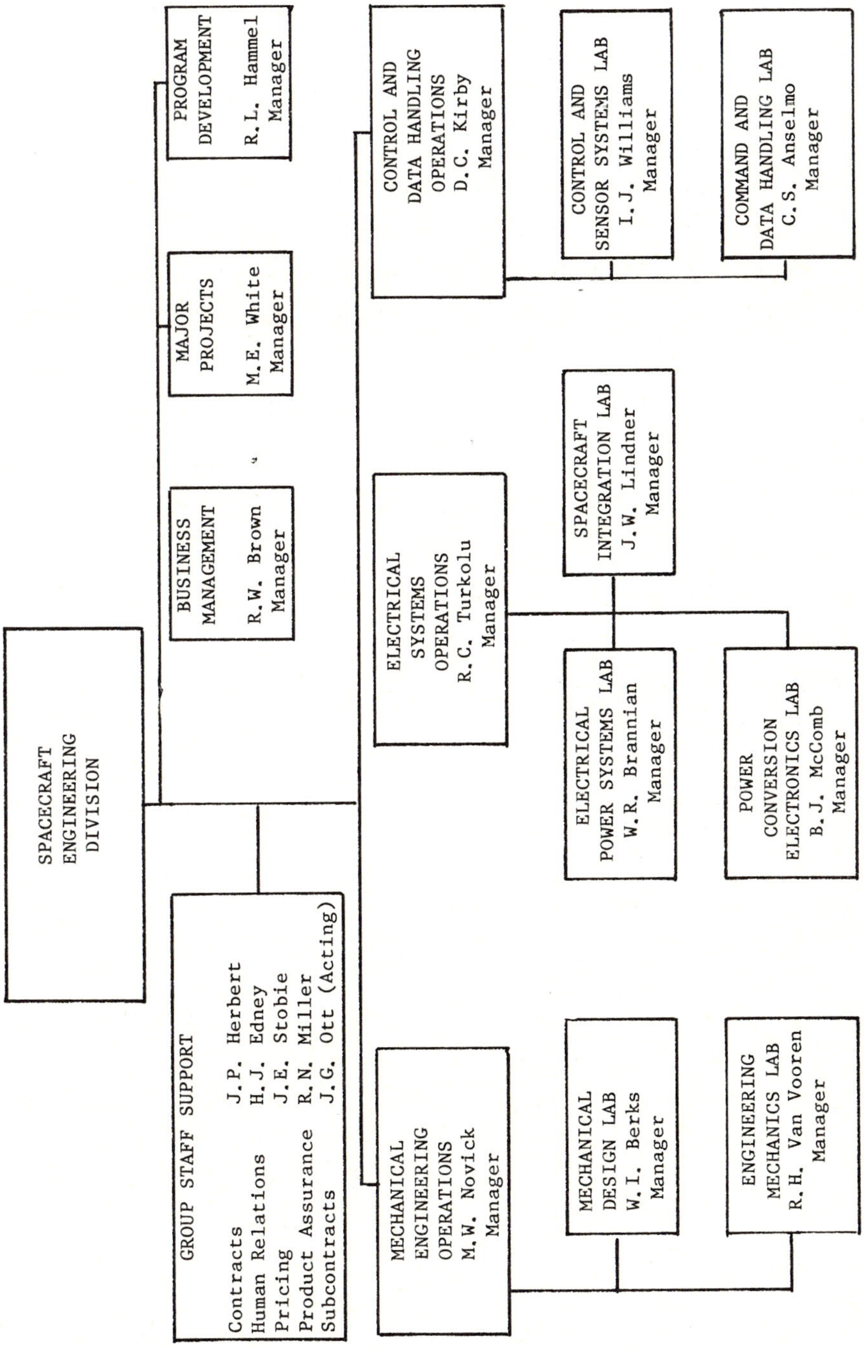
SPACECRAFT ENGINEERING DIVISION
GROUP STAFF SUPPORT
Contracts J.P. Herbert
Human Relations H.J. Edney
Pricing J.E. Stobie
Product Assurance R.N. Miller
Subcontracts J.G. Ott (Acting)
BUSINESS MANAGEMENT
R.W. Brown Manager
MAJOR PROJECTS
M.E. White Manager
PROGRAM DEVELOPMENT
R.L. Hammel Manager
MECHANICAL ENGINEERING OPERATIONS
M.W. Novick Manager
MECHANICAL DESIGN LAB
W.I. Berks Manager
ENGINEERING MECHANICS LAB
R.H. Van Vooren Manager
ELECTRICAL SYSTEMS OPERATIONS
R.C. Turkolu Manager
ELECTRICAL POWER SYSTEMS LAB
W.R. Brannian Manager
POWER CONVERSION ELECTRONICS LAB
B.J. McComb Manager
SPACECRAFT INTEGRATION LAB
J.W. Lindner Manager
CONTROL AND DATA HANDLING OPERATIONS
D.C. Kirby Manager
CONTROL AND SENSOR SYSTEMS LAB
I.J. Williams Manager
COMMAND AND DATA HANDLING LAB
C.S. Anselmo Manager

Figure 5

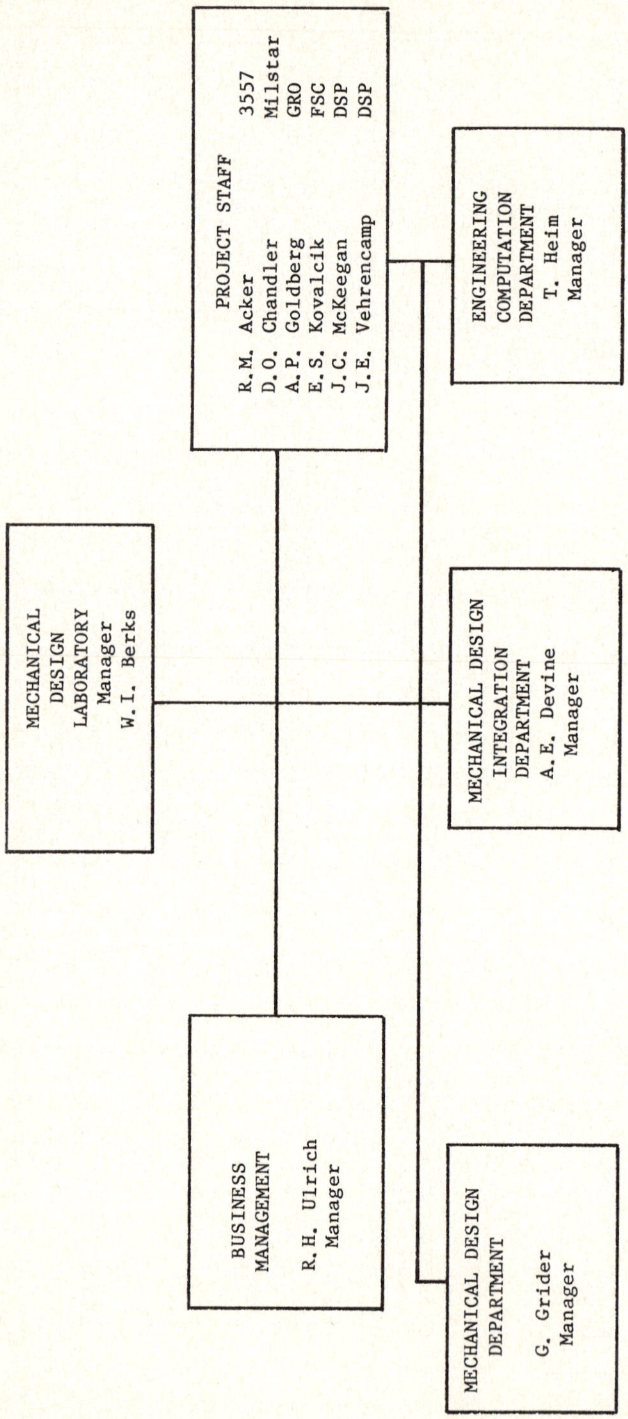
PROJECT STAFF
R.M. Acker 3557
D.O. Chandler Milstar
A.P. Goldberg GRO
E.S. Kovalcik FSC
J.C. McKeegan DSP
J.E. Vehrencamp DSP
ENGINEERING COMPUTATION DEPARTMENT
T. Heim
Manager
MECHANICAL DESIGN LABORATORY
Manager
W.I. Berks
MECHANICAL DESIGN INTEGRATION DEPARTMENT
A.E. Devine
Manager
BUSINESS MANAGEMENT
R.H. Ulrich
Manager
MECHANICAL DESIGN DEPARTMENT
G. Grider
Manager

Figure 6

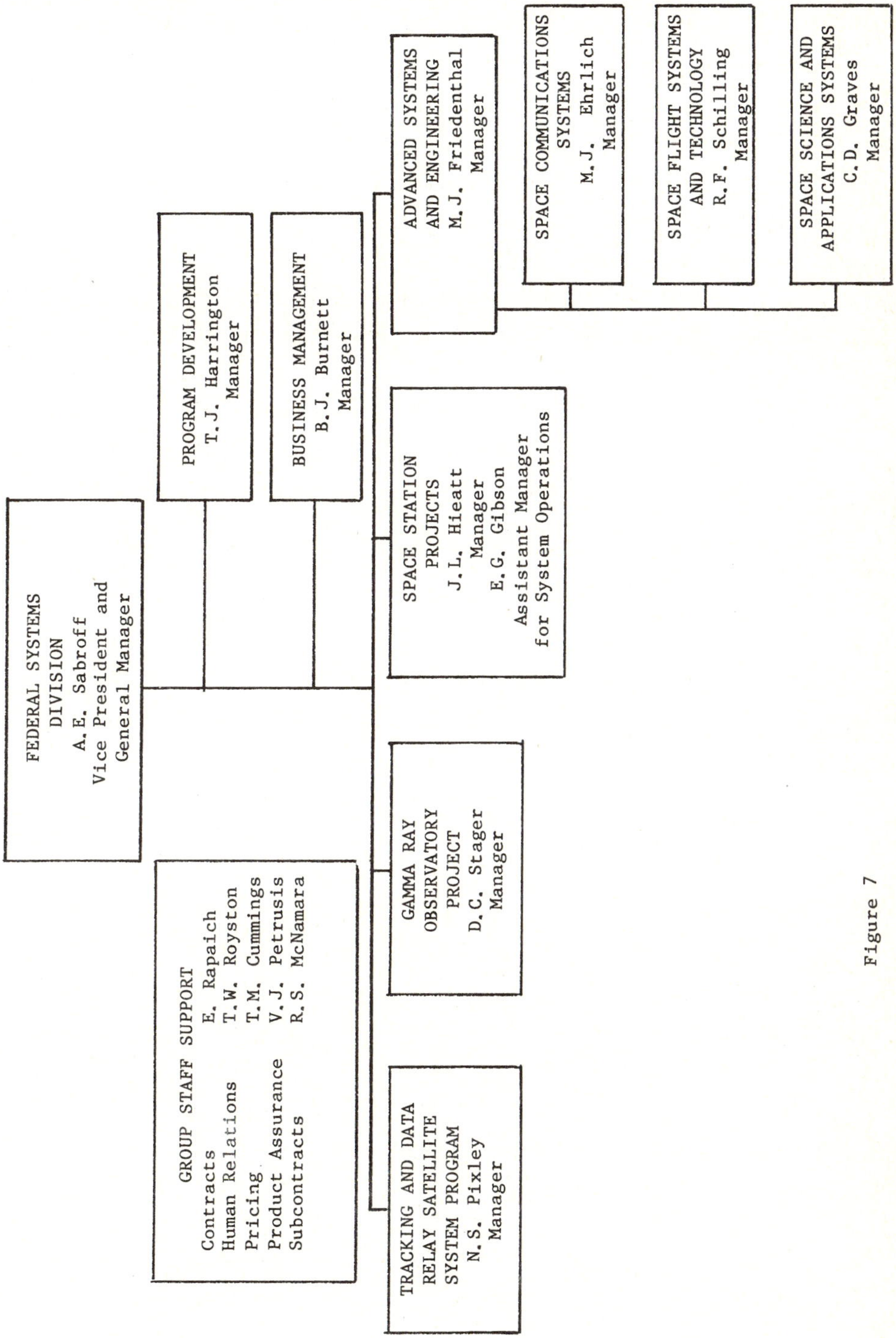
FEDERAL SYSTEMS DIVISION
A.E. Sabroff
Vice President and General Manager
PROGRAM DEVELOPMENT
T.J. Harrington
Manager
BUSINESS MANAGEMENT
B.J. Burnett
Manager
GROUP STAFF SUPPORT
Contracts E. Rapaich
Human Relations T.W. Royston
Pricing T.M. Cummings
Product Assurance V.J. Petrusis
Subcontracts R.S. McNamara
TRACKING AND DATA RELAY SATELLITE SYSTEM PROGRAM
N.S. Pixley
Manager
GAMMA RAY OBSERVATORY PROJECT
D.C. Stager
Manager
SPACE STATION PROJECTS
J.L. Hieatt
Manager
E.G. Gibson
Assistant Manager for System Operations
ADVANCED SYSTEMS AND ENGINEERING
M.J. Friedenthal
Manager
SPACE COMMUNICATIONS SYSTEMS
M.J. Ehrlich
Manager
SPACE FLIGHT SYSTEMS AND TECHNOLOGY
R.F. Schilling
Manager
SPACE SCIENCE AND APPLICATIONS SYSTEMS
C.D. Graves
Manager

Figure 7

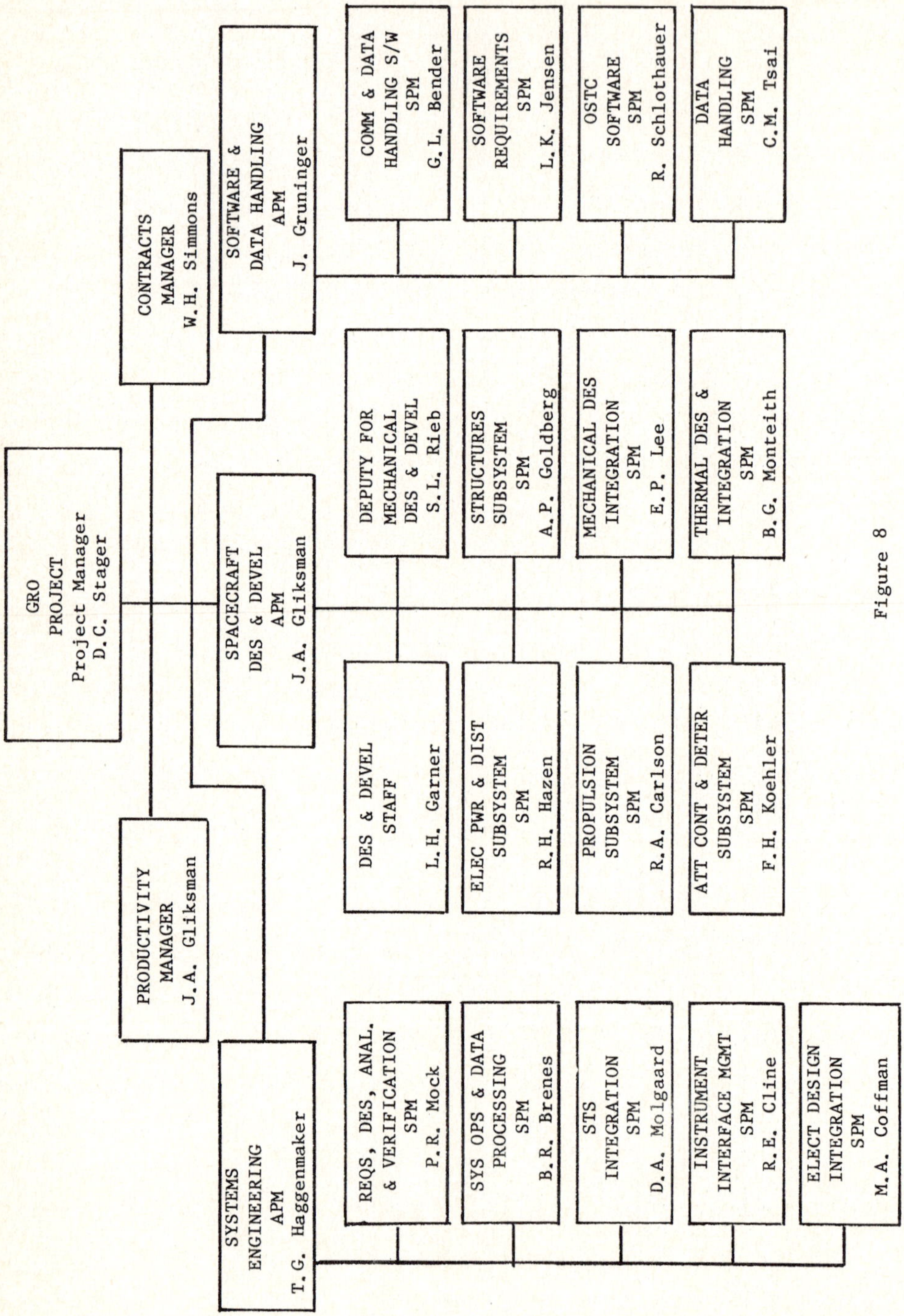
GRO
PROJECT
Project Manager
D.C. Stager
CONTRACTS
MANAGER
W.H. Simmons
PRODUCTIVITY
MANAGER
J.A. Gliksman
SOFTWARE &
DATA HANDLING
APM
J. Gruninger
SPACECRAFT
DES & DEVEL
APM
J.A. Gliksman
SYSTEMS
ENGINEERING
APM
T.G. Haggenmaker
COMM & DATA
HANDLING S/W
SPM
G.L. Bender
SOFTWARE
REQUIREMENTS
SPM
L.K. Jensen
OSTC
SOFTWARE
SPM
R. Schlothauer
DATA
HANDLING
SPM
C.M. Tsai
DEPUTY FOR
MECHANICAL
DES & DEVEL
S.L. Rieb
STRUCTURES
SUBSYSTEM
SPM
A.P. Goldberg
MECHANICAL DES
INTEGRATION
SPM
E.P. Lee
THERMAL DES &
INTEGRATION
SPM
B.G. Monteith
DES & DEVEL
STAFF
L.H. Garner
ELEC PWR & DIST
SUBSYSTEM
SPM
R.H. Hazen
PROPULSION
SUBSYSTEM
SPM
R.A. Carlson
ATT CONT & DETER
SUBSYSTEM
SPM
F.H. Koehler
REQS, DES, ANAL.
& VERIFICATION
SPM
P.R. Mock
SYS OPS & DATA
PROCESSING
SPM
B.R. Brenes
STS
INTEGRATION
SPM
D.A. Molgaard
INSTRUMENT
INTERFACE MGMT
SPM
R.E. Cline
ELECT DESIGN
INTEGRATION
SPM
M.A. Coffman

Figure 8

aspects: First, it is concerned with requirements for various components of the spacecraft. The systems engineer must allocate these requirements to the people responsible for the equipment design and construction. This involves trade-offs and interface decisions. The Systems Engineer must understand and agree with the plans for conduct of the design and development program. Secondly, the Systems Engineers are responsible for integration of the design and the hardware. They serve as a focal point for interface control and must assure that the configuration is adequately documented and controlled. They must assure that the requirements allocations are traceable and they keep track of the design progress from a performance measurement standpoint. Finally, systems Engineering is responsible for verification of the design and its performace. They approve the verification plans, make sure that the tests are adequate and assure that discrepancies are properly resolved.

The intersection of projects and functions are subprojects. From a functional standpoint, the subproject is a grouping of people responsible for a single technical discipline asigned to a particular project. The example previously given was the structures subproject responsible for GRO. Other subprojects are responsible for the design of the attitude control system, spacecraft reliability, etc. People who work in the functional organization are assigned to the subproject for the duration of the program and receive project direction from the Project Manager. However, the functional manager is responsible for their assignment and responsible for the adequacy of the performance of these people on the program. The subproject manager applies their talents to the project and is dedicated to the project and its proper conduct.

One of the important organizational aspects of matrix management is personnel selection and assignment. Key personnel in the project organization are selected by the Product Division Manager. There are a variety of techniques used to determine who is best suited for these key jobs. In general, both the Product Division Managers and the Functional Division Managers keep track of people who are suitable for such assignments. A number of precursor jobs are often important to the selection process. For instance, the person who conducts the study and writes the proposal is generally a prime candidate for the Project Manager. The functional divisions make it a normal process to identify those people whose next job should be on a project. Quite often people will identify themselves as candidates for project positions. This process is encouraged within the functional organizations and the human relations organization by a process of open posting of job openings.

From the functional organization standpoint the selection of the people to go on to a project has a number of different dimensions. Since the exact start up date of most projects is unknown until the project is awarded, the functional manager must respond rapidly to unexpected requirements for people.

It goes without saying that good people are always busy, and a good functional manager will make sure that he has only enough people on board to conduct the business at hand. This requires the functional manager to be constantly examining the assignment of his people to make sure that there are an adequate number of qualified people available to be rotated or moved into the new projects as they start up. Clearly this rotation or replacement of people must be worked carefully with the project managers to make sure that the projects are well served and that new people are rotated into the projects in a timely manner.

PROJECT MANAGEMENT

A number of the techniques used for project management are peculiar to the aerospace industry and perhaps to TRW specifically. The first of these is the principle of a formalized work breakdown structure; that is, a tiered arrangement of project tasks. Each of the work elements is characterized by a plan which indicates what the task is, when it will be accomplished, and what resources will be applied to the task. This plan is accompanied by authorization from the manager of that portion of the project and is also approved by the functional manager responsible for work conduct. In effect, the task plan and the project work authorization represent a contract between the functional organization and the project organization for the task at hand. This may seem unnecessarily cumbersome, but the process has clear benefits. First, it should be observed that effective work requires good plans. The formal work breakdown structure assists in the understanding of the scope of the work. The construction of the work breakdown structure helps to identify problems and provides a framework for discussing the plans. Periodic reports against the plan provide a framework for discussing progress on the program and for identifying problems.

A second specialized technique used in the aerospace industry is concerned with schedule control. Schedule complexity for a development program is almost universally underestimated. Many of the tasks on a spacecraft development program are interrelated and the exact nature of interaction is imperfectly understood. Problems occur which affect tasks underway and also require that new tasks be initiated. A good scheduling system allows for dynamic reprogramming and provides feedback about schedule performance. The scheduling system should provide a forum for a discussion of progress as well as help in planning future actions. The most commonly used technique for schedule control is milestone charting (Gantt charts). It is normal for the subproject to report milestone progress on a monthly basis. However, this technique provides little visibility into the interactive nature of the schedules. In a number of cases the milestones must be supplemented by PERT diagrams or constraint networks to indicate the nature of the interaction. PERT networks are particularly useful in facilitating communication on schedule matters.

Reporting and design review on a spacecraft development program are critical. Typically, the Project Manager reviews subproject status on a monthly basis. This provides the subproject manager with an opportunity to indicate his status, report on any problems he has, and forecast the need for resources for the remainder of the program. These program status reviews are also attended by functional managers and provide an opportunity for discussion between project and function of both progress and assignment of resources.

A normal portion of the spacecraft development program is a series of design reviews. These are structured as key milestones and consist of presentation of engineering data to a select design review committee. They also provide for participation of interfacing areas. Design reviews are held at least twice and sometimes three times during the program at unit, subsystem, and the system levels. The design review package provides formal documentation of the design at the time of review.

Most programs also employ a weekly systems engineering meeting or design meeting in order to provide a forum for data interchange and problem discussion. Sometimes these meetings are attended by customer representatives as well as system and subsystem engineers.

ORGANIZATIONAL DEVELOPMENT/TEAM BUILDING

As indicated earlier, the TRW organization is composed of both project and functional arms. In order to form an effective project team it is important that both of these organizational arms address the task of building the project team. To the functional organization it is extremely important that the management team be dedicated to support the project in terms of resource assignment and maintenance of technology base. A strong functional organization at the laboratory level assures that personnel within the laboratory are adequate in numbers, experience, and technical base and thus capable of supporting the project as required. In general, the functional organizations at TRW pay considerable attention to internal working relations and to the development of technology which, in their opinion, will be required for upcoming projects. Having adopted organization objectives consistent with their role of project support, they are in a position to assign personnel to the project and make sure the assigned personnel represent their function in an exemplary manner. In the project organization it is extremely important that the project manager and assistant project managers communicate well and are able to resolve problems within the project framework. Since the project managers and assistant project managers are generally aggressive, accomplishment oriented people and, in the TRW culture, of high technical caliber, the project team tends to force and direct all activities in their sphere of influence. It is common practice to colocate all project personnel (including subproject personnel assigned to the project) in the same building in order to facilitate the direct interaction of the project team.

It is common in a technical organization to spend a great deal of effort working technical problems and very little effort addressing how the problems are being worked. As pointed out earlier, organizational teams at TRW are formed at several levels; typically functional teams at Laboratory, Operations and Division level and Project and Subproject teams. Conscious attention is paid to working relations in each of these teams. A typical technique used to examine working relations is the offsite meeting or team retreat. Often these meetings are designed or facilitated by a moderator with training in industrial psychology and experience in interpersonal relations.

One of the common techniques used in building effective teams is a two step process in which the team members first focus on organizational mission. Having defined succinctly why this organization exists, they then focus on the boundaries between their domain and adjacent organizations and teams. The conscious examination of messages, both information and attitude, flowing across their boundaries provide the team with valuable insight into their effectivity. Gainful effort can then be applied to shaping both the organizational domain and the message flow to improve effectively.

The preceding paragraphs described some of the techniques used to form effective organizational teams. Clearly the formation of strong project and functional organizations can be of assistance in crystalizing responsibilites. However, the process can also run the risk of creating fiefdoms which are in conflict for responsibility or credit. The matrix tends to avoid this latter problem by providing a system of checks and balances. In the TRW Space and Technology group success is indicated by proper spacecraft operation in orbit and continued business growth. Both the functional and project arms are evaluated against these criteria. Since the structure has both functional and project representation, organizational conflict tends to be around ways and means. The matrix therefore provides for controlled organizational conflict.

One of the common misconceptions of the functional side of the matrix is as a supplier of people, i.e., a body shop. Many project managers approach their task by asking the functional organization for particular people of known and high technical caliber. Once assigned, they expect the functional organization to disappear but be ready to reabsorb the people when the project is done. Of course the risk with this model is that the project capabilities are restricted to the project manager and his chosen few. Furthermore, the best people are always in multiple demand. The body shop model might work for one project but cannot be expected to handle multiple projects. Proper employment of the matrix avoids this problem by providing the functional manager with responsibility not only for people but for their function.

In the spacecraft development business, problem identification and resolution is extremely important. Not only are the spacecraft very complex and expensive, but once launched they are inaccesssible. There's no such thing as a spacecraft recall. (In view of the recent successful in-orbit repair of the Solar Max Spacecraft by the shuttle crew, this comment should be qualified. However, the author believes, in view of the retrieval and repair expense, that is better to design and build the spacecraft right in the first place than to repair after launch.) The author believes that every problem is known or at least suspected somewhere in the organization prior to launch. The management challenge is to allow problems to surface. The matrix provides very effective avenues for problem identification and resolution. The functional organization oversees all projects and is in a unique position to assure that problems which are common between projects are avoided.

Several other techniques are used at TRW to help in problem identification. Classically many problems occur at the interfaces between equipment and between organizations. At TRW there are separate organizations whose task is to track interfaces. For instance, one such organization is responsible for keeping track of all wires and signals flowing between components. Another organization tracks mechanical interfaces, etc.

Once a problem is identified, the project organization generally takes responsibility for finding a solution. Typically, a person is assigned to work each significant problem and if the problem is severe, an entire task force may be fielded.

A technique which is very useful in providing ideas that may help in problem solution is nominal group technique (NGT). The technique is based on the theory that group effort is more effective than individual effort. A typical NGT group consists of approximately 20 people. In form, the NGT session consists of rapid fire voicing of ideas in sequence around the room. The ideas are recorded and after listing, the group expands, clarifies and combines the ideas. The group then prioritizes them. With such a list, the manager can apply the ideas to the problem at hand.

CONCLUSIONS/PROJECTIONS

The preceding discussion has focused on spacecraft development as an interdisicplinary management example. The spacecraft development progam was characterized and some of the management techniques used at TRW were described. The chief technique is the matrix organization with its capacity for both project and functional focus. Some of the TRW management and organizational development techniques were described. These techniques have proven effective at TRW and they may be of use in the management of other interdisciplinary pursuits.

No current discussion of high-technology management techniques would be complete without reference to improved productivity. Certainly the central thrust of management development at TRW is toward greater productivity. Although the specifics of the TRW productivity program are beyond the scope of this discussion, it is felt that the matrix organization is ideally suited to pursue productivity enhancement. The functional organization arm is dedicated to effective and efficient operation by function. Although measurement of high technology productivity has been illusive, the functional organizations are ideally situated to solve this problem and produce quantifiable and observable results. The project organization at TRW has taken the initiative to incorporate productivity improvement elements in each of the spacecraft programs; some with contractural provisions. This productivity thrust provides a new and vital dimension to an already exciting and challenging management field.

MANAGING HIGH TECHNOLOGY
An Interdisciplinary Perspective
B.W. Mar, W.T. Newell, and B.O. Saxberg (Editors)

ORGANIZED RESEARCH UNITS IN ACADEME REVISITED

Robert S. Friedman and Renee C. Friedman

The Pennsylvania State University
207 Old Main
University Park, PA 16802

ABSTRACT

Linkages between universities and government in the education and research processes are of long standing. The societal forces that helped forge these partnerships persist. Our nation's need for sophisticated technological and scientific innovations will likely increase in the ensuing decades. Fulfilling policy goals with respect to the prevention and cure of disease, national security, energy self-sufficiency, exportable food supplies and industrial, social and economic well-being will continue to require the assistance of our universities in the recruitment and training of large numbers of research scientists and engineers.

Organized research units (ORUs) represent one of the major devices employed at leading research universities for the conduct of research and the development and maintenance of reservoirs of talented young research scientists and engineers. (The term organized research units refers to the broad range of nondepartmental structures often called centers, institutes, laboratories, bureaus, or clinics at the host institutions.) Among the study objectives were: an exploration of the current structure, organization, and mode of operation of ORUs as understood by the major actors in the system; an examination of the dynamics of the relationship between departments and ORUs in the resource allocation process in the core university and medical school; and an appraisal of the nature of the research conducted by ORUs and its utility in securing initial job opportunities for doctoral students and postdoctoral scientists and fellows.

OVERVIEW

Among the quieter, but no less radical, alterations in the academic topography during the past century have been the incorporation and proliferation of organized research units (ORUs) (Friedman, 1977; 1978; Friedman and Friedman, 1982; Ikenberry and Friedman, 1972; Rossi, 1964). The concept evolved, in part, because of the perceived need for organizations parallel to, but apart from departments and priorities of society. Almost one hundred years ago the government recognized the utility of such arrangements in meeting national goals. The structural antecedents of contemporary ORUs, the

Agricultural Experiment Stations, grew out of federal legislation, the Hatch Act of 1887, designed to foster the science of agriculture (Friedman,, 1977; 1978; 1984; Friedman and Friedman, 1982).

Land grant colleges and universities were especially poised for their arrival. This institutional set was created to promote the "liberal and practical education of the industrial classes in all the pursuits and professions of life" and to emphasize the branches of learning related to agriculture and the mechanical arts (Friedman and Friedman, 1982). The American educational response--linking applied, problem-focused research to the teaching function--contrasted sharply with the procedure adopted by the German universities. As Ben-David (1976) noted:

> ...the integration of research and teaching, which had been the main source of strength of German science throughout the middle of the nineteenth century, became a problem for the German system by the turn of the century. As a result, teaching was institutionally separated from research, first within the university through the concentration of research in 'institutes' that were personal research establishments of professors virtually separated from the university, and subsequently--in 1911, through the founding of the Kaiser Wilhelm Gesellschaft--by establishing pure research institutions without any teaching functions at all (Ben-David, 1976, p. 103).

From observation to experimentation, from singular activity to teamwork, from little science to big science--the ascendancy of the research mission within universities and societal dependence on its scientific and technological products helped accelerate the establishment of ORUs on American campuses. Additionally, professionalization, the Academic Revolution about which Jencks and Riesman (1968) wrote, emphasized why individual faculty advocate the ORU mode. Expectation of research productivity and the judgment of professional competence and subsequent career advancement based on the quantity and quality of that research output made ORU association particularly appealing (Friedman and Friedman, 1982; Ikenberry and Friedman, 1972).

ORUs promised something for everybody. Administrators sometimes welcomed them as devices for institutional development, including the recruitment and retention of faculty; increasing coordination and communication among departments and programs; strengthening graduate education and research programs; resolving internal conflicts; renovating and reforming existing departments; creating special areas of academic emphasis and specialization; and enhancing institutional visibility (Friedman, 1977; 1978; Friedman and Friedman, 1982; Ikenberry and Friedman, 1972).

Faculty recognized the accessibility to research resources and career satisfaction obtained through ORU affiliation. Unlike departments where reasonable parity must be maintained, ORUs offered perquisites such as travel monies, twelve-month contracts, research time, secretarial assistance, space and equipment, and graduate assistants. And for those who aspired to executive leadership, the ORU directorship was not without the elements of power.

Sponsors, particularly the federal government, derived benefits from the ORU mode. Given the enormity of demands encumbered by a technologically oriented society committed to democratic values, the federal bureaucracy cannot attract, retain, and support on a continuing basis the manpower, equipment and physical plant needed to cope with rapidly changing needs and requirements (Friedman, 1971; 1977; 1978; Friedman and Friedman, 1982). For industry as well, the ORU seemed ideally suited for task fulfillment of limited duration.

ORUs were championed for delivering what departments could not. In essence, they became the tabulae rasae of academe. They could provide a focus for a group of faculty from diverse disciplines who were concerned about the same subject. They were purportedly better suited to respond to new constituencies and enabled a temporary restructuring of the university in ways not possible in departments. Unbound by numerical conventions and prescribed personnel mixes like departments, ORUs might be able to assemble new and different configurations of professional talent and might be expanded and contracted at will. Lastly, because of their appeal to problem-oriented sponsors, ORUs had the potential to generate sufficient resources to make them financially self-sustaining (Friedman and Friedman, 1982; Ikenberry and Friedman, 1972). In sum, inter- and multidisciplinarity, transience, elasticity and their cost-free nature were important selling points for ORUs. Obviously not every ORU established is sold on the basis of any or all of these features, but these are what the conventional wisdom tells us generates support for their creation.

Data from our NSF funded study, The Role of University Organized Research Units in Academic Science,[1] offer clues as to the degree to which each of these rationalizations for ORU mode prevail in major research universities and their medical schools in the United States and suggest to us some guidelines for policy with respect to ORUs in academe (Friedman and Friedman, 1982).

MULTIDISCIPLINARITY AND INTERDISCIPLINARITY

Scholars have defined terms such as multidisciplinarity, interdisciplinarity, cross-disciplinarity and transdisciplinarity according to the type of coordination and degree of collaboration among disciplines. Germane to our analysis is Alpert's distinction between multidisciplinary and interdisciplinary styles (1967). He proposed the term multidisciplinary to describe a center or laboratory in which individual scholars from different disciplines or departments share common facilities, common research approaches or a common environment. An interdisciplinary program requires insights from experts in a number of disciplines and demands an interactive joint effort to each a solution.

Although multidisciplinarity and interdisciplinarity are prime justifications for the existence of ORUs, in the real world they do not appear as frequently as might be expected. As Table 1 illustrates, only about one-third of ORU directors sampled reported faculty interactions with more than one department. ORU directors were asked about their research mode. Three models were identified: (1) faculty and staff from different disciplines work together on a single project (interdisciplinary); (2) faculty and staff from different disciplines work separately on different aspects of a

TABLE 1
Departmental Representation by ORU Groupings

Number of Departments	ORU Groupings									
	Biol. Sci.	Med. Sci. (CORE)[a]	Agr. Sci.	Phy. Sci.[b]	Engin-eering	Soc. Sci.[c]	Humani-ties	Profes-sional[d]	Med Sci. (MED)	Row Average
1	68.6%	50.0%	73.7%	62.5%	68.1%	66.4%	59.5%	79.3%	66.9%	67.4%
2	12.4	12.5	5.3	22.1	16.9	9.3	21.5	18.3	13.1	14.2
3	9.8	25.0	0.0	3.8	5.0	3.0	2.5	0.0	5.2	4.2
4	3.9	0.0	5.3	2.9	0.0	6.8	0.0	0.0	0.0	2.9
5	5.2	0.0	5.3	2.9	2.5	3.2	0.0	1.2	5.2	3.3
6	0.0	0.0	0.0	3.8	3.7	1.7	0.0	0.0	4.4	2.2
7	0.0	0.0	0.0	1.0	0.0	1.7	0.0	0.0	0.7	0.8
8	0.0	12.5	5.3	0.0	0.0	6.3	0.0	1.2	0.7	2.4
9	0.0	0.0	0.0	0.0	1.2	0.8	0.0	0.0	0.0	0.4
10 or more	0.0	0.0	5.3	1.0	2.5	0.8	16.5	0.0	2.8	2.0
Total	99.9	100	100.2	100	99.9	100	100	100	100	99.8
% Reporting	83.8	48.0	58.3	87.5	86.2	91.9	98.3	91.7	88.1	87.4

SOURCE: Friedman and Friedman 1982.

NOTE: Groupings are based on disciplinary specialty of ORU directors in conformity with NSF fields of science. In the few instances in which current assignment diverges from fields of training, the research thrust of the organization is the basis of group assignment.

[a]Composed of those with medical, veterinary, and dental training who operate in the core milieu.

[b]NSF atmospheric and geological sciences, mathematics, and computer sciences classifications are subsumed under physical sciences.

[c]Psychology is merged with the NSF social sciences list.

[d]Composed of those who hold professional degrees in law, business, education, and social work.

total program (multidisciplinary); and (3) faculty and staff from a single discipline work together but have assistance available from consultants from other disciplines as required (unidisciplinary). A fourth category was included to recognize different emphases in different projects. As shown in Table 2, engineering-directed ORUs were the only ones in which a majority reported an interdisciplinary work style. In all other intellectual categories the multidisciplinary pattern was the most frequently reported. Further, among all ORUs more than 20 percent acknowledge a unidisciplinary work style.

Numerous organizational, professional, and personal factors account for the reluctance of faculty to venture beyond disciplinary confines. There is an informal hierarchy of disciplines and fields in academe in which the more prestigious theoretical basic disciplines regard scholarly interaction with colleagues in the applied, less theoretical fields as an unrewarding experience. Lack of uniform policies with respect to overhead retention is a further disincentive to interdisciplinary and multidisciplinary activity. Consequently, faculty in more privileged colleges and departments in such institutions are likely to be discouraged from relationships with colleagues from less favored units. Additionally, by virtue of its control over the reward system and access to tenure, the department remains the single most influential force in shaping the career patterns of faculty. Multidisciplinary research may be exciting and innovative but it usually represents a radical departure from accepted practice and norms. The scholar who straddles many fields runs the risk of being acceptable to none. Favorable review from one's peers is what the academic game is all about.

Even with departmental approbation, most academics are reluctant to leave the nest. Without delving into the psychosocial nature of identity, it should be recognized that the rites of passage we undergo during the socialization process in order to internalize the values, attitudes, and norms of the peer group to which we aspire preclude much wandering off the reservation (Bragg, 1976; Erickson, 1968). Our discipline and/or profession furnishes us with a self-identity, a common language, and a set of research tools with which we define and are defined by the bigger world. Almost without exception, the professorial titles of ORU directors--the universally understood message--were telegraphed first on the identification section of the survey forms (Friedman and Friedman, 1982). Further, the rate at which new knowledge is generated and its concomitant, intellectual obsolescence, tends to make us more, not less xenophobic.

Lastly, ORUs may take on a disciplinary bias in spite of best intentions:

> ...Mores and values in personnel selection may cause professional staff members to value candidates with views and competencies in harmony with their own, and thus the institutes may evolve a more homogeneous staff over time. Different disciplines define the same problem differently and tend to approach tasks with different methodological tools and procedures. The process of defining tasks and resolving these "differences" very likely screens in certain disciplines and screens out others. In short, one could hypothesize that the same forces which have esca-

TABLE 2

Nature of Research Mode by ORU Groupings

Research Mode	ORU Groupings									
	Biol. Sci.	Med. Sci (CORE)[a]	Agr. Sci.	Phy. Sci.[b]	Engin-eering	Soc. Sci.	Humani-ties	Profes-sional[c]	Med Sci (MED)	Row Average
Faculty and staff from different disciplines work together on a single project.	23.5%	20.0%	26.2%	21.0%	58.6%	26.6%	16.0%	31.3%	22.8%	28.0%
Faculty and staff from different disciplines work separately on different aspects of a total program.	57.9	60.0	47.7	54.6	29.0	37.3	49.4	38.5	54.3	45.1
Faculty and staff from a single discipline work together but have assistance available.	17.5	20.0	26.2	24.5	11.2	30.0	2.5	30.2	20.3	22.9
Combinations of the above research modes.	1.1	0.0	0.0	0.0	1.2	6.0	32.1	0.0	2.6	4.0
TOTAL	100	100	100.1	100.1	100	99.9	100	100	100	100
% Reporting	100	60.0	100	96.6	90.6	96.3	100	100	75.4	92.1

SOURCE: Friedman and Friedman 1982

NOTE: Groupings are based on disciplinary specialty of ORU directors in conformity with NSF fields of science. In the few instances in which current assignment diverges from fields of training, the research thrust of the organization is the basis of group assignment.

[a]Composed of those with medical, veterinary, and dental training who operate in the core millieu.

[b]NSF atmospheric and geological sciences, mathematics, and computer sciences classifications are subsumed under physical sciences.

[c]Psychology is merged with the NSF social sciences list.

[d]Composed of those who hold professional degrees in law, business, education, and social work.

> lated the power and attractiveness of academic departments in recent decades have exerted a continuing pressure toward disciplinary homogeneity in institutes and centers as well (Ikenberry, and Friedman, 1972, p 60).

Uniform overhead policy for all colleges and departments would solve one of the organizational problems to which we have alluded. Administrative fiat, however, will not generate radical changes in academic mores. Faculty members tend to be internally motivated rather than outwardly directed.

Crane's typology of knowledge acquisition--revolutionary, cumulative, random--explains much about the individual versus collective approach to research (Crane, 1972). When scholars become dependent upon the methodologies of other disciplines, they either acquire that knowledge through self-education and training or seek outside help. As the social sciences progressed from normative to empirical research, they styled themselves after their colleagues in the physical and life sciences (Friedman, 1977; Friedman, 1978; Friedman and Friedman, 1982; Ikenberry and Friedman, 1972; Merton, 1973; Rossi, 1964). The escalating cost of equipment, the increasing intricacy of scientific problems, and the needed expertise amplified the team effort that characterized much of the scientific milieu. The key word is dependency, because of the immediacy with which scientists must validate emerging theories and concepts and the relevancy of those discoveries to the unraveling of more complex scientific mysteries. "Scientific truths discovered in one age are essential for scientific progress in another ... in this sense science has always been a cumulative, team activity" (Weinberg, 1970, p. 1046). The same urgency does not pervade the humanities. The shelf life of an idea, in contrast with the sciences, is infinite. Therefore, scholars in the humanities may draw at random upon any idea they wish to pursue. The notion of the humanist as the solitary toiler is deeply ingrained and reinforced during the apprenticeship.

Interaction can be encouraged by reducing the traditional physical segregation of faculty by discipline. Horizontal integration permits the natural affinity that exists between scholars from different departments who share common research interests to ripen. Unless faculty from diverse disciplines know each other personally and professionally beforehand and can synchronize their work styles, no organization can bond the research group together quickly or easily (Mar, et al., 1976).

So long as multidisciplinary work remains in its infancy, a continued disciplinary predominance makes sense in terms of quality control. At the current stage of our knowledge, it would be unwise for civil engineers to make primary judgments about the work of sociologists. However, if there are to be major thrusts in dealing with problems that extend beyond disciplinary confines, there must be some readjustment of the traditional criteria of evaluation. The reward system must be altered so that equal attention is paid to applications of theory that have payoffs (Friedman and Friedman, 1976). Socialization patterns must be adjusted. Engineering is the applied expression of high paradigm disciplines. The ability of academic engineers to engage in interdisciplinary research may stem, in part, from such a mind-set, a preoccupation with problem solving, nurtured and reinforced during the socialization process.

TRANSIENCE

ORUs are frequently touted as compensatory mechanisms for the rigidity and scholasticism attributed to departments. They presumably homestead the territories not yet annexed to departments. They serve as vehicles for high risk experimentation. And above all, they are viewed as temporary, endowing neither the university nor the department with *de jure* or *de facto* permanent staff. Based on the circumstantial evidence presented in Table 3, the verdict is inconclusive. If the yardstick of comparison is departmental longevity, then ORUs are "the new kids in town." Less than 11 percent of the core university ORUs and 4 percent of the medical school ORUs were established prior to 1940. But if the standard is a promise of termination no later than puberty, then core university ORUs are "graying" in academe along with departments. Only about one-third of them were born after 1970. By contrast, more than 50 percent of the medical school ORUs have not reached high school age.

Medical schools have promoted the concept of program project grants, the so-called centers without walls. Operating like matrix organizations, the staffs needed for heterogeneous work tasks are drawn from departments, but the members maintain their primary allegiances to their chairpersons and colleagues back home. Such research delivery models do not disrupt the status quo, widen the division between clinicians and bench scientists, or require major rearrangements of the administrative structure. Further, they self-destruct when the one- to five-year grants terminate (Friedman and Friedman, 1982). Perhaps the reason that medical schools can effect this mode with greater ease than the core universities is because there are fewer layers in the authority structure to advise and consent. In addition, courtesy and salaried joint appointments, commonplace in medical schools, facilitate personnel exchanges.

These aggregate data celebrate the survivors and mask the casualties. ORUs, to a large extent creations of a demanding society, are subject to the whims of a vacillating public not always cetain about its needs and priorities. A previous study of ORUs chronicled their fortunes and misfortunes in academe during the 1960s and 1970s (Friedman, 1977). The greatest number of obituary notices appeared among those units in education, government, labor, law, and social sciences. In effect, those fields linked to public policy and derivatives of social science disciplines display the greatest volatility and are, therefore, in the most precarious circumstance (Friedman, 1977; 1978; Friedman and Friedman, 1982). Because of their general reliance on software and limited space needs, they are also easier and cheaper to assemble and disassemble than their counterparts in engineering, physical sciences and life sciences.

In the current academic milieu reallocations and contractions in programs prompt several policy caveats with regard to the creation, continuance, and severance of ORUs. Clearly those activities that enhance institutional prestige, complement programmatic excellence, and do not replicate research in progress elsewhere should not receive automatic termination notices because the clock is running out.

Organizational longevity must be viewed within a broader context. The hidden agenda item--what faculty and administrators fear most--is the diminution of institutional reputation. Like all complex

TABLE 3

ORU Groupings by Decades of Establishment (Core University and Medical School)

Groupings	Pre-1940	1940-1950	1950-1960	1960-1970	1970-1981	Total % By Field
Biological Sciences	5.0	2.2	58.3	22.3	12.2	12.6
Medical Sciences (CORE)[a]	3.7	11.1	7.4	18.5	59.3	2.4
Agricultural Sciences	14.6	20.8	2.1	37.5	25.0	4.3
Physical Sciences[b]	8.6	4.6	17.7	44.0	25.1	15.9
Engineering	21.5	3.0	6.7	35.6	33.3	12.2
Social Sciences[c]	12.2	14.6	4.9	25.5	42.9	34.9
Humanities	3.2	0.0	33.3	25.4	38.1	5.7
Professional[d]	7.5	0.0	8.3	45.9	38.4	12.0
Total (CORE) % by Decades	10.7	7.6	15.8	32.0	33.9	100.0
Medical Sciences (MED)	3.7	13.8	2.6	26.4	53.5	100.0
Total (MED) % by Decades	3.7	13.8	2.6	26.4	53.5	100.0

SOURCE: Friedman and Friedman 1982

NOTE: Groupings are based on disciplinary specialty of ORU directors in conformity with NSF fields of science. In the few instances in which current assignment diverges from fields of training, the research thrust of the organization is the basis of group assignment.

[a]Composed of those with medical, veterinary, and dental training who operate in the core milieu.

[b]NSF atmospheric and geological sciences, mathematics, and computer sciences classifications are subsumed under physical sciences.

[c]Psychology is merged with the NSF social sciences list.

[d]Composed of those who hold professional degrees in law, business, education, and social work.

organizations, universities have diffuse and sometimes conflicting goals and missions. One goal, however, is paramount and universal among the leading research universities: the maximization of prestige. Prestige is not an abstraction but rather the cornerstone of economically motivated behavior. The preeminent universities attract the renowned scholars, the intellectually outstanding graduate and undergraduate students, and the funding that assures continued supremacy.

In times of scarce resources, two opposing strategies hold the most economic promise. Institutions can adopt a market utility approach whereby they abandon any serious attempt to be more than a regional or local university, pander to local interests, and actively seek the research grants and contracts that the other camp disdains. The prestige model derives its revenue by sustaining its national and international reputation, catering to a cosmopolitan faculty and student body, and enjoining only those research activities that accentuate its greatness (Garvin, 1980). It is the latter group of institutions upon which this research project focused. Therefore, the availablity of external funding cannot be the sole criterion for the establishment of peripheral organizations or their endurance.

ELASTICITY

ORUs may be exclusively faculty, a combination of faculty, graduate students, and if appropriate, postdoctoral scientists and fellows; or any meld of the above with or without staff assistance. Further, the associated faculty may be full- or part-time, on temporary or regular appointment (Friedman and Friedman, 1982, p. 129). The accordion effect, the ability to expand and contract at will, is one of the alleged characteristics of ORUs.

While one swallow does not a summer make, the FY 1980 mean staff configurations presented in Table 4 lead to a conclusion: claims about the elasticity of ORUs are exaggerated. Humanities, Professional, and Medical Sciences (MED) demonstrate the permanent nature of ORU employment for faculty and staff. Biological Sciences, Social Sciences, and Physical Sciences augment their regularly appointed faculty cohorts with temporary additions. Agricultural Sciences and Engineering also offer temporary faculty appointments, but relative to their total staff size, these add-ons make ripples not waves. The professional, nonprofessioral staff lines are static except in Engineering, and this may be influenced by the shortage of available trained personnel (Friedman and Friedman, 1982, p. 148).

Other insights into the structure of research can be gleaned from the data. Rossi's distinction between the consortium and research firm model is demonstrated in the ratios of professional staff to faculty:

> A consortium is a relatively loose confederation of faculty members held together primarily by their interest in access to the resources held by the collective unity involved and who have developed a very low level of division of labor among themselves. ... Such research centers have little organizational capability to pursue a unified line of research. ... A consortium is a coalition of several very independent researchers who guard their individual autonomies quite carefully.

TABLE 4
Regular and Temporary Mean ORU Staff Configurations FY 1980

		Regular ORU Appointments		Temporary ORU Appointments	
Groupings	% Reporting	Fully Budgeted	Partially Budgeted	Fully Budgeted	Partially Budgeted
Biological Sciences	92				
Rank and Budgeted in Dept.		---	11.93	---	0.84
Rank Without Budget in Dept.		1.60	1.33	3.12	0.32
No Rank		2.93	0.52	0.35	0.40
Medical Sciences (CORE)[a]	36				
Rank and Budgeted in Dept.		---	7.66	---	2.17
Rank Without Budget in Dept.		2.00	0.00	0.00	0.00
No Rank		1.33	0.00	0.00	0.00
Agricultural Sciences	54				
Rank and Budgeted in Dept.		---	53.14	---	4.17
Rank Without Budget in Dept.		1.49	0.74	0.37	0.00
No Rank		6.17	0.51	0.43	0.00
Physical Sciences[b]	81				
Rank and Budgeted in Dept.		---	5.98	---	1.40
Rank Without Budget in Dept.		1.17	0.74	0.56	0.08
No Rank		7.10	3.51	0.52	0.53
Engineering	75				
Rank and Budgeted in Dept.		---	16.63	---	1.77
Rank Without Budget in Dept.		1.21	0.49	0.10	0.63
No Rank		1.48	0.29	2.53	3.29

Table 4 - continued

Groupings	% Reporting	Regular ORU Appointments: Fully Budgeted	Regular ORU Appointments: Partially Budgeted	Temporary ORU Appointments: Fully Budgeted	Temporary ORU Appointments: Partially Budgeted
Social Sciences[c]	81				
Rank and Budgeted in Dept.		---	3.75	---	1.61
Rank Without Budget in Dept.		1.04	0.53	0.10	0.11
No Rank		2.65	2.10	0.44	0.59
Humanities	80				
Rank and Budgeted in Dept.		---	6.20	---	0.88
Rank Without Budget in Dept.		0.44	0.23	0.03	0.64
No Rank		0.34	0.41	0.00	0.06
Professional[d]	86				
Rank and Budgeted in Dept.		---	3.23	---	0.27
Rank Without Budget in Dept.		0.61	0.11	0.07	0.00
No Rank		0.51	0.16	0.39	0.95
Medical Sciences (MED)	74				
Rank and Budgeted in Dept.		---	4.70	---	0.30
Rank Without Budget in Dept.		1.57	0.84	0.07	0.03
No Rank		1.84	0.39	1.06	0.11

SOURCE: Friedman and Friedman 1982

[a]Composed of those with medical, veterinary, and dental training who operate in the core milieu.

[b]NSF atmospheric and geological sciences, mathematics, and computer sciences classifications are subsumed under physical sciences.

[c]Psychology is merged with the NSF social sciences list.

[d]Composed of those who hold professional degrees in law, business, education, and social work.

TABLE 5
Funding Means, Ratios, and Rankings by ORU Groupings or FY 1980
(Dollars in Thousands)

Groupings	% Reporting	Internal Mean Funding	External Mean Funding	Ratio	Ranking
Biological Sciences	73.5%	$ 490	$1,369	2.8:1	5
Medical Sciences (CORE)[a]	88.0	247	854	3.5:1	3
Agricultural Sciences	93.8	1,422	1,959	1.4:1	7.5
Physical Sciences[b]	86.4	494	1,425	2.9:1	4
Engineering	97.8	386	1,474	3.8:1	2
Soc. Sciences[c]	75.1	345	476	1.4:1	7.5
Humanities	68.3	356	80	0.2:1	9
Professional[d]	78.2	115	174	1.5:1	6
Medical Sciences (MED)	69.0	210	957	4.6:1	1
Total	78.3	$ 383	$ 929	2.4:1	

SOURCE: Friedman and Friedman 1982

[a]Composed of those with medical, veterinary, and dental training who operate in the core milieu.

[b]NSF atmospheric and geological sciences, mathematics, and computer sciences classifications are subsumed under physical sciences

[c]Psychology is merged with the NSF social sciences list.

[d]Composed of those who hold professional degrees in law, business, education,and social work.

cost about twice the figures presented. Only a small portion, principally in the Medical Science (MED) category, are institutionally unfunded. They are program project grants described earlier.

In fairness, sponsors sometimes demand that universities pledge resources as a sign of commitment to the projects. Grants from government agencies like NSF contain cost-sharing provisions.

In policy terms institutions must recognize that ORUs do not provide a free lunch. They generate costs as well as benefits and must fit the primary mission of the institution. Indeed, while a strong case can be made for some return to units of overhead realized from external grants and contracts, institutions should be cautious in developing such schemes. The capacity to generate external funding through entrepreneurship needs to be tempered by the basic goals and values of an institution. Contracts, for example, to perform service functions may generate large revenues for ORUs but if they do not entail new research thrusts, they could be antithetical to the welfare of the institution.

SUMMARY DISCUSSION

Our intent is not to scuttle ORUs but rather to encourage truth in packaging. No, they do not often function like an academic United Nations welding together disciplinarians who speak different languages and chart directions with different intellectual compasses. Multidisciplinary research does take place in the university setting and the ORU can be an accommodating device for that purpose, but it is not admistrative fiat or structure, _per se_, that makes it happen. One might also ask what is wrong with disciplinary research, especially if it represents a critical mass of expertise and enables the group to focus on targeted research without departmental interruption or the hassles of departmental politics. Further, in the modern complex university where, for example, economists and psychologists can be geographically dispersed in many colleges and departments, the ORU may serve as a recall from the diaspora, a form of disciplinary self-renewal.

Not every innovation a university attempts succeeds or should. ORUs are seen as temporary incubators for new ideas, emerging technologies and high risk experimentation. Once their purposes have been accomplished, ORUs are supposed to self-destruct. That is not the way it usually happens. "Almost all organizations build in self-maintenance mechanisms and find ways to extend their life spans beyond their original mandates or social utility. ORU's exhibit the same ingenuity and will to survive" (Friedman and Friedman, 1982, p. 76). Like departments, ORUs develop constituencies that will rally to the defense of the organization if it is threatened with extinction. ORU directors and professorial staff can easily become accustomed to the good life, the perquisites that are meagerly rationed out in departments. Furthermore, legitimacy and status are not accorded to ephemeral organizations.

Policy-centered ORUs--transportation, water resources, environmental health, population--do tend to be somewhat elastic, not by design but by necessity. Universities mirror the fads and fashions of society. Funding insecurity that results from these frequent and erratic shifts in research priorities plague policy-research centers. In addition, the outcomes of policy research may be redistributive in function, benefitting the have-nots at the expense of the haves (Friedman and Friedman, 1982). Even in those areas where the ORU leadership must accommodate feast and famine, the lack of research continuity and the inability to make long-term personnel commitments are antithetical to the university culture.

ORUs are not without cost in the financial or literal sense. College campuses are relatively quiet now, but that was not the case during the Vietnam War when prime targets were centers and institutes engaged in classified research. Such federally-funded research and development centers (FFRDCs) were not only free of charge but also contributed managerial fees to the host universities. Nevertheless, the price paid was enormous. As the data show, ORUs cost real dollars as well, but that is not without redeeming qualities.

Despite all of these reservations, ORUs can lay claim to numerous accomplishments. Yes, they do cost money but they act as magnets in attracting funds from the private and public sectors that, under most circumstances, would not otherwise flow into the departmental coffers. Rather than diminishing the research focus of the department, ORUs affirm the institutional commitment to societal needs and run interference so that the departmentally-based faculty may continue to engage in esoteric or fundamental research (Friedman and Friedman, 1984). Even if the university does not recoup its investment, publicly-supported institutions, particularly, do owe public service to the state and nation.

Educationally, ORUs have much to commend them. They offer educational opportunities to graduate students that can blend ivory tower disciplinary training with real world needs and problems. They provide greater opportunities for students to become aware of methods of studying large complex problems having a broad ecology of interacting factors (Jencks and Riesman, 1968). They serve as temporary residences for postdoctoral fellows and other professionals seeking permanent positions in academic and nonacademic sectors. In medical schools they act as a bridge between basic science and clinical departments and introduce medical students to the notion and importance of research. ORUs can assemble more and better research equipment that is available in departments. ORUs make available specialized personnel such as computer specialists and other difficult to employ within departments. ORUs are sources of dissertation support for graduate students and employment agencies for undergraduate and graduate students as well as faculty (Friedman and Friedman, 1982). ORUs are an idea whose time has come.

FOOTNOTE

1. The project focused on the role played by organized research units (ORUs) at 78 leading research universities and their medical schools. The 78 institutions included all of the 51 Research Universities I and 27 of the 47 Research Universities II (1976 Carnegie Classification). These universities and their medical schools accounted for nearly half of over 6,000 ORU listings in the sixth edition (1979) of the <u>Research Centers Directory</u> and more than 76 percent of federally

financed R&D expenditures at colleges and universities in FY 1979. Interviews and mailed questionnaires to four respondent categories--university research administrators, ORU directors, department chairpersons, postdoctoral scientist/fellows--constituted the principal data sources. Medical school data were collected and analyzed separately. The average useable questionnaire response rate varied by category: 84 percent of the university research administrators, 65 percent of the ORU directors, 50 percent of the department chairpersons, and 66 percent of the postdoctoral scientists/fellows completed their respective survey forms. For complete methodological detail, see chapter 2 of the aforementioned study.

REFERENCES

Abler, R. 1975. Memorandum no. 8. In The academic administration of research: a descriptive analysis. Henry Sams, principal investigator, pp. 136-139. University Park: The Pennsylvania State University Graduate School.

Alpert, D. 1967. The role and structure of interdisciplinary and multidisciplinary research centers. Address to the Council of Graduate Schools in the U.S., December 5 at Washington, D.C. Mimeographed.

Ben-David, J. 1976. Center of learning: Britain, France, Germany, United States. New York: McGraw-Hill.

Bragg, A.K. 1976. The socialization process in higher education. ERIC/AAHE research report no. 7. Washington, D.C.: American Association for Higher Education.

Bucher, R. and Strauss, A. 1961. Professionals in process. The American Journal of Sociology 66(January):325-334.

Carnegie Council on Policy Studies in Higher Education. 1976. A classification of institutions of higher education, revised ed. Berkeley: Carnegie Council.

Crane, D. 1972. Invisible colleges: diffusion of knowledge in scientific communities. Chicago: University of Chicago Press.

Deutsch, K.W., Platt, J. and Senghaas, D. 1971. Conditions favoring major advances in social science. Science 171(February): 450-459.

Erikson, E.H. 1968. Identity, youth and crisis. New York: W.W. Norton.

Friedman, R.C. 1977. The continuing saga of institutes and centers. Report no. 28. University Park: The Pennsylvania State University Center for the Study of Higher Education.

Friedman, R.C. 1978. Centers and institutes: the state of the art. In Policy research centers director, eds. S. Nagel and M. Need, pp. 60-74. Urbana, Illinois: University of Illinois Policy Studies Organization.

Friedman, R.C. and Friedman, R.S. 1976. Social and behavioral sciences in the 1970s. In Land-grant universities and their continuing challenge, ed. G.L. Anderson, pp. 160-177. Lansing: Michigan State University Press.

Friedman, R.S. 1971. Professionalism: expertise and policy making. New York: General Learning Press.

Friedman, R.S. and Friedman, R.C.. 1982. The role of university organized research units in academic science. A report to the National Science Foundation NTIS PB 82-253394. Springfield, Virginia: National Technical Information Service.

Friedman, R.S. and Friedman, R.C. 1984. Managing the organized research unit. Educational Record 65(Winter):27-30.

Garvin, D.A. 1980. The economics of university behavior. New York: Academic Press.

Ikenberry, S.O. and Friedman, R.C. 1972. Beyond academic departments. San Francisco: Jossey-Bass.

Jencks, D. and Riesman, D. 1968. The academic revolution. Garden City, New York: Doubleday.

Kuhn, T.S. 1970. The structure of scientific revolutions, 2nd ed. Chicago: University of Chicago Press.

Mar, B.W., Newell, W.T. and Saxberg, B.O. 1976. Interdisciplinary research in the university setting. Environmental Science and Technology 10(July):650-653

Merton, R.K. 1973. The sociology of science. Introduction by W.W. Storer. Chicago: The university of Chicago Press.

National Science Foundation. 1980. Federal funds for research and development: fiscal years 1978, 1979 and 1980, vol. 28. NSF 80-315. Washington, D.C.: U.S. Government Printing Office.

Palmer, A.M. 1979. Research centers directory, 6th ed. Detroit: Gale Research.

Pelz, D.C. and Andrews, F.M. 1966. Scientists in organizations: productive climates for research and development. New York: John Wiley.

Pettigrew, A.M. 1979. On studying organizational cultures. Administrative science quarterly 14(December):570-581.

Rossi, P.H. 1964. Researchers, scholars and policy makers: the politics of large-scale research. Daedalus 93:1142-1161.

Rossi, P.H. 1976. Assessing organizational capacity for education research and development in an academic institution. Educational Researcher 5(April):3-10.

Weinberg, A.M. 1970. Scientific teams and scientific laboratories. Daedalus 10(Fall):1056-1075

MANAGING HIGH TECHNOLOGY
An Interdisciplinary Perspective
B.W. Mar, W.T. Newell, and B.O. Saxberg (Editors)

UNIVERSITY INTERDISCIPLINARY ENGINEERING CENTERS: NEW DIRECTIONS--PERSISTENT ORGANIZATIONAL PROBLEMS

Graham S. Toft and F. Tomlinson Sparrow

Institute for Interdisciplinary Engineering Studies,
Purdue University
West Lafayette, Indiana 47907

INTRODUCTION

Irving Shapiro (1980) once described the reason for an organization's existence as follows: "The dual test of an organization is how well it performs its chosen functions and how well it matches the values of the society which charters it." In the jargon of organization theory, Shapiro is referring to organizational efficiency, and organizational effectiveness, respectively--or, in Peter Drucker's more colorful phrases, "doing things right" and "doing the right things."

This paper is about both organizational efficiency and organizational effectiveness of university interdisciplinary engineering centers. It is in part a summary of a National Science Foundation (NSF) sponsored workshop held at Purdue University in November 1983 on "Making Interdisciplinary Engineering Centers Work." It is also part independent comment and synthesis of ideas by the authors.

With respect to organizational efficiency, the purpose of this paper is to offer practical guidelines for dealing with the perennial problems associated with organizing and managing interdisciplinary engineering centers in hostile university settings which include distrust by department heads, a laissez faire attitude on the part of university administrators and a questioning of the fundamental contributions of interdisciplinary research by a majority of the faculty. The NSF conferees concluded that contingency theory probably best describes the organization and management of such centers and that, therefore, no single model or best management plan can be followed. A portion of the text is devoted to summarizing focal internal and external management strategies that can be pieced together to increase the probability of success in a particular university setting.

With respect to organizational effectiveness, the paper begins with a discussion of why interdisciplinary engineering centers survive and how their core mission is currently in flux as the United States government turns its interest to the role that such centers might play in advancing industrial technology in a competitive international climate.

THE SURVIVAL OF UNIVERSITY INTERDISCIPLINARY ENGINEERING CENTERS

By and large, interdisciplinary engineering centers appear to have "done the right things" and for that reason, despite the several barriers discussed in the next section, they survive and several prosper in the United States. The reasons for their organizational effectiveness appear to be related to change and to constancy.

First, interdisciplinary engineering centers have been responsive to changing national imperatives. In the Sputnik era, emphasis was on the engineering sciences and a general drawing back from the traditional view of engineers as designers. The generics of materials science, engineering mechanics, engineering physics, etc., were stressed in all engineering disciplines. These centers became the meeting place for formulating common integrative themes for all fields of engineering. In the seventies, the driving force became nonmarket goods and social concerns (product safety, environmental quality, mobility, technology assessment, etc.). Interdisciplinary engineering centers have become sociotechnical think tanks. Today, as national concern over tougher international competition, trade imbalance and slippage in technology leadership heightens in the United States, we see interdisciplinary engineering focuses shifting toward automated manufacturing, the new biology, photonics and the materials sciences all with an emphasis on commercial applications. To that end, the NSF budget plan for Fiscal Year 85 proposes the funding of campus-based interdisciplinary centers in areas of engineering that reflect industry needs. The mid-eighties sees interdisciplinary engineering centers in flux as they shift to a new niche of advancing industrial technology. The challenge for university interdisciplinary engineering centers appears to be to facilitate the rapid flow-through of scientific discoveries into proven technology for adoption in American industry (Figure 1). Robert M. White, President of the National Academy of Engineering states (1983):

> There is, however, between the traditional role of the National Science Foundation [that is, to fund basic research]--and engineering practice, which is not regarded as being central to the NSF mission--a class of engineering activity, multidisciplinary in nature, directly addressing major national technological problems that can represent new and innovative areas of interest for the National Science Foundation in both substance and process. These new engineering activities represent fruitful opportunities for the National Science Foundation initiatives.

Second, despite their everchanging substantive focus, interdisciplinary engineering centers display a constant stream of second order benefits to their universities, industry and society at large. These include:

> High economic returns appear to occur where disciplines can coalesce around pressing complex problems; i.e., "economies of scope" can be achieved where complex problems are attacked from different disciplinary perspectives.
>
> Advances in one discipline rapidly spillover as new research methodologies in another, such as improved instrumentation, innovations in research techniques or totally new paradigms.
>
> Interdisciplinary engineering research has led to new subdisciplinary fields, such as biomedical engineering, environmental engineering, and automated manufacturing.

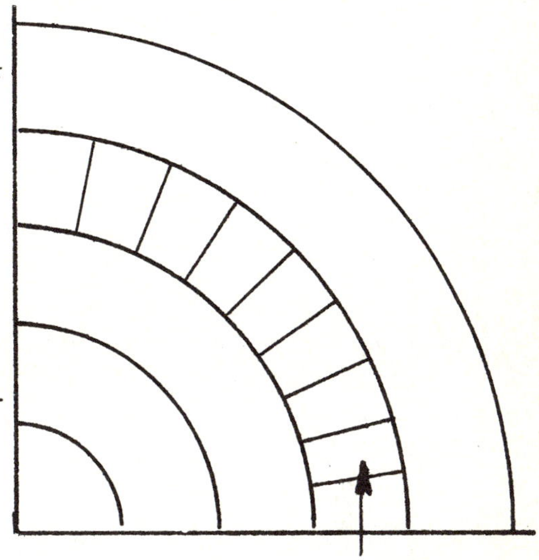

Figure 1
The Scope of University Interdisciplinary Engineering Centers in Industrial Technology Advance for the 80's and 90's.

Interdisciplinary engineering centers serve as a training ground for industrial researchers and R&D managers.

Interdisciplinary engineering centers facilitate technology transfer by acting as an intermediary between recent scientific discoveries and commercialization through prototype development and the dissemination of generic applied research.

While these reasons for being are noteworthy, interdisciplinary engineering centers face considerable threats in a university setting. These are discussed in the next section.

BARRIERS TO SUCCESS

Paradoxically, these very attributes of interdisciplinary engineering activity that enhance survival are also the sources of organizational conflict and barriers to success in a university setting. In short, there is disharmony between organizational effectiveness and organizational efficiency. At least six barriers to efficient operations in a university setting can be identified.

1. Variable Demand Requiring Flexible Operations. Because interdisciplinary engineering centers are highly responsive to fast-moving national imperatives they are also plagued by an unstable revenue stream and variable work program. This makes planning for space, equipment, and

personnel difficult. The need for a loose structure and flexible operating mode conflicts with the traditional administrative style of the university based on predictable income, permanent disciplinary research laboratories and long-term contracts with staff, suppliers, and sponsors.

As pointed out in the first section, problems come and go in society, and a country's engineering research resources must be flexible enough to mount attacks on those problems and, once solved, to shift direction. This kind of adaptability directly contrasts with the management of disciplinary research in general, where the accumulation of knowledge is crucial and continuity of research must be continued regardless of societal needs. This implies that interdisciplinary research projects have a very distinct life cycle that places special demands and restrictions on the use of university resources. The closing out of an interdisciplinary center, for example, is not to be equated with failure: indeed the opposite may be true.

2. Faculty Resources. Interdisciplinary engineering dollars will not necessarily buy faculty housed in traditional departments. Senior faculty are often comfortably tenured and see little need to sell their services to interdisciplinary projects. At the other extreme, junior faculty are preoccupied with positioning themselves for tenure, such as publishing in the right disciplinary journals. Premature commitment by junior faculty to interdisciplinary projects can mean slippage to the minor leagues.

3. Professional Research Resources. Basic research operations in the university have relied heavily on post docs and graduate students. This has led to university personnel policies that treat research professionals as second class citizens. Productive interdisciplinary engineering research depends upon mature experienced research professionals who can work as coequals with discipline-based faculty. Interdisciplinary engineering centers particularly benefit from nontraditional engineering scholars who can be distinguished from traditional engineering faculty by some of the following characteristics: enjoy messy real world problems; more inclined to synthesis (putting the pieces together) than analysis (busting the problem apart); use trial and error learning; use an action research mode that involves regular dialogue with client/user; decision (normative) science orienation as contrasted with natural (positivistic) science orientation; skilled in quasi-experimental design and pilot testing as opposed to clinical laboratory testing. Attractive job classifications for such nontraditional engineering scholars are difficult to create in a university.

4. Conflict with Traditional Research Culture of University. The traditional research values of the university, namely, academic freedom, pursuit of knowlege, freedom of information, and solo research, are violated to some degree in applied engineering endeavors. In the latter case the research agenda is demand (user) driven, timing is more

critical, and the findings may be proprietary in nature. Secrecy and suppression of information are antithetical to the culture of the university and generally a faculty member is far more anxious to have a paper published than a patent registered. This poses major dilemmas for industrial and defense sponsors of applied engineering research. Much of the difficulties in collaborating with industry is directly or indirectly associated with the control of information and obtaining patents. From industry's perspective technical know-how equals trade secret.

5. External Constituencies. With the decline of R&D by federal mission agencies and without a RANN type organization in NSF, interdisciplinary engineering appears to have had insufficient advocacy in Washington. New interdisciplinary engineering centers proposed by NSF in the 1985 budget could restore the balance between applied and basic research and promote excellence in engineering as a means to improved industrial competitiveness. Nevertheless, there is probably need for stronger representation in Washington on behalf of interdisciplinary engineering teaching and research. This could come from some established university association such as the Association of Land Grant Colleges.

6. Working with Industry and Government. While the notion of partnerships between university, industry, and government is popular, viable collaboration is extremely difficult due to disparate organizational cultures, missions, and modes of operation. Success appears to hinge on finding win-win situations. For example, in the creation of an innovations (or incubator) center, government is in the position to provide seed capital through research contracts, public venture funds and tax inducements to high risk investors. Government could also provide initial paid-in operating and/or capital grants as NSF did with the Industry/University Cooperative Research Centers (IUCR). On the other hand, university proximity to an innovations center provides consulting opportunities for faculty, part-time university opportunities for struggling entrepreneurs, and joint research contracts between fledgling technology companies and university R&D laboratories.

This general discussion of barriers points to the importance of deliberative organization design for each proposed interdisciplinary engineering center. Thought must be given to internal and external strategies outlined in the next section.

INTERNAL MANAGEMENT STRATEGIES

Focal internal management strategies are required to address at least three core organization design problems: the Downside Risk Problem, the Incentives Problem and the Intra-University Constituencies Problem.

Downside Risk Problem

Goal: Preparedness for the end stage of the funding/topic life cycle.

Strategies:

Portfolio. Look for funding/topic diversification in the portfolio of center activities unless concentration in a few areas with few funding sources does provide a market niche.

Structure. Avoid the tenure issue: one way to do this is to organize programs rather than centers; maintain a flexible structure using, for example, matrix management.

Human Resources. Minimize full-time staff; maximize use of professors, graduate students and post doctoral researchers with home bases in disciplinary departments.

Organizational Dynamics. Encourage project leadership to keep the topic alive, evolving and adaptable to new funding opportunities.

Financial Resources. Accumulate stabilization funds to compensate for the vagaries of interdisciplinary funding. University accountants and granting authorities may frown upon such rainy day funds.

Incentives Problem

Goal: Rewards to ensure both faculty and professional staff job satisfaction and quality of work life.

Strategies:

Portfolio. Diversify topics to provide job enrichment for professional staff; allow professional staff freedom to secure own funding, then recruit faculty as subcontractors.

Structure. Tailor reward system to meet needs of each participant, e.g., junior faculty rewarded with resources to generalize and publish results in disciplinary journals; senior faculty rewarded with resources to initiate new basic research topics and superior facilities and equipment; professionals presented with growth opportunities and multiyear contracts.

Organizational Dynamics. Stress intrinsic rewards such a intellectual stimulation and creative atmosphere through effective team building.

Financial Resources. Provide for a capital budget to ensure state-of-the-art facilities and equipment.

Intra-university Constituencies Problem

Goal: Healthy relationships with department heads and central administration.

Strategies:

Portfolio. Organize a research agenda encompassing several key departments.

Structure. Encourage department heads to participate in governing and setting research agenda in return for receiving funds to release faculty to participate in interdisciplinary endeavors.

Human Resources. Nurture a strong advocate of interdisciplinary centers within the central administration.

Organizational Dynamics. Stress communication within university, e.g., newsletters, annual reports, open houses; strive for social/physical proximity with departmental research, i.e. ideally, interdisciplinary research facilities should not be too distant from departmental laboratories.

Financial Resources. Ensure that university accounting practices allow departments to receive full budget credit from interdisciplinary research fund transfers.

EXTERNAL MANAGEMENT STRATEGIES

Focal external management strategies are increasing in importance as interdisciplinary engineering centers interface more with industry and governments, and as the intellectual property, patent management, product liability problems become more acute.

The Intellectual Property Problem

Goal: Rationally managed and marketable patents, with a view to holding patents for societal benefit rather than for the protection of trade secrets.

Strategies: Create an independent technology marketing activity to handle patents, licenses and royalties. This may be a part of the university organization but should be run as an enterprise with break-even or profit goals.

The Conflict of Interest Problem

Goal: Minimize conflict of interest, wherein enterpreneurial activity of faculty is divorced from traditional teaching, research and community service responsibilities.

Strategies: Create an enterprise/innovations center through which academic enterpreneurs can refine marketable ideas, write business plans and form companies. This center should be at "arms length" from the university with little or no financial or legal connections.

Product Liability Problem

Goal: Minimize likelihood of product liability litigation against the university.

Strategies: Create arms length patent and licensing corporation; change federal law to protect the university where good faith is present; find deep pocket (major insurer) to stand between university and final consumer.

The National Laboratories - University Collaboration Problem

Goal: Develop collaborative, as opposed to competitive, relationships with national laboratories.

Strategies: Seek more collaborative interdisciplinary research proposals between universities and national laboratories; encourage university senior faculty to be more involved on national advisory panels.

Federal Government Contract Problem

Goal: In addition to securing university interdisciplinary engineering centers funded under the proposed NSF budget, federal procurement policies and procedures giving greater recognition to the downside risk problem and to the heavy reliance on contracts, as opposed to grants, in interdisciplinary research.

Strategies (suggestions for federal action):

IR&D Provision. (Private defense contractors are allowed a certain percentage of their contract volume for independent research and development (IR&D). This is to encourage high-risk, exploratory investigations, just the kind universities are expected to perform. With a few exceptions, e.g., the Johns Hopkins Applied Physics Laboratory, universities are disallowed IR&D for both defense and nondefense contracts.) Provide IR&D to interdisciplinary engineering centers to smooth some of the irregularites in revenue stream and to provide improved continuity in frontier work.

Equipment Provision. (Until the late sixties it was common for United States government contracts to provide for necessary new equipment. More recent curtailment of this practice leaves interdisciplinary engineering laboratories at a disadvantage in providing the most up-to-date equipment.) Provide for state-of-the-art equipment in contracts and grants to improve quality of work, and to act as an incentive to draw disciplinary specialists into interdisciplinary ventures.

Overhead Charges for Capital. (In government contracts, universities write off buildings on a 50-year life cycle, instead of the customary 20 or 30 years. Capital equipment must be carried on a straight line depreciation of 15 years.) Allow for realistic depreciation rates especially considering the short life of high technology equipment.

OMB Circular A21. Re-examine this circular with a view toward rewarding results rather than effort.

CONCLUSIONS

Because of the pace and everchanging external environment in which university interdisciplinary engineering centers must operate, their survival, health and growth appears to have depended on a flexible management style and adaptive organizational structure. Their past responsiveness to changing national imperatives suggests little reason to fear that the next challenge, that of advancing industrial technology, will not be met.

Despite their apparent organizational effectiveness, university interdisciplinary engineering centers continue to face longstanding organization and management problems. These barriers to their efficiency can be condensed under three mega-problem headings: human resources management; university support; and industry, university, government endeavors.

First, the human resources mega-problem stems from the need for an appropriate mix of tenured, discipline-oriented faculty and problem-user-oriented research professionals. Both senior and junior tenured faculty are difficult to attract, while university personnel policies generally make research professionals second class citizens based on the postdoc syndrome of basic research laboratories.

Second, while interdisciplinary engineering research will be supported in the main from outside the university, a modest level of funding coupled with sympathetic treatment from university bureaucrats regarding policy compliance provides increased stability, continuity and needed flexibility.

Third, while industry, university, government partnerships sound promising, ideological and practical considerations are numerous and deep-seated. These include conflict of interest, overkill of the extension service rule, timely publication, and patent rights.

A variety of management strategies and alternative organizational arrangements appear to have worked for different institutional settings. Contingency theory, i.e., it all depends, best describes the management approach. The essence of the design and selection of management strategies is a recognition that universities are organized along disciplinary lines. The disciplinary characteristics of the particular university must be well understood, e.g., where the disciplinary power centers lie such as with department heads or deans, to ensure that productive interdisciplinary centers can survive the hostile environment.

From discussions at the NSF Workshop in "Making Interdisciplinary Centers Work," one management strategy that became conspicuously important in all institutional settings was that of building constituencies both inside and outside the university. University administrators, basic researchers, federal policy makers and mission agencies must come to recognize that centers for interdisciplinary engineering research are a part of an advanced industrial society and serve at its pleasure through processes of mutual adjustment. Internal dialogue with the central administration, department heads, and interested senior and junior faculty should be routine. In external relations, regular contact with federal mission agencies is important, as are diligent and persistent searches for joint ventures with industry.

REFERENCES

Shapiro, I.S. 1980. Business and the public policy process. In Business and public policy, ed. J.T. Dunlop. Cambridge, MA: Harvard University Press.

White, R.M. 1983. Presentation to July 1983 meeting of the National Science Board.

MANAGING HIGH TECHNOLOGY
An Interdisciplinary Perspective
B.W. Mar, W.T. Newell, and B.O. Saxberg (Editors)

FORTY INTERDISCIPLINARY RESEARCH PROJECTS: MULTIPLE SKILLS AND PEER REVIEW

Alan L. Porter and Frederick A. Rossini

School of Industrial and Systems Engineering
Georgia Institute of Technology
Atlanta, GA (USA) 30332

This comparative study examines forty research projects sponsored by the National Science Foundation (NSF). These predominantly interdisciplinary (IDR) projects include basic, applied, and policy research. In the course of this study, we have compiled a bibliography (Chubin et al., 1984) and developed a new conceptual framework that emphasizes intellectual skills, instead of disciplines, as basic units (Porter and Rossini, 1984; Rossini and Porter, this volume). We have also compiled a number of observations on the support and conduct of interdisciplinary research (Porter et al., 1984). This paper concentrates on two sets of empirical results: evidence that the multiple skills framework describes complex research projects in a useful way, and evidence that peer review of proposals discriminates against IDR.

The forty projects are drawn from five NSF programs: Environmental Geosciences; Neurobiology; Archeology; Earthquake Hazard Mitigation, Societal Response Program; and Science and Technology to Aid the Handicapped. Our information came from Carter (1982) discussion with the program managers, Chubin et al. (1984) perusal of each of the proposals, Cole et al. (1981) access to sanitized peer reviews of proposals, Harnad (1982) phone interviews with the principal investigators (PI's), and Linstone and Turoff (1975) examination of some of the research products (many of the projects were still underway).

The project sample was purposive: we asked program managers to identify projects they perceived as interdisciplinary (plus a few problem-oriented, disciplinary projects for comparison). As such, they cannot be assumed representative of all NSF projects. To examine some twenty hypotheses, we split the sample, program by program, with each investigator handling one subsample. This provided a reliability check on operationalizing the skill measures and two checks on the hypotheses.

MULTIPLE RESEARCH SKILLS

Table 1 illustrates some general characteristics by type of project. Applied research shows the greatest range of skills and diversity of disciplines within projects. These projects are also least likely to have continuing financial support and enduring teams.

We also compared twenty-four projects housed in academic departments with sixteen the others. These did not differ in terms of how interdisciplinary they were[1], team size, or how many skills were involved. They did have fewer participants from other disciplines. This suggests that researchers may acquire needed research skills, rather than recruit collaborators who possess them (see Rossini and Porter, this volume).

A most basic question concerning the notion of intellectual skills is whether these research skills can be identified. Figures 1-3 indicate that they can. PI's related well to our nominations of substantive areas (S) and techniques (T) based on our review of their proposals. The three projects show different skill distribution patterns. Note that the laboratory project (Figure 1) incorporates diverse skills, yet these are largely possessed by each individual. High skill density (Table 1) also corresponds with high communication density, suggesting that exchange of information may relate to skills. Figure 2 illustrates a pattern of relatively unique skills in a long-term collaboration. Figure 3 is an ad hoc team performing applied research. One can package the skills needed for IDR in quite different ways.

Note also that participant backgrounds do not specify skill contributions in a neat fashion. Those possessing a given skill may not employ it on a project. Conversely, others may acquire and use a skill during the project duration. Disciplinary training is not a sure guide to research contribution.

Let us now consider four basic hypotheses pertaining to research skills. The first suggests:

> At least one substantive area and one technique can always be identified.

This was supported in 35 of the 40 projects. One contained no S's: it reflected support to purchase equipment to be shared among independent researchers. Four contained no T's. We believe this reflects a coding problem, a need to disaggregate the substantive concerns. In certain policy studies, it was hard to classify a given skill (e.g., integrative modeling) as S or T.

The second hypothesis holds that:

> At least one S or one T will demand high expertise (unless the project represents cross-paradigm synthesis).

This assumes that researchers can distinguish between expert level (frontier, state-of-the-art knowledge) and technician level (journeyman, routine application of well-known information). In general they could, yielding the following distribution:

Median Research Skills Identified Per Project

Skill	Expert Level	Journeyman Level
Substantive Expertise	2.7	0.6
Technical Expertise	1.7	1.4

Table 1
Contrasts Among Basic, Applied & Policy Research

Variable Examined	Whole Sample			
	Basic Research	Applied Research	Policy Research	Statistical Significance
Number of Projects	17	13	10	
Range of Skills*	1.6	2.7	1.8	.003
% of Team from Outside PI's Grand Disciplinary Category**	7.3	38.8	10.5	.0001
Projects Representing Continuation Support, NSF or Other	88.2	61.5	30.0	.01
Ad Hoc Team Configurations (Rather than Quasi-permanent (lab) or Single Investigator)	41.2	92.3	60.0	.08
Skill Density (Degree to which each team member possesses all requisite substantive and technique expertise)	52.3	41.6	62.2	.03

Note: Values are means. Statistical significance indicates the probability level for F test of the one way Analysis of Variance or for Chi Square.

* Coded 1 = Skills lie within an established research area; 2 = Skills within grand category; 3 = Skills span grand categories.

**Grand Disciplinary categories represented are engineering, life sciences, physical sciences, professional fields, and social sciences.

Substantive Area) or T (Technique)	Skill	Level of Expertise 1 = Journeyman 2 = Expert	Team Members (by Discipline) 1) Prehistorian	2) Anthropologist	3) Prehistorian	4) Geologist	5) Archeologist	6) Archeologist	7) Illustrator	8) Botanist	9) Geologist	10) Botanist	11) Paleontologist	12) Archeologist (Ph.D. Student)	13) Archeologist (Ph.D. Student)	14) Chemist
T	Archeological Survey	2	X	X	X		X	X						X	X	
S	Geological Mapping	2			X	X					X					
T	Pollen Analysis	1								X						
S	Neolithic Adaption	2	X	X	X		X	X								
T	Statistical Analysis	1		X												
T	Botanical Analysis	1										X				
S	Faunal Analyses	2											X			
T	Excavation	2	X	X	X		X	X						X	X	
T	Artifact Analyses	2	X	X			X	X						X	X	X

Figure 2
S&T Matrix for a Large Project (Archeology)

			Team Members (by Discipline)											
S (Substantive) or T (Technique)	Skill	Level of Expertise 1 = Journeyman 2 = Expert	1) Artificial Languages Expert	2) Technician	3) Rehabilitation Specialist	4) Linquist	5) Business Administration	6) Psychologist	7) Electrical Engineer	8) Center for Disabled	9) Electrical Engineer	10) Software Engineer	11) Hardware Engineer	12) Psycholinguist
T	Interview Disabled Consumers	1			X					X				
T	Interview Others	1			X					X				
S	Voice Output	2	X			X	X	X		X	X			
T	Prototype Construction	2									X	X	X	
S	Evaluation	2	X		X							X		
T	Simulation	1		X							X			

Figure 3
S&T Matrix for a Several Phase Science & Technology
to Aid the Handicapped Project

Expert level skills range from 1 (on three projects) to 16 (on a single project) with a median of 4.3, supporting the hypothesis. Of the roughly six discrete skills distinguished per project, about half are substantive in nature. Journeyman level substantive expertise is rarely noted. Perhaps such knowledge is taken as general background. The only statistically significant anomaly between the coders was in the frequency of journeyman techniques noted. Further refinement of the coding is needed.

The third hypothesis asserts:

> Complexity in terms of S&T can be measured; it will not correlate simply with number of disciplines.

None of the twenty project subsamples showed a significant correlation. For the whole sample, the correlation between the number of disciplines was nonsignificant ($\acute{r}$ = -.02 for core team; .07 for whole project team). This implies that counting disciplines does not capture the intellectual complexity of a research endeavor.

A more intricate hypothesis holds that:

> Problems requiring substantive expertise in two or more areas are more likely to need collaboration than problems requiring technique expertise in a similar number of areas.

A test was whether number of team members correlated with number of S's more highly than with number of T's. It does, in each sample of 20 projects, by an average of .25. It appears that we can begin to dissect research collaboration in terms of the type of skills involved.

PEER REVIEW

One of the great concerns for interdisciplinary research projects is that they not fall between the cracks of organizations designed to foster disciplinary research (Porter et al., 1980). Peer review processes raise such concerns, both with regard to choosing proposals to receive funding (cf., Cole et al., 1981; Carter et al., 1982) and refereeing papers for publications (cf., Harnad, 1982). We were fortunate to receive access to sanitized proposal review data on 37 of the projects.

The peer reviews of these NSF projects showed notable differences on several dimensions. Reviewers favored academic researchers. We examined average ratings by the type of organization with which the PI was associated. No significant differences were seen among academic departments and centers, nor among nonacademic units such as quasi-public organizations or large/small consulting firms. However, the difference between academic and nonacademic units was highly significant ($t = 4.29$, $p = .0001$):

PI Organization	Reviews (N)	Rating (Mean)
Academic	161	1.63
Non-academic	86	2.18

(Ratings scaled from 1 = excellent to 5 = poor; range is constricted in that these reviews all pertain to funded proposals.)

Interestingly, there are no such differences by type of reviewers. Academic and nonacademic reviewers do not give significantly different ratings, nor do researchers and research users.

Several other characteristics relate to how favorable reviews tend to be. Unfortunately, they are confounded with the academic/non-academic distinction, and with each other. Basic research proposals are rated more highly than applied or policy research proposals. Science proposals rate higher than engineering ones. Proposals in certain NSF programs average better ratings than those in other programs. Essentially, we can only say that these factors associate with strong reviews, not which, if any, of them affect ratings received. We also cannot say that scientific merit is the same for these projects. A rather remarkable finding is that these factors appear to outweigh differences among individual proposals. For example, in a multiple linear regression, differences among ratings are better accounted for by NSF program ($F = 16.17$, $p = .001$) than by individual project differences ($F = 2.06$, $p = .15$). That implies that the funding cutoff varies by program.

We hypothesized that:

> Peer ratings will be lower for more interdisciplinary proposals.

Unfortunately, our sample contains mostly interdisciplinary projects so that it is not possible to contrast ratings of IDR projects with those received by clearly disciplinary projects. Also, all the proposals for which we have reviews are funded. Certain comments made by reviewers do suggest discomfort with cross-disciplinary work. For instance, one economist faulted a proposal for including noneconomic aspects. On occasion, a reviewer would indicate reluctance to grade a proposal in other than his or her domain of expertise.

We were able to observe the degree of correlation between peer rating and how interdisciplinary a project was (recall Footnote 1). The more interdisciplinary the project, the poorer the rating ($r = .29$; $rho = .23$).

Our most intriguing finding offers a clue as to why interdisciplinary proposals are down-rated. It is reasonable for a reviewer of proposed research to favor that which is more familiar personally. In such a case, one is apt to understand better what is planned; may know the researchers personally or by reputation, hence appreciate their expertise; and could be more secure in making strong recommendations. We were able to classify reviews by the similarity of the disciplinary affiliation of the reviewer and the PI. Table 2 indicates a marked tendency of reviewers to rate proposals from PI's from their own discipline more favorably.

Table 2
Peer Review Ratings by Disciplinary Similiarity of PI and Reviewer

Disciplinary Match[1]			
Degree of Match	Reviews (N)	Ratings (Mean)	Standard Deviation
Same	76	1.69	.85
Similar	52	1.68	.80
Different	67	2.21	1.00
[F = 7.61, p = .0007]			
General Disciplinary Category Match[2]			
Same	112	1.73	.86
Different	82	2.07	.98
[F = 6.70, p = .01]			

[1]Our judgment based on information available on PI and Reviewer.

[2]Mechanical comparison of the general disciplinary grouping (physical science, life science, social science, engineering, or professional fields -- as per National Research Council codes) of the PI and Reviewer.

This suggests a grave concern in reviewing IDR proposals. The obvious strategy of composing a review team by including one reviewer who knows one aspect, another who is expert in a second area included in the proposed research, and so on, implies a set of reviewers each unfamiliar with much of the work. According to our present findings, that means an expected down-rating of the proposal in comparison to one of equivalent merit, but for which individual reviewers can more fully grasp the entire research strategy.

If IDR is to prosper, such hurdles must not be ignored. To give such proposals an even chance, one has several choices. One is to lower rating standards, but that is difficult to justify within programs. Given NSF's policy of situating applied research projects in the basic research programs, this is problematic. Cross-program review often entails additional jeopardies (e.g., even more reviewers with only partial expertise or multiple panel reviews). Another possibility is to seek reviewers who are expert across the breadth of the proposed work. They may be difficult to secure for frontier research and there is danger of in-breeding in cases where only a small number of potential reviewers share interests fully. Panels have an advantage in allowing discussion to help each other understand unfamiliar aspects of a proposal under review. We have support for such an uncertainty reduction strategy in remarks of some panelists that they "changed their mind after discussion." However, panels are costly and must be constituted to address sets of proposals. Novel IDR is likely to fall on the fringe of panel expertise, hence face an expected poorer rating (scientific merits equal).

We suggest that incorporation of feedback in the review process could help remedy the problem of restricted reviewer expertise. Extending the present (sometimes used) provision of an opportunity for the PI to respond to reviewer concerns before a final funding decision, we advocate a Delphi process (cf., Linstone and Turoff, 1975). Mail reviewers would comment on a given proposal, but not provide a numerical rating. These comments would be provided anonymously to the PI and to the other reviewers. The PI and reviewers would be given an opportunity to explain the proposed research more satisfactorily. Only after receiving these amplifications would reviewers make the numerical rating. Such a process should improve reviewers' perspectives on the full project and resolve concerns due to unfamiliarity with certain aspects.

CONCLUSIONS

Effective interdisciplinary research means solving complex problems by bringing to bear multiple research skills. This empirical study demonstrates that the STRAP framework, in particular the notion of substantive and technique skills, is understandable to researchers. The framework also generates testable hypotheses concerning ways to structure multiskill research. In so doing, it opens new dimensions for research management. New problem solving strategies are suggested. For instance, the single senior investigator (lab) model for IDR might find useful application in certain applied or policy areas. Conversely, the ad hoc team of individuals with distinct skills might enrich basic science IDR. Focusing on intellectual skills in lieu of disciplines is also more conducive to study of nonacademic, problem-focused research. As formulated, skills are addressed inductively. It will be useful to compile skill sets and to compare these with existing intellectual classifications (e.g., bibliometric categories, subdisciplinary groupings).

As discussed, peer review is crucial to research. We have shown evidence of inequalities in peer ratings of proposals. In particular, the demonstration of substantially lower ratings when the reviewer's discipline differs from that of the PI, suggests that IDR cannot be reviewed the same way as disciplinary projects.

ACKNOWLEDGEMENTS

We gratefully acknowledge research support from the Office of Interdisciplinary Research, National Science Foundation (Grant OIR-8209893). We thank our colleagues in the study, D.E. Chubin and E. Kotzakidou; our helpful project monitor, R.R. Goulet; the NSF program managers who shared insights and files; and the researchers who told us how projects really were performed. The views expressed are those of the authors and do not necessarily reflect the views of NSF or Georgia Tech.

FOOTNOTES

1. Degree of interdisciplinarity is based on a factor analysis. This yielded a factor composed of five variables: PI judgment of how interdisciplinary the project was, our judgment of how interdisciplinary the project was, number of disciplines on the team, range of skills represented, and the percentage of the team from outside the PI's grand disciplinary category (see Table 1 note).

REFERENCES

Carter, G.M. 1982. What we know and do not know about the NIH peer review system. Santa Monica, CA: The Rand Corporation, N-1878-RC/NIH.

Chubin, D.E., Porter, A.L., and Rossini, F.A. 1984. Interdisciplinary research (problem-focused, multi-skilled research) -- National Science Foundation Experiences, vol. 2 Bibliographical analysis. Georgia Institute of Technology.

Cole, S., Cole, J.R., and Simon, G.A. 1981. Chance and consensus in peer review. Science 214:881-886.

Harnad, S., ed. 1982. The behavioral and brain sciences, Special Symposium Issue: Peer Commentary on Peer Review 5(2).

Linstone, H.A. and Turoff, M., eds. 1975. The Delphi method: techniques and applications. Reading, MA: Addison-Wesley.

Porter, and Rossini, F.A. 1984. Interdisciplinary research redefined: multi-skill, problem-focused Research in the STRAP framework. R&D Management 14(2):105-111.

Porter, A.L., Rossini, F.A., and Chubin, D.E. 1984. Interdisciplinary research (problem focused, multi-skilled research)-- National Science Foundation experiences. Georgia Institute of Technology.

Porter, A.L., Fossini, F.A., Chubin, D.E. Connolly, T. 1980. Between disciplines. Science 209:966.

Rossini, F.A., and Porter, A.L. 1984. Interdisciplinary research without disciplines, this volume.

MANAGING HIGH TECHNOLOGY
An Interdisciplinary Perspective
B.W. Mar, W.T. Newell, and B.O. Saxberg (Editors)
Elsevier Science Publishers B.V. (North-Holland), 1985

A SURVEY OF INTERDISCIPLINARY APPLIED RESEARCH IN THE U.S. FEDERAL GOVERNMENT

Albert H. Teich and Jill H. Pace

American Association for the Advancement of Science
1776 Massachusetts Ave., N.W.
Washington D.C. 20036

INTRODUCTION

The purpose of this paper is to examine the support of interdisciplinary research by the U.S. federal government (in the civilian sector) by reviewing statistics on the R&D budgets of key federal agencies. Under the proposed budget for fiscal year 1985, a total of $54.3 billion in new budget authority would be allocated to R&D. This represents slightly less than half of the U.S. national R&D effort; the balance of the nation's R&D funding comes principally from industry, and to a much smaller extent, from state and local governments, and from universities and other nonprofit institutions.

By far the largest share of federal R&D funding is devoted to the military. Of the proposed FY 1985 total of $54.3 billion, about 70 percent ($37.9 billion) would be spent on military R&D through the Department of Defense (DOD) and the Department of Energy (DOE). Military R&D is a very specialized activity. Much of it is classified and detailed information on even the unclassified portions is not readily available. This paper therefore excludes military R&D and concentrates on the remaining $16.4 billion of civilian R&D.

Distinguishing between disciplinary and interdisciplinary research in this statistical aggregate is not a straightforward task. In most R&D activities, the federal government is primarily concerned with the solution of a particular problem or set of problems and only secondarily concerned with whether the R&D is carried out in a disciplinary or interdisciplinary mode. The disciplinary or interdisciplinary character is of importance principally to the performer of the R&D. In gathering R&D statistics the federal government classifies basic and applied research by fields of science. It does not classify development by field.

We would contend that the concept of disciplinarity is mainly applicable to basic research, where the primary thrust of the researcher is to contribute to the advancement of knowledge and the primary audience for the results is the community of researchers working in the same or related areas. In most applied research, and especially in development, the main defining characteristic is the practical application to which the R&D is intended to contribute rather than the field or fields within which the work is conducted.

Most federal agencies that support R&D are organized according to the problems with which they are concerned rather than the disciplines which they support. Only such agencies as the National Science Foundation (NSF), and such parts of agencies as the high energy physics or basic energy sciences programs of DOE, whose primary mission is to support basic research in academic or quasi-academic institutions, are organized by discipline. The other agencies are interdisciplinary by virtue of their organizational nature--although some (such as DOE) may encompass a broader range of disciplines in their activities than others (such as the National Oceanic and Atmospheric Administration, NOAA).

As a means of simplifying our problem of surveying interdisciplinary research, therefore, we have chosen to examine the civilian applied research and development component of federal R&D as a rough first approximation of federal interdisciplinary research.

As Table 1 shows, federal support of nondefense applied R&D totals $9.01 billion under the proposed FY 1985 budget. Under this heading falls an extremely diverse set of activities ranging from studies of alternative peanut drying techniques to such high technology programs as magnetic fusion. These R&D efforts have their roots in various areas of government responsibility: for example, providing for public health and safety by controlling environmental pollution, supporting the nation's agricultural enterprise, or assisting states, localities, and individuals to protect themselves against natural hazards, such as earthquakes, floods, or tornadoes. Although performer statistics are not available for the proposed FY 1985 budget, comparable figures from NSF (for FY 1984) indicate that about 26 percent of federal civilian applied R&D is conducted in government laboratories, 35 percent is conducted in industry, 15 percent in universities and colleges, and the balance in federally funded R&D centers (FFRDCs), state and local governments, and other nonprofit institutions.

FORMS OF FEDERAL CIVILIAN APPLIED R&D

Federal support for applied research and development stems from a variety of different motives. A useful way of understanding applied R&D is to develop a classification scheme in terms of these motives, and apply the scheme to the overall federal R&D program as represented in the current (FY 1985) budget. For purposes of this analysis, we have adapted a taxonomy developed by Nelson and Langlois (1983) in their article, "Industrial Innovation Policy: Lessons from American History."

Nelson and Langlois performed a series of case studies of innovations in different industries and examined the implications of their findings for federal support of civilian R&D. They suggest that there are four general kinds of civilian R&D programs, each supported by government for a different set of reasons.

1. programs attendant on government procurement or some other well-defined public objective

2. programs to support research on generic technologies --research in the gray area between basic scientific research and applied R&D

3. programs to support applied R&D in the service of well-defined clientele demands

4. programs that insist on "picking winners" in commercial applied R&D.

Table 1
R&D in the Proposed FY 1985 Budget by Agency*
(budget authority in millions, current dollars)

	Basic Research	Applied Research	Development	Facilities	TOTAL
DOD - military	$ 930	$ 2,326	$ 31,690	$ 390	$ 35,336
DOE - AEDA	4	694	1,323	+ 501	2,522
DOE - other	936	1,019	800	489	3,244
NASA	827	1,088	1,425	127	3,466
HHS	2,913	1,676	348	15	4,952
(NIH)	(2,738)	(1,261)	(342)	(14)	(4,356)
NSF	1,330	78	0	18	1,427
Agriculture	420	450	29	289	26
Interior	97	248	20	*	364
Transportation	**	83	407	11	502
EPA	25	173	81	0	280
Commerce	18	202	51	12	283
NRC	0	168	0	0	168
VA	15	157	24	28	224
AID	9	140	115	5	269
Labor	7	11	1	0	19
Education	14	64	26	3	107
TVA	6	3	18	*	27
Smithsonian	66	0	0	2	68
HUD	0	8	13	0	20
Corps of Engineers	1	19	14	0	34
Justice	3	13	28	0	44
FEMA	0	6	0	0	6
Treasury	2	8	4	*	14
TOTAL	$ 7,623	$ 8,635	$36,416	+ $ 1,628	$54,301
(Civilian R&D)	($6,689)	($ 5,615)	($ 3,403)	($ 737)	($16,444)

* Amounts represent authors' estimates based on data from "Special Analysis K," supporting OMB documents and agency budget justifications. Detail may not add to totals due to rounding.

** Less than $0.5 million.

\+ Denotes nondefense applied research and development treated in this analysis.

Our approach employs essentially these four categories, but views them as cells in a 2 by 3 matrix, reflecting two distinct dimensions: (1) the nature of the R&D itself, and (2) the intended user of the R&D (Figure 1). With respect to the nature of the R&D, we can distinguish programs in terms of their generality or technological specificity. In order to simplify, we treat this as a dichotomy, although we recognize that it is really a spectrum ranging from activities that are close to basic research on one end, to highly targeted product development efforts on the other end. With respect to intended users, we consider three categories: programs for which government itself is the user, programs for which a specific outside interest group (or clientele) is the user, and programs which seek to serve a broader segment of society, beyond the government or a specific interest group, which we have labeled "other."

		Intended User		
		Government	Interest Group	Other
Nature of R&D	General	A	B	C
	Specific	A	B	D

A. Procurement-related
B. Clientele-oriented
C. Generic Research
D. Favored Technology

Figure 1
Matrix of Federal Applied R&D Programs

Within the first two user categories the nature of the R&D is less important for our purposes than the identity of the user. That is, if government or a specific client-group is the intended user, then the degree of generality or specificity of the R&D itself is not all that relevant to our analysis. Thus, in our scheme, both cells in the left-hand column are considered together as procurement-related R&D, equivalent to Nelson and Langlois's first category. Similarly, both cells in the middle column are regarded as clientele-oriented applied R&D. When the user community is less well-defined however, the degree of generality or specificity of the R&D becomes more important, since, as Nelson and Langlois point out, it influences potential private sector applications. The more general type of R&D in this category is equivalent to Nelson and Langlois's generic technology type, while the more specific is, in their terms, picking winners, or in ours, favored technology.

Procurement-related R&D

In procurement-related civilian R&D the federal government itself is the primary client. Included in this category are applied R&D programs supporting regulatory missions, such as those of FDA or EPA; space R&D in NASA apart from that closely linked to aeronautical and space applications; and R&D in areas serving important public needs where no market exists, such as R&D on improved air traffic control in FAA. The rationale for these programs grows out of the government's own needs--i.e., the need to pursue its missions most effectively, and in a broad sense, the government itself makes the market for the results.

Generic Research

The generic category includes R&D on problems relevant to the commercial marketplace but somewhat removed from specific, immediate applications. As Nelson and Langlois put it,

> much of it is the sort nonpatentable and nonspecific knowledge--broad design concepts, properties of materials, and testing concepts--that is generally shared among scientists and does not pose a strong threat to proprietary interests (1983, p. 816).

Examples would include much of NASA's work on aeronautical R&D or NBS's research on computer software standardization. The work supports industry needs but covers broad areas which few individual firms have resources or motivation to explore on their own. Nelson and Langlois include both basic and applied research in this category. Our usage excludes basic research, which is classified separately in the budget analyses which we are using, but we recognize that the line is a fuzzy one and, as we discuss below, good arguments can be made for dropping the distinction entirely.

Clientele-Oriented R&D

The primary defining characteristic of clientele-oriented R&D is the participation of a specific interest group in priority-setting, conduct, and use of the R&D. The agricultural research and extension system, with its decentralized structure and traditional linkages with farmers and farm organizations, represents the best developed example. As Nelson notes (1982, p.464),

> ...the agricultural sciences have in general managed to define their niche appropriately. The research they do lies in between the basic academic sciences such as chemistry and biology on the one side and on the other the research that goes on in public experiment stations and private companies to develop, for example, better seeds or fertilizers. Both sides influence the kind of research that is done and monitor quality and efficacy.

Another example might be the mining research activities of the Bureau of Mines. Whatever criticisms might be made of this type of federal R&D program, and there are many, it is generally regarded as an effective means of employing federal resources to serve specific needs in particular areas of the private sector.

Favored Technology

The fourth category, which we term favored technology R&D (picking winners, in Nelson and Langlois's terms), involves programs in which federal agencies participate in the selection of specific commercial technological products or devices, often large and expensive, for R&D and demonstration. Generally, these programs are geared to bring to the market a technology that would not reach commercialization (at least within the desired time-frame) without such assistance. Noteworthy examples include the abortive supersonic transport program, HUD's Operation Breakthrough, and many of DOE's more recent energy demonstration projects.

These categories are, of course, not mutually exclusive or entirely unambiguous. Certain federal R&D programs might well fall into more than one category, depending on one's point of view or on how the program description is written. Nevertheless, we believe that the categories are sufficiently distinct to allow use of them for at least a roughcut analysis of the federal R&D budget. The next section attempts such an analysis.

CIVILIAN APPLIED R&D IN THE FEDERAL GOVERNMENT: AGENCY PROGRAMS

Introduction

Of the $9 billion in federal nondefense applied R&D in the FY 1985 budget, more than $8 billion is accounted for by nine agencies. The R&D programs of these agencies are described in the budget justification documents submitted by the agencies to the Congress as part of the budget approval cycle. For this analysis, we have reviewed these budget documents, extracted from them brief descriptions of the relevant R&D programs, and applied our taxonomy to these programs. This section of the paper includes these program descriptions and the basis of our classification by category.

Department of Health and Human Services

National Institutes of Health. The National Institutes of Health (NIH) is the principal biomedical and behavioral research agency of the federal government, and most of its support is classified as basic research. Nevertheless, more than one-third of NIH-supported R&D is considered by NIH to be applied research and development. Much of this falls under the NIH category of clinical research, in which the fundamental knowledge developed by basic research is used to classify pathological conditions and identify methods for intervention, prevention and amelioration of those conditions. Such studies include clinical investigations in humans and preclinical research to develop disease models, diagnostic tests, and prosthetic devices. We have classified NIH's applied research as generic.

Food and Drug Administration. The Food and Drug Administration (FDA) supports and conducts a variety of regulatory and research activities directed toward consumer protection, to ensure the safety and purity of foods, human and animal drugs, biological products, and therapeutic devices, and to reduce unnecessary human exposure to man-made radiation. Virtually all of this R&D is procurement-related in the sense that it serves FDA's regulatory activities. The agency supports a limited amount of extramural research (e.g., for developing standards); almost all research directly related to regulation is performed in-house. FDA's applied R&D activities include research on human drugs, as part of the drug review and approval process. FDA also supports development of orphan products, which include drugs, foods for medical uses, and medical devices with limited applications, which are not likely to be developed by the private sector. In our classification scheme, this activity falls into the category of favored technology development.

Centers for Disease Control. The Centers for Disease Control (CDC) is involved in research directed toward the prevention of illness, disability and premature death. It is the lead agency for responding to health emergencies, including natural disasters, technological hazards, and disease outbreaks. CDC also works in the areas of occupational safety and health and family planning. We have classified the entire CDC applied research and development program as generic research.

Alcohol, Drug Abuse and Mental Health Administation. The Alcohol, Drug Abuse and Mental Health Administration (ADAMHA) focuses its R&D activities on the prevention and treatment of problems related to alcohol and drug abuse, and mental and emotional illnesses. Its research programs are administered by its three institutes, the National Institute of Mental Health (NIMH), the National Institute on Drug Abuse (NIDA), and the National Institute on Alcohol Abuse and Alcoholism (NIAAA). ADAMHA research areas include social and behavioral research on crime and delinquency, minority mental health problems, rape services development, and biological and biochemical research on the genetics of alcoholism transmission, and on health effects related to the genetics of alcoholism transmission, and on health effects related to the use of various drugs. The functions of these programs are analogous to those of NIH and CDC and we have classified them as generic research.

National Aeronautics and Space Administration

Applied research and development programs in the National Aeronautics and Space Administration (NASA) totalled $2,512 million in the FY 1985 budget. Most of this total is devoted to what we would classify as procurement-related programs, that is, programs for which the government itself is the client for the R&D. This category includes all space systems activities and the weather and land satellite programs.

NASA's applied R&D also includes some generic technology efforts. The largest of these by far is the aeronautical research program. This program supports basic research in aeronautical disciplines, and more applied systems research applicable to general classes of advanced aircraft. The program works closely with aeronautical manufacturers "to ensure that the technology is compatible with practical design considerations and can be successfully transferred into use." Since the 1940's, this has been NASA's policy with respect to aeronautics; that is, that NASA would perform basic research and generic applied research, as well as enough development to demonstrate the viability of a new technology, and the private sector would then be responsible for product development, as well as any additional, product-specific R&D it chose to undertake.

Department of Energy

The Department of Energy's (DOE's) applied research and development efforts are among the largest and most controversial in the federal government. These efforts have been substantially reoriented during the past several years. The Reagan Administration's energy policy has consistently emphasized nuclear R&D programs while calling for reductions in programs for fossil, solar, environmental and conservation R&D. The result of the administration's cutbacks and redistribution of funds, combined with Congressional termination of support for the Clinch River Breeder Reactor project, has been to leave most of DOE's programs--which three years ago had included many "favored technology" programs--focused on longer-term generic R&D activities. These generic programs, and the few remaining exceptions which fall into other categories in our scheme, are described below.

Solar and Geothermal Energy. Applied solar energy research, all counted as generic research, is conducted in many program areas, including: solar building energy systems, photovoltaic energy systems, solar thermal energy systems, biofuels energy systems, wind energy systems, and ocean energy systems.

The two major geothermal energy R&D programs are geopressured resources and geothermal technology development. Both are classified as generic research.

Nuclear Fission and Fusion. The civilian radioactive waste R&D program emphasizes spent fuel storage R&D, particularly at-reactor storage technologies and alternative disposal concepts research, examining the feasibility of disposing of nuclear waste in stable geologic formations in the deep ocean floor.

The FY 1985 breeder R&D program "will be aimed at evolving and confirming those concepts that are best suited for the market and are viable alternatives to other energy sources." Following termination of the Clinch River Breeder Reactor project, the restructured breeder program is intended to resolve the "technological and economic uncertainties which are a prerequisite for the private sector to demonstrate and deploy the breeder." The revised program will not include a new federally funded demonstration project. Most of the nuclear programs would now be categorized as generic R&D, although some, such as the advanced nuclear systems activities, are procurement-related, and some, such as the water-cooled breeder development project, are considered favored technology.

According to DOE, "the proper role of the Federal Government in fusion is to demonstrate scientific and engineering feasibility." DOE's Magnetic Fusion Energy program is viewed as a three-phase effort in "developing the technical data base required for the private sector to determine the commercial potential of fusion." The two major magnetic confinement concepts being pursued as part of this base program are closed magnetic toroidal systems (particularly tokamaks) and open magnetic mirror systems.

Electric Energy Systems and Storage. The systems program supports research aimed at increasing the efficiency, reliability and adaptability of electric power transmission and distribution systems. Such work, affecting both the cost of power to consumers and the security of energy systems, "requires a sustaining and longer-range view than any individual utility, group of utilities, or research arm of the utilities, has thus far deployed; hence, the Federal presence, " according to DOE. The energy storage program focuses primarily on electrochemistry as applied to battery and fuel cell development and on developing new storage applications for more cost-effective use of energy sources. These program justifications strongly suggest a generic research classification for all of the applied R&D in this area.

Fossil Energy R&D. The DOE FY 1985 budget for fossil energy R&D ($139.6 million) is intended to focus on "long-term, high-risk R&D that the private sector lacks the incentive to finance." Within the coal program recent emphasis has been placed on acid rain-related research, particularly control technology. The gas program is intended to aid the private sector in developing cost-effective diagnostic and extraction technologies to produce gas more efficiently and economically from unconventional gas resources. Research is also conducted on the resource potential of methane hydrates and deep source gas. Although the big ticket synfuels demonstration plants which were terminated by the Reagan Administration were certainly favored technology programs, the current efforts, examples of which were described above, seem to be fairly described as generic research.

Energy Conservation. Conservation R&D programs operate in the areas of buildings and community systems, industrial energy conservation, transportation energy conservation and multisector programs. The program tries to encourage industry to switch from petroleum to more abundant energy sources through various generic research projects.

Department of Commerce

National Oceanic and Atmospheric Administration (NOAA). NOAA's research programs fall under several broad categories: ocean and coastal programs, marine fishery resource programs, atmospheric programs, and satellite and environmental data and information services. Most applied R&D in NOAA falls into the category of procurement related, tied to NOAA's weather service mission, satellite activities, and similar efforts. We have classified the fishery resource programs as clientele-oriented, because of their direct role in serving the fishing industry.

National Bureau of Standards. The National Bureau of Standards (NBS) develops and maintains a wide range of standards for measurements to support domestic and international standards processes, promotes technological innovation, and supports related applied research and development. With the exception of some procurement related work in support of radiation regulations and computer standards, all NBS's applied R&D falls in the generic R&D category.

Department of Agriculture

The U.S. Department of Agriculture's (USDA's) research programs are divided among three major bureaus, the Agricultural Research Service, the Cooperative State Research Service, and the Forest Service, and several smaller programs.

USDA's applied R&D falls into our category of clientele-oriented

R&D, since the user community plays such a central role in the formulation of federal priorities in this area. However, some work which addresses broad resource conservation and use patterns, and other R&D which is not crop- or animal-specific, could be classified as generic technology R&D.

Agricultural Research Service. The Agricultural Research Service (ARS) is the department's in-house research agency, supporting basic and applied research and development at 145 separate field locations on plant and animal production; use and improvement of soil, water and air resources; processing, storage, distribution, food safety and consumer services; and human nutrition.

Cooperative State Research Service. The Cooperative State Research Service (CSRS) supports agricultural research at state agricultural experiment stations and other eligible institutions, such as the 1890 land grant colleges and Tuskegee Institute. It also supports competitive basic research grants and special research grants.

Forest Service. Forest Service research is intended to improve forest and range land productivity, and includes efforts in forest genetics, silviculture, integrated pest management, forest engineering, hydrology and wildlife biology. Specific efforts are related to protecting wood from decay, developing processes to treat the lignin-bearing wastes from pulp and paper mills, predicting the effects of surface mining and reclamation on streamflow and water quality, and other resource management activities.

Department of the Interior

As the principal conservation agency in the United States, the Department of the Interior is responsible for most nationally owned public lands and natural resources. The Department conducts and coordinates an array of research activities related to the management and use of these resources.

Some research relates to the environments required for the existence of various species of fish and wildlife, and the effects of management practices on them. These procurement-related, mission-oriented research efforts include activities to identify and evaluate the increasing array of agricultural and industrial chemicals and other environmental pollutants that affect fish populations and their habitats; programs to assess the effects of forest and range management practices on wildlife species; and development and evaluaton of control techniques for migratory bird management.

The U.S. Geological Survey supports geologic research and hazards investigations, minerals research and water resources investigations. Most of these are generic technology R&D; some procurement-related R&D is performed in support of federal regulatory activities.

The Bureau of Mines is responsible for conducting research involving mineral reserves and the production, consumption and recycling of mineral materials. R&D programs include clientele-oriented programs aimed at improving the technology for extracting minerals from the ground and for processing and using minerals and mineral products; and procurement-related R&D to assess the minerals potential of public lands, and to improve mineral material performance and critical and strategic mineral substitutes.

Other Agencies

Three other federal agencies with significant applied R&D programs were included in our survey. R&D in support of the regulatory programs of EPA and NRC was, on a first approximation, classified as all procurement-related, while NSF's applied research activities (mainly industrial innovation and cooperative industry/university activities) were classified as generic R&D. Applied R&D programs of other federal civilian agencies were not classified.

ANALYSIS

The foregoing review of federal applied R&D programs is a very broad brush treatment. Describing $9 billion worth of R&D in a relatively few pages and classifying programs on the basis of budget justification documents forces one to work at a high level of aggregation and ignore many important details. Yet, crude as it may be, there is a good deal to be learned from this exercise. The categories we have chosen are helpful in better understanding the nature of federal applied R&D programs. The best way to see this is to examine the overall federal applied R&D effort by category, cross-cutting the agency analyses of the previous section. Table 2 summarizes the results of the agency analyses of the previous section. As can be seen in the table, of the $9.01 billion in nondefense applied R&D, we were able to classify programs representing all but $1.5 billion. Among the $7.5 billion worth of R&D programs that were categorized, 58 percent were generic research, 34 percent were procurement-related, 7 percent were clientele oriented, and less than 1 percent were favored technology programs.

Procurement-Related R&D

The largest procurement-related federal applied R&D program (outside of defense) is that of NASA. The mission of space exploration is unique in many ways and--apart from specific applications such as communication--there is little alternative to a central role for the federal government if the nation wishes to pursue this mission. As in the military, the government's role makes it the direct client for the R&D it sponsors and leaves little doubt about the necessity for federal sponsorship of the R&D.

Similarly, the nature of NOAA's mission--especially that portion relating to weather forecasting--and those elements of the Interior Department's mission relating to resource conservation make portions of their R&D programs rather clear-cut procurement-related efforts. The technologies (with a couple of exceptions) are distinct from those of NASA or the military, but the nature of the government role is similar, in that it is clearly defined by the fact that the agency itself is the intended user of the results.

Other procurement-related programs, about $500 million, fall in the regulatory domain, including the programs of EPA, NRC, and FDA. Here, too, the roles seem rather clear-cut. If EPA is required to set National Ambient Air Quality Standards based on an understanding of the health and environmental effects of air pollution and an ability to monitor air quality, then it must conduct R&D to gain the

Table 2
FY 1985 Nondefense Applied Research and Development, by Type and Agency
(budget authority in millions, current $)

	Proposed FY 1985 Budget by Category					
	Total	Procurement-related	Generic	Clientele-oriented	Favored Techn.	Not Classif.
DOE	$1,819	$ 65	$1,457		$ 68	$ 229
NASA	2,513	1,753	760			
HHS	2,024	78	1,885		3	58
NSF	78		78			
Agriculture	479		8	$ 471		
Interior	268	141	80	13		34
NRC	168	168				
EPA	254	254				
Commerce	253	143	77	20		13
Other	1,162					1,162
TOTAL	$9,018	$2,602	$4,345	$ 504	$ 71	$1,496

understanding and the ability. If EPA does not sponsor this R&D, there seems little prospect of having it done elsewhere. This seems inherent in the concept of procurement-related applied R&D.

Generic Research

Generic research is by far the largest category in federal applied R&D. Nearly $1.5 billion in DOE programs, nearly $2 billion in HHS and over $750 million in NASA R&D comprise the bulk of this category, with smaller amounts from NSF, Commerce, Interior and Agriculture. The programs may be distinguished from procurement-related programs, in concept at least, by virtue of the fact that the results are primarily intended for use in the private sector rather than by government. In most cases, this distinction seem reasonably clear-cut and not difficult to make in practice. For example, it applies to DOE's energy technology programs, since the results of the R&D are to be applied in energy production which takes place mainly (although not exclusively) in the private sector.

There are, however, a couple of ambiguities in the generic research notion. In some respects, the line between basic research and generic applied research seems somewhat arbitrary. Most members of the scientific community, including those involved in the performance of the work, would probably classify as basic research much of what the official NIH and NBS agency figures show as applied (and we have classified as generic). What seems to define the applied character of this work is the fact that it is carried out with greater awareness of its potential applications and the choice of problems and research directions seems directed more on the basis of the value of potential applications than strictly on the scientific importance of the problem. Decisions of this type generally reflect a series of compromises anyway, and it is not clear that there is an unambiguous place to draw the line or that it would be particularly useful to do so (National Science Foundation, 1979).

Another troublesome area is the distinction between generic and clientele-oriented R&D. If federally sponsored generic research is to be useful, then an industry or set of industries must exist capable of employing and benefiting from its results. The way in which the health care industry (medical device manufacturers, pharmaceutical firms, practicing physicians, etc.) puts to work the products of NIH research is an example. At the same time, relationships between the government R&D sponsor and the firms in the private sector must exist through which the firms, as users, may influence the course of the R&D. When such relationships stop yielding generic research and start producing clientele-oriented work is not at all self-evident, and we have some doubts about the utility of the distinction, as discussed in the next section.

One final point about the federal effort in generic applied research: When the numbers are assembled in the manner we have pulled them together, perhaps the most striking feature of the generic research picture is the extent to which federal efforts are concentrated in the areas of health, energy and aeronautics, and how small the federal effort relevant to other areas of the U.S. economy is by comparison. The reason certainly is rooted in the way government responds to problems, but if the overall health of the U.S. economy is regarded as the problem to which generic research should be addressed, the pattern of funding might look rather different.

Clientele-Oriented Research

Slightly more than $500 million in federal applied R&D has been classified as clientele-oriented. Most of this is in the Department of Agriculture, with smaller pieces belonging to Interior (Bureau of Mines) and Commerce (National Marine Fisheries Service). As noted above, the category is somewhat troublesome. We have put into it those programs more or less implied by the Nelson and Langlois definition from which we adapted the category--programs that seem to be tied closely to a specific industry interest group--but we remain a bit uncertain as to whether many generic research programs in DOE (e.g., nuclear and fossil fuels), NIH, and NASA (aeronautics) might not have fit here as well.

One difference, perhaps, is that clientele-oriented efforts may go beyond the bounds of generic research to more specific problems and technologies that in other realms might be regarded as proprietary in nature. The structure of the industry and the types of relations that exist between it and the federal R&D agency may make the proprietary character less important. On the whole though, it would seem that not much would be lost in collapsing the categories if the analysis were to be repeated.

Favored Technology R&D

Only a very small portion of federal applied R&D, about $70 million, appeared to fit the favored technology category. This was less than one percent of total nondefense applied R&D. Given the amount of policy attention this category has received, the small size of the effort was somewhat surprising. In large measure, it seems, the Reagan Administration policies aimed at eliminating such efforts have succeeded--either in terminating the programs or at least in disguising their budget presentations.

Nearly all of the favored technology programs were in DOE, where they represent a remnant of the formerly substantial set of DOE energy demonstration programs in synfuels, solar energy, and other areas. The only other agency represented in the category was HHS, whose orphan drugs and devices program in FDA is aimed at bringing to market drugs and medical devices for which demand would otherwise not be sufficient to make the R&D economically viable. It is possible that some addition favored technology efforts are lurking in the guise of generic research in agencies discussed above or elsewhere in agencies (like DOT) whose budgets were not scrutinized in this analysis. It seem highly unlikely, however, that enough exists to change the overall conclusion that the category is a very small and specialized component of federal applied R&D.

DISCUSSION

The implication of the analysis from which we derived our category scheme is that certain types of appplied R&D are appropriate for the federal government to be funding and others are not. Procurement-related R&D has a special rationale because of the fact that the government is the client. The other three categories are private sector oriented. Generic research is generally accepted as an effective means for government to support industrial innovation. Clientele-oriented R&D raises some questions, but they are mostly concerned with the public interest versus private interest dimensions of the programs rather than the appropriateness of the federal role. Favored technology programs, however, are the <u>bêtes noires</u> of civilian applied R&D. They involve federal bureaucrats in making decisions that the market can make better on its own, and often, we are told, these programs fail or produce expensive white elephants.

Our analysis suggests that federal policymakers may have learned some lessons with respect to favored technologies. In any case, there are few such efforts left to kick around any more. Those that survive in DOE are generally scaled-down versions of projects that have been in construction for some time. The only other current examples are in FDA's orphan drugs and devices program which seem small-scale, relatively noncontroversial, and based on a sound rationale.

Although it requires something of an analytical leap to conclude much from the contrast between these two very different types of favored technology programs, we would speculate that there is a direct relationship between the size of a given favored technology project and its controversiality. The mediating factor is politics. Favored technology projects attempt to substitute administrative judgments for the economics of the marketplace. In relatively small projects, this can be done by experts on the basis of fairly sound criteria. Once a project exceeds a certain size, however, it becomes both a vested interest and an irresistable target for others in the agency, other agencies, interest groups, and, of course, the Congress, who have alternative schemes, different judgments about economic potential, or who simply want a piece of the action. This changes the overall character of the effort and makes it much more vulnerable to misdirection.

In any case, the favored technology component of federal applied R&D probably does not merit a great deal of additional policy attention at this time with respect to ongoing programs. More important might be a study of organizational models other than favored technology projects as a means of facilitating the commercialization of technologies such as magnetic fusion power which industry by itself may not be able to bring to fruition.

The issue with regard to generic applied research is somewhat different. As noted above, our analysis reveals that federal generic research efforts are concentrated in a relatively small number of areas. While the areas are important ones, they are not necessarily those of greatest need with respect to the support of U.S. industry or enhancement of the nation's economy.

There is growing recognition among policy makers and members of the scientific and industrial communities that support of generic research is one potentially important means by which the federal government may be able to contribute to industrial development. A more careful look at the nature of existing federal generic research programs, analyzing in greater depth some of the areas we have sketched out here, might be useful in developing policy directions for a potential expanded or redirected federal generic research effort. One approach might be to refine and apply in more general fashion the methods used experimentally in a recent Congressional Budget Office report to assess the relevance of federal R&D to various industries (1984, pp. 66-69).

We do not wish to imply that there is necessarily something wrong with existing generic research efforts. Indeed, in such areas as health and aeronautics these efforts have been enormously successful. The issue is one of perspective. Taking an industrial development perspective, one might start by evaluating the need for generic applied research in different industrial sectors (for example, by such criteria as the overall contribution of the sector to the economy, the structure of the industry, the ripeness of the technologies applicable to the sector, etc.) and then examining the distribution of federal resources and capabilities relative to these needs. The results might suggest a distribution of effort not too different from that which already exists. On the other hand, they might reveal some opportunities for significant shifts in the character and emphasis of federal generic research. In any case, they are likely to be thought-provoking.

ACKNOWLEDGEMENTS

This paper is adapted from work performed for the division of Policy Research and Analysis, U.S. National Science Foundation. The support of the Foundation is gratefully acknowledged.

NOTE

The discussions of agency R&D programs are based on FY 1985 budget justification documents prepared by the agencies for submission to the Congress in February 1984. These are not cited individually here. While the documents are not generally available to the public, most are reprinted in full in the authorization and appropiations hearings published by the relevant Congressional committees. Additional information on agency R&D programs may be found in AAAS Report IX: Research and Development, FY 1985, published by the American Association for the Advancement of Science on behalf of the Intersociety Working Group (Washington, D.C.: March 1984).

REFERENCES

National Science Foundation. 1979. Categories of scientific research (papers presented at an NSF seminar, Washington, D.C., December 8, 1979)

Nelson, R.R., and Langlois, R.N. 1983. Industrial innovation policy: lessons from American history. Science 219:814-818.

Nelson, R.R., ed. 1982. Government and technical progress: a cross-industry analysis. Elmsford, NY: Pergamon Press, Inc.

U.S. Congress, Congressional Budget Office. 1984. Federal support for R&D and innovation. Washington, D.C.: Congressional Budget Office, April.

MANAGING HIGH TECHNOLOGY
An Interdisciplinary Perspective
B.W. Mar, W.T. Newell, and B.O. Saxberg (Editors)

THINKING INTERDISCIPLINARILY - ACTING TRADITIONALLY

Gyorgy Darvas[1]

Institute for Research Organization of the
Hungarian Academy of Sciences
H-1051 Budapest, Munnich Ferenc utca 18

Covering nearly 100 thousand square kilometers and having a population of not quite 11 million of whom about 35,000 persons are performing scientific research, Hungary is a relatively small country in Eastern-Central Europe. Her potential could be compared to that of a minor state in the U.S.

Since research and technological development are among the decisive factors of the country's progress, science policy has been playing a more and more important role in Hungary's general policy over the past decades. Accordingly, as the government had realized the increasing socio-economic weight of science, not only political decisions of ever higher levels were made on science, but also modern scientific approach was gaining ground in them--thus, among others, interdisciplinarity. And while the level of decision making passed towards the topmost one, i.e., government level, also the organs and institutions of science policy management were taking shape, forming a uniform system in our days. It is apparent that in the management hierarchy from bottom up, there is less and less responsibility distinguished by the aspect of disciplinarity.

The spread of modern attitudes in science management is reflected not only in the development of the institutional system, but also in project- and program-like planning, as well as in interdisciplinary target setting. Planning is a central element of Hungarian politics. It prevails also in science policy. This is why a modern approach in the science policy or science management can be observed first of all in the planning system of science, in other words on the manner in which the political sphere expresses its expectations with science. Respecting this situation, I can state that the management of Hungarian science has shifted towards interdisciplinary progress over the past two or three decades.

Following some top-level political decisions, a National Long-term Scientific Research Plan (valid for approximately 15 years) was adopted for the first time in 1972; this plan has been updated several times since then, partly through topic modification necessitated by internal progress in science and new economic requirements and partly by extending the time horizon. At present, the plan includes 19 national level priority research tasks their achievement engaging 30 to 40 percent of the country's research capacity. (It would be inexpedient to engage the entire research potential in planned topics, since it must be kept open at all times, a part of it being reserved for coping with emerging problems in basic research. A further proportion, amounting to 20 to 30 percent, is engaged in performing the projects and departmental tasks of various national science management bodies and ministries. All these efforts together represent the topical orientation of Hungary's scientific research and technological development, along with the main proportions of utilizing the R&D potential.) The interdisciplinary approach to the projects outlined in the plans could be clearly indicated by listing the titles of the 19 research and development tasks highlighted at the national level. The reason I refrain here from giving that list is not that I want to save the reader from boredom, but that the plan is considered outdated in its present form. Instead of this, the frameworks in which today's science management policy is thinking to further develop the National Long-term Scientific Research Plan are dealt with below.

According to present conceptions the updated Plan has to include so-called main research aims, indicating the long-term objectives of scientific activity; these aims have to serve as fundaments for the medium-term program plans.

The research aims can be formulated as follows:

1. Conducting highly significant fundamental research in the natural sciences with special regard to the key spheres in the technological development of the industry, agriculture and infrastructure

2. Research of high importance in economics, political and law sciences, sociology, demography, history and philology

3. Research of the devices and integrated systems in computer technology telecommunication and automation, as well as the systems of information processing

4. Research aimed at exploration, effective exploitation and rational application of natural resources and mineral raw materials

5. Research aimed at achieving power economy, as well as rational consumption of energy and resources

6. Searching new materials and technologies

7. Research aimed at biological and medical achievements having major social and public health effects, as well as maintaining the health and working ability of the population

8. Research in agrotechniques (and biology, technology and chemistry paving the way for them) with the aim of increasing yields and rendering food production more economical

9. Environment protection research with special regard to protecting the ecological systems

10. Research related to the development of the public administration of Hungarian society, economy, and management

11. Research aimed at establishing cultural progress of society, along with the development of general culture, pedagogy and public education.

The National Medium-term Research and Development Plan for 1981-1985 is one of the most important and in several aspects (contrary to the image harbored on Hungary's centralized planning all over the world) novel characteristic product of R&D planning. The plan is made up of three main chapters: science policy objectives, titles of research topics, and financial means available for achieving the objectives. It is the chapter devoted to research topics we are interested in now. All of them being complex research and development activities aimed at achieving important and concrete social or economic (industrial or agricultural) objectives, the programs have been selected to meet four main criteria, one of them denominated as follows: "Performance of the program demands coordinated cooperation of several disciplines and branches of application." Accordingly, the Plan includes at present the following 19 programs, most of them being of interdisciplinary character.

1. Medium-term research and development tasks of the power economy
2. R&D tasks for machine production technology
3. Development of general purpose component and subassembly manufacture for machine production
4. R&D of microelectric components, technologies, and basic materials.
5. R&D of telecommunication equipment, with special regard to the establishment of systems
6. R&D of computer software systems.
7. Development of original pharmaceutical research
8. R&D of pesticides
9. Development of grain production
10. Meat and milk production based on mass fodder and byproducts
11. R&D tasks associated with the safety of nuclear power stations
12. Regional environment protection of Lake Balaton
13. Research of structural materials and material technology
14. R&D of the industrial technology of protein production
15. Research, development and implementation of diagnostical and medical, therapeutical products in human biology
16. The organizational system of Hungary's economy
17. Research aimed at developing the system of public education

18. Research aimed at establishing a general national social policy concept

19. Research for setting the foundation of settlement development control

Let us take a look at the means available for management to put the above-mentioned interdisciplinary research concepts into practice. First of all, what about research workers?

Researchers are developed in several steps, the first being graduation at a university. A lot of graduates choosing the researcher's life remain at universities. Following the several hundred years old German pattern, Hungarian universities consist of traditional "Cathedras" organized to teach a single discipline. Although the number of chairs or departments (cathedras) has increased with the number of disciplines, the organizational structure (and disciplinary isolation) is still the same. In order to take a doctor's degree in, e.g., medicine, a student has to read anatomy, biochemistry, pharmacology, urology, etc. at the respective departments, and take part at clinical practice at urologial, gynecological, and cardiological, and so on, disciplinary separated clinics. And the same applies to other faculties. There are no interdisciplinary organizational units at the universities, and none of them offer separate interdisciplinary education. Graduates are given diplomas valid for one (or two) of the disciplines.

And what is postgraduate education like? Scientific degrees exceeding the doctor's degree obtainable at the universities are awarded by a national Scientific Qualification Committee (candidate's and doctor's in a given science degree: Candidatus Scientarium and Doctor Scientarium). The Committee is an independent body made up of elected members. Scientific degrees are awarded in the following 30 disciplines:

- Linguistics
- Literature science
- Musicology, etnography and folklore
- Philosophy
- Economic science
- History
- History of fine art
- Law and administrative sciences
- Mechanical engineering, Metallurgy
- General biology
- Experimental biology
- Psychology
- Defense science
- Sociology
- Pedagogy
- Geography and meteorology
- Mathematics
- Physics and astronomy
- Agronomy
- Animal husbandry
- Agricultural economics
- Theory of medicine
- Clinical medicine
- Organic chemistry
- Power technology
- Telecommunication engineering
- Architecture and transport
- Mining and geology
- Inorganic chemistry
- Political science

No quarter is given there: on writing a dissertation to apply for a degree to be granted, every scientist must choose the topic from among the disciplines listed here. The aim, i.e. taking a scientific degree, more or less sets the possible line of action: the applicants selecting a peripheral, i.e. interdisciplinary, topic lose much of their chances from the outset because the 30 specialized subcommittees are manned by the representatives of the respective disciplines, and we all know what professional favoritism is like. In principle, it is possible to request the nomination of an interdisciplinary qualification committee, but practical experience shows that it seldom happens and even then a scientific degree is only awarded in one of the above-mentioned 30 disciplines, in the one most closely associated with the topic.

The Hungarian Academy of Sciences is the highest scientific forum of the country. A scientist having won exceptional fame in his/her special field of study during his/her scientific career and having earned a doctor's degree may be elected by the existing members of the Academy by secret voting. The Academy of Sciences has 10 scientific sections:

1. Linguistics and Literary Studies Section
2. Philosophy and Historical Sciences Section
3. Mathematics and Physical Sciences Section
4. Agricultural Sciences Section
5. Medical Sciences Section
6. Technical Sciences Section
7. Chemical Sciences Section
8. Biological Sciences Section
9. Economics and Law Section
10. Geological and Mining Sciences Section

Within that system, the Academy has assigned the individual disciplines to the respective Sections but the 117 disciplines registered with the academy fail to cover equally the actual branches of science studied in Hungary. For example, while the biological sciences are divided to 16 subbranches, mathematics and physics represent a single discipline each of the 117 ones listed, although it is well known also here in the U.S. what a large number of excellent physicists and mathematicians--active in different subdisciplines--are given by Hungary to the world's science. New members are elected by and among the members of one of the ten Sections. This means that everybody having qualified for membership candidacy must fit into one category or another. This paradoxical situation has been realized also by the members of the "Learned Society." After 157 years of existence the Academy passed a resolution at the 1982 General Assembly to reform the system of candidacy by including a paragraph to facilitate interdisciplinary preparation of the application for candidacy. In the following I quote from Resolution No. 26/1982 of the Academy's Presidium: "On the final version of the resolutions passed at the General Assembly of 1982:"

3.2 Interdisciplinary preparation for candidacy.

3.2.1 In quite exceptional cases, candidation for corresponding membership may be prepared through the participation of an interdisciplinary commission.

3.2.2 Interdisciplinary preparation is feasible if several sections are authorized to nearly the same degree to judge scientific achievement of a candidate.

3.2.3 Interdisciplinary preparation of candidacy may be initiated by those authorized for nomination by making an explicit recommendation for it, provided that all sections concerned are represented among them by at least a member each.

3.2.4 The recommendation shall state why interdisciplinary preparation is deemed fit, along with the list of sections having right to pass judgment.

3.2.5 Recommendation shall be filed with either the President of the Academy, or simultaneously, the Heads of the respective sections. In the latter case, the Heads of sections shall inform the president of the recommendation.

3.2.6 In order to prepare a recommendation for candidacy specified under Para. 3.2.3, an interim intersectional (interdisciplinary) commission shall be established. Such an action calls for a decree made to that effect by the Presidium. Consequently, the procedure must be initiated in due time, so as to facilitate publication of the recommendation in the Academy's periodical, along with the statement of the fact that an interdisciplinary procedure is on hand, should the Presidium pass a decision in the affirmative.

3.2.7 An intersectional commission consists of 20 to 24 members, elected from among the sections concerned by the members of those sections with due regard to the scientific activity of the candidate in question and aiming at a proportionate representation. The Commission shall be headed by Vice-president of the Academy having jurisdiction over the discipline concerned. Should several Vice-presidents be responsible for the case at the same time, the Head of the Commission shall be nominated by the President.

3.2.8 Recommendation of a candidate for corresponding membership is decided upon by taking a secret ballot at the intersectional commission. The sections concerned shall be informed on the result of the ballot by the Head of the Commission in writing.

3.2.9 With the knowledge of the recommendation coming from the intersectional commission, the respective sections shall hold meetings and take ballots separately on the issue of candidacy. Depending on the outcome of those ballots, and with respect to the statement of the intersectional commission, the respective sections shall recommend the candidate for acceptance to the Presidium in accordance with the general rules.

3.2.10 With the knowledge of the statements (ballots) of the respective sections and the intersectional commission the Presidium shall decide on the submission of the resolution to the General Assembly in accordance with the general rules."

By quoting this decree at great length my aim was to highlight the grotesque character of the procedure because a procedure in need of such a tedious and complicated regulation cannot indicate anything else but something out of place in the natural functioning of the institution concerned. The Presidium of the Academy was of the opinion that they had succeeded in updating the method of selecting new members. But instead of regarding the problem as a natural part of selecting members, the Presidium has specified a separate discriminative procedure for interdisciplinary scientists.

The reader might think that the case has been settled in this manner, or the paradoxical situation been remedied somehow. Well, the point of the story is still to come. At the General Assembly of 7 May 1982, two scientists were elected corresponding members by "interdisciplinary procedure" (along with 32 others, who were elected by the regular procedure). Thereafter, the Presidium passed a resolution on 25 May 1982, "assigning Corresponding Member Gy.I. to the Chemical Sciences Section and Corresponding Member G.E. to the Biological Sciences Section". So the attempt was utterly in vain: everybody must toe the line!

The example quoted here is an eloquent testimony to the cumbersome actions of a time-honored body of excellent scientists trying to break away from old customs and adapt their proceeding to novel conditions of scientific life.

The Academy runs many scientific committees, having a total membership of almost 3,000. This does of course not mean quite that many scientists, since some members are on several committees. I have made inquiries as to what interdisciplinary attitude is reflected in the composition of the committees. Following from their intersectional character, the 11 committees adjunct to the Presidium naturally are of interdisciplinary composition. With the 10 scientific sections a total of 133 committees are associated. Of them, 44 (33.1 percent) are nominally or by function of interdisciplinary character. However, on taking a look at their actual composition it turns out that in many of them over half of the members had been selected from a single discipline each. I consider the committees interdisciplinary if at least half of the members come from more than one discipline, characteristic of the function of the committee concerned. Only 20 of such committees have been found (representing 15.0 percent of all committees). And only a few of the latter overlap with the committees that can be regarded as interdisciplinary by their functions. On one hand it is a consequence of their above mentioned homogeneity. On the other hand, some of the committees that seemed to be monodisciplinary at first sight, proved to be of interdisciplinary composition. It was interesting to see that in the final outcome the committees had turned out to be of less interdisciplinary composition than could be expected in view of their functions, destination and titles. Total number of members amounted to 2,747 (considering the multiplications of membership). Of them, 588 (21.4 percent) were regarded as interdisciplinary members, i.e. belonging to a discipline other than represented by the highest number of members in a given committee. Of all scientific sections the lowest ratio of interdisciplinary members (11.4 percent) was found in the committees of the Medical Sciences Section, which was quite a surprise. The highest ratios (31.7 percent and 30.3 percent, respectively) were found for the Geological and Mining Sciences Section's and the Mathematics and Physical Sciences Section's Committees. (For details see Table 1.)

Let us now see on what kind of institutional basis the interdisciplinary research programs could be performed. Having surveyed the universities, let us size up the situation at the research institutes functioning under the auspices of the Hungarian Academy of Sciences. About one-third of the independent (full time) research institutes come under this category. Regarding the number of research workers, this represents approximately 12 percent of the country's research potential, and regarding the expenditures, approximately 14 percent. As it stands, this network of institutes is about 30 years old, and has not been fundamentally modified

Table 1

Scientific Committees of the Hungarian Academy of Science

I - 11 Intersectional Committees operating directly under the Presidium

II - Composition of the Committees of the Scientific Sections of the Academy of Sciences

Scientific Sections	1	2	3	4	5	6	7	8	9	10	Total	
Number of Committees	14	17	11	12	15	23	7	12	11	11	133	
Number of committees regarded as inter-disciplinary by name, function or activity	4	2	3	3	6	6	3	6	6	5	44 /33.1%/	
Number of committes regarded as inter-disciplinary by composition	4	1	3	1	1	2	0	2	4	2	20 /15.0%/	
Number of committee members	289	42	231	236	290	528	176	287	282	186	2747	Corr = 0.60
Number of inter-disciplinary members	50	53	70	48	33	94	25	61	95	59	588	n = 21.7
Percentage of inter-	17.3	21.9	30.3	20.3	11.4	17.8	14.2	21.3	33.7	31.7	21.4%	Corr. = 0.38 n = 7.18

as yet. The network of institutes supervised by the Academy is basically of disciplinary structure. Irrespective of the internal division of the institutes their denomination tells much of the story.

1. Institute of Mathematics
2. Institute of Nuclear Research
3. Research Institute for Technical Physics
4. Central Research Institute for Physics
5. Central Research Institute for Chemistry
6. Research Institute for Technical Chemistry
7. Isotope Institute
8. Astronomical Observatory
9. Geodetical and Geophysical Research Institute
10. Geographical Research Institute
11. Limnology Research Institute for Balaton
12. Ecological and Botanical Research Institute
13. Szeged Biological Research Center
14. Natural Science Research Laboratories
15. Research Laboratory for Mining Chemistry
16. Computer and Automation Research Institute
17. Institute of Experimental Medicine
18. Agricultural Research Institute
19. Research Institute for Plant Protection
20. Research Institute for Veterinary Medicine
21. Research Institute for Soil Science and Agricultural Chemistry
22. Institute of Philosophy
23. Institute for Psychology
24. Institute of Sociology
25. Intitute for Law and Administrative Sciences
26. Institute for World Economics
27. Institute of Economics
28. Research Institute for Industrial Economics
29. Archeological Institute
30. Institute of Historical Research
31. Etnographical Research Group
32. Institute of Literary Studies
33. Institute of Linguistics
34. Institute for Musicology
35. Research Institute for Art History
36. Regional Research Ventures Center
37. Institute for Research Organization

Under such organizational conditions, preparations and execution of some interdisciplinary program would demand extensive organizational efforts on behalf of research workers of the highest goodwill and interdisciplinary openness. The task usually falls to the leader of the research topic in question, in spite of his (or her) usual insufficient skill in R&D management. Most of the difficulties arise from having to cooperate with researchers working at other research units; such undertakings are known to be energy intensive and time-consuming. There are only a handful of major research centers (first of all those listed under Points 13 and 4) at which intramural interdisciplinary cooperation is readily feasible.

Although these conditions bear age-old traditions, modern ways of thinking are spreading among research workers. We have information on the growing number of interdisciplinary research topics, and more and more research projects are known to be converted from individual ones into integrated interdisciplinary ventures. This progress has been strongly promoted by the topical planning of research.

It is interesting to note, however, that in Hungary the requirement of thinking interdisciplinarily has not been produced by the science itself, but it has been implanted externally (partly from abroad, and partly in the form of societal expectations). I am far from taking sides in the debate going on among science historians on externalism and internalism. On the contrary, in my view external and internal factors play equally determining roles in the progress of science. But only with regard to the science as a whole. The above case study dealing with the implementation of interdisciplinary thinking in Hungarian science indicates that occasionally any of the two factors mentioned before can gain the ascendancy over the others. But I am convinced that in my country interdisciplinarity could have rooted externally only after it had arisen somewhere else in the world as a result of self-development of science. But also that self-development must have been influenced by external requirements and expectations.

Interdisciplinarity has spread to various degrees in the individual countries. Such comparisons have already been demonstrated in papers of mine presented at previous conferences. The data acquired for further countries are being processed along with their comparison.

FOOTNOTE

1. At the author's request, portions of this paper were unedited.

REFERENCES

Since all sources of data referred to have been taken from official documents and handbooks printed in Hungarian, a language hardly understandable to the nonHungarian reader, I refrain from listing them here. On request, the references are readily available.

MANAGING HIGH TECHNOLOGY
An Interdisciplinary Perspective
B.W. Mar, W.T. Newell, and B.O. Saxberg (Editors)

THE USE OF INTERDISCIPLINARY RESEARCH IN ASSESSING THE ENVIRONMENTAL IMPACT OF EMERGING TECHNOLOGIES

Kan Chen,[1] Paul Appasamy,[1] Michael Donahue[2] and John Reuss[3]

[1]Program in Technology Assessment; East Engineering Building
Ann Arbor, michigan 48109
[2]Great Lakes Commission; Ann Arbor, Michigan
[3]U.S. Environmental Protection Agency; Washington, D.C.

INTRODUCTION

Many of the participants at this conference have stressed the need for an interdisciplinary approach to the management of high technology. Although the focus of much of the discussion on high technologies has been on improvements in productivity and economic growth, it is also important to consider the potential environmental and social impact of these technologies. Despite the clean image of high technology industries in comparison with the traditional smokestack industries, there are significant environmental and social issues to be considered. There have been serious problems with groundwater contamination, air pollution, and hazardous waste disposal caused, at least in part, by the microelectronics industry in the Santa Clara Valley in California (LaDou, 1984). Similarly, there is considerable environmental concern regarding the deliberate release of genetically engineered microorganisms developed by the biotechnology industry (Office of Technology Assessment, 1982). There have also been serious social concerns raised regarding the impact of automation on employment and the ethical aspects of engineering and patenting new forms of life. Thus, both the microelectronics industry and the biotechnology industry, often referred to as the key elements of a high tech strategy, require close attention in terms of their environmental and social impact. The race towards rapid commercialization of these high technology industries must be managed carefully to avoid the mistakes made with earlier technologies--particularly chemicals and nuclear power. An assessment of the risks balanced against the economic benefits of these technologies should be done using an interdisciplinary perspective. The first step, however, is the identification of potential environmental problems caused by these and other emerging technologies. The focus of this paper is on the process of anticipating environmental problems in a systematic manner.

Since environmental problems are complex and multifaceted, they require the attention of many different disciplines. The United States Congress emphasized the interdisciplinary nature of environmental problems in the National Environmental Policy Act. Federal agencies were directed to utilize a systematic, interdisciplinary approach which will insure the integrated use of the natural and social sciences and the environmental design arts in the preparation of environmental impact statements (National Environmental Policy Act of 1969, Public Law 91-190). It seems reasonable, therefore, that the development of a system to anticipate emerging environmental problems be based on an interdisciplinary approach as well.

The Environmental Problems Anticipatory System (EPAS) was developed as part of a cooperative research effort between the University of Michigan and the U.S. Environmental Protection Agency. The purpose of the research was to develop a systematic procedure for the anticipation of emerging environmental problems. The research effort was conducted over a three year period during which the system was designed, refined, and tested. The major focus of EPAS is an interdisciplinary approach involving forward looking people from different sectors to identify and analyze emerging environmental problems. This paper discusses the interdisciplinary framework of EPAS, and some initial experiences in the development and use of the system.

CONCEPTUAL FRAMEWORK

Environmental problems are caused by driving forces in the economic, social, and political domains, as well as from developments in science and technology. There is a complex set of interrelationships among these four domains which result in the emergence of environmental problems. This concept was tested using three case studies. The case studies involved brief environmental assessments of indoor air pollution; recombinant DNA technology; and electric vehicles. These assessments were given to experts in each of the four domains. They concluded that each of the three problems could have been identified if developments and trends in the four domains had been monitored in a collective fashion; i.e. that an interdisciplinary approach is essential in anticipating emerging environmental problems.

In order to anticipate environmental problems in a systematic manner, it is necessary to monitor the four domains. The interdisciplinary information generated by monitoring should be amenable to analysis. It is necessary, therefore, to reduce broad concerns to specific indicators that can be integrated in a workshop format. Using appropriate criteria, this information can then be used to identify and select specific problems for further study. The selected problems are then scoped through a mini or microassessment process[1] to determine if the problem merits further attention. Thus, EPAS essentially involves four subsystems as shown in Figure 1:

<u>Monitoring</u> the scientific, economic, social, and political domains to detect events and trends that might coalesce into environmental problems

<u>Analyzing</u> these quantitative and qualitative clues in a workshop format with the involvement of representatives from government, industry, public interest groups and the scientific community to identify a candidate list of emerging environmental problems

<u>Selecting</u> a manageable set of environmental problems from the candidate list using a set of appropriate criteria by involving the same group of people at the workshop

<u>Scoping</u> the selected problems through an assessment process with the help of experts, and a peer review process

Implicit in the design of EPAS are three major concepts derived from futures research:

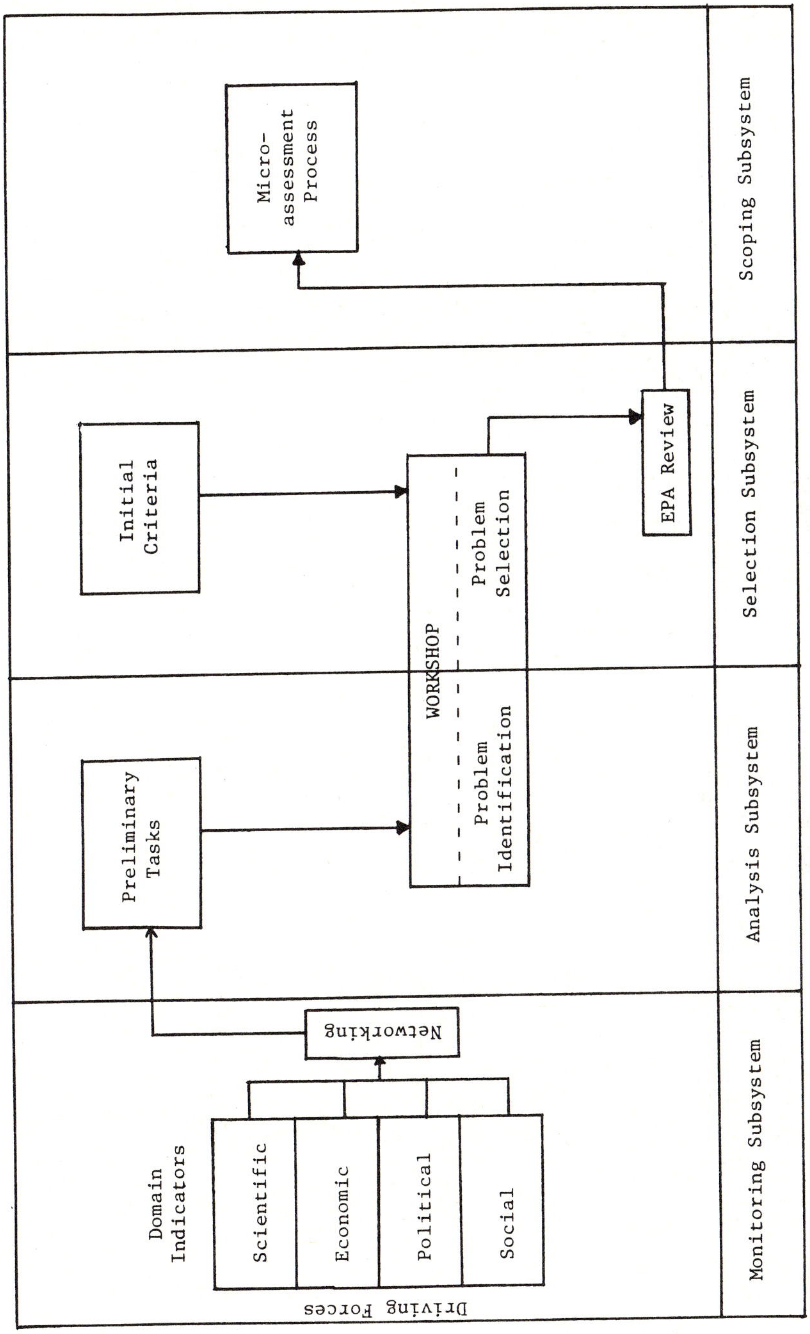

Figure 1
Environmental Problems Anticipatory System
(EPAS)

1. The Systematic Use of Expertise

2. Use of a Qualitative Interdisciplinary Model

3. Use of Indicators in Forecasting

THE SYSTEMATIC USE OF EXPERTISE

Throughout the EPAS system, there is substantial reliance on experts. In the monitoring subsystem, experts are used in the identification of driving forces and are involved in the network. In the analysis and selection subsystem, experts attend a workshop to identify and select problems. Finally, in the scoping subsystem, experts prepare assessments and participate in the peer review process. Thus, the realiance on experts in every part of EPAS is a key aspect of the methodology.

The rationale for such heavy reliance on experts is based on the belief that the insight of experts is more dependable than any theoretical model which presumes to provide accurate solutions. Olaf Helmer describes the need for expert judgment as follows:

> While model-building is an extremely systematic expedient to promote the understanding and control of our environment, reliance on the use of expert judgment, though often unsystematic, is more than expedient; it is an absolute necessity. Expert opinion must be called on whenever it becomes necessary to choose among several alternative courses of action in the absence of an accepted body of theoretical knowledge that would clearly single out one course as the preferred alternative... While in some cases the expert judgment that is available may be derived from explicit application of existing theories, more often it may be highly intuitive in character and based on insights that, although no less reliable, may have thus far defied articulation within a theoretical framework. Much of this potential expertise now remains untapped in practical policy considerations (Helmer, 1983).

The complexity of environmental problems suggested the need to rely on experts rather than computer models, resulting in an expert-oriented methodology in the design of EPAS.

QUALITATIVE INTERDISCIPLINARY MODEL

The conceptual framework was developed using a qualitative model whose hypothesis is that driving forces in the four domains play a causal role in the emergence of environmental problems. This approach differs to some extent from other attempts to identify and prioritize environmental issues. The Conservation Foundation recently published a report, entitled, "In a Deluge of Problems, Where are Worst Threats?" The report listed the results of six surveys to define the major environmental problems confronting society (Odell, 1983). Most of the studies were done using experts in a conference or workshop format or through mail surveys. None of the other studies required the experts to use a common interdisciplinary base of information for their predictions. The major emphasis of EPAS is that the workshop participants integrate the interdisciplinary information prior to making their predictions. This process is aided by using lists of indicators, background papers and presentations. The conceptual model visualizes the group of workshop participants as processors and analyzers. They receive common information which they process using their background and experience. They also analyze the information in an interdisciplinary setting guided by structured group techniques. The result of the interaction is a list of emerging environmental problems. Furthermore, EPAS has an assessment or scoping component that provides a substantive analysis of the importance of the selected problems. The scoping process also provides feedback to the monitoring subsystem on driving forces and indicators.

USE OF INDICATORS IN FORECASTING

There is substantial literature in the social sciences on the use of indicators. The National Science Foundation regularly publishes a list of Science Indicators. The Bureau of the Census provides a list of Social Indicators. There are many publications such as "The Survey of the Current Business" of the U.S. Department of Commerce that provide Economic Indicators. The Council of Environmental Quality published a report on Environmental Indicators. EPAS proposes to combine selective indicator information from these different domains in an interdisciplinary framework. In EPAS, the term environmental indicators was defined as the range of qualitative clues and quantitative measures which facilitiate the anticipation and visualization of potential environmental problems.

Thus, the use of indicators and driving forces in EPAS is a development which we hope will improve the quality of forecasting in the environmental field. Having discussed the theoretical basis for EPAS, we can now proceed to discuss each of the subsystems of EPAS in some detail.

SYSTEM DESCRIPTION

Monitoring Subsystem

As mentioned earlier, the purpose of monitoring is to identify the trends and developments in each of the domains which coalesce in an interrelated manner to create environmental problems. Thus, it is necessary to monitor each of the four domains: economic, social, political, and, science and technology. In EPAS, we sugggest that this be done through a networking process. Such a process should include traditional literature review as well as personal contacts by telephone, interviews and informal discussions with key experts, decision makers, regulators, and business people.

The review of the literature should include sources of information in the four domains that provide qualitative or quantiative clues regarding the emergence of environmental problems. The personal contacts should be done with representatives in government (federal, regional, state, local), the scientific and research community, the private sector, interested groups, and possibly international agencies or organizations.

In testing the monitoring subsystem we used three techniques.

First, we interviewed planning and environmental staffs of numerous public agencies and private sector organizations. These individuals helped us in the development of the system and also agreed to participate in an information network. Next, we conducted review of literature sources suggested by these individuals and by our domain experts and prepared lists of indicators in the four domains. Last, we also conducted telephone and mail surveys of selected individuals to obtain their views on emerging environmental problems. The information generated by these efforts were used as the basis for further analysis.

Analysis Subsystem

The next step is to analyze the information provided by the monitoring subsystem using an interdisciplinary approach. Numbers and graphs supplied by the monitoring subsystem do not spell out environmental problems automatically, but provide rich clues for problem identification by knowledgeable and forward-looking people.

The primary mechanism to carry out the analysis is an interactive workshop that will involve 30-40 participants selected from the diverse group of people identified in the networking process. However, the workshop should be preceded by three preliminary tasks. First, a list of indicators of environmental problems needs to be culled from the monitoring information. Second, a survey of participants and resource people should be conducted to learn about their concerns about environmental issues, background and knowledge bases. The survey could be done by telephone interview or through questionnaires. The concise summarizing of the results will provide a useful digest on the diversity of opinions and analysis that could be used in the design of the workshop. Lastly, some analysts and practitioners should be commissioned to prepare background papers to assist the staff and project participants in understanding the issues and alternatives.

The objective of the workshop is two-fold:

1. To assess, evaluate and improve on a set of qualitative and quantitative indicators developed earlier for use in predicting future problems by using structured group techniques

2. To identify as many possible future environmental problems as possible from a set of stimuli (e.g. indicators, presentations, discussion) using structured group techniques. These techniques are used to facilitate group interactions in a creative yet focused manner, and to avoid dominance of interaction by a few individuals

In 1981 and 1982 two workshops were conducted at the University of Michigan. The first workshop was essentially a pilot one using EPA officials and project staff. The second workshop involved several representatives from the environmental staffs of private corporations. Examples of environmental problems that emerged from each of the workshops are given in Tables 1 and 2.

Table 1
Examples of Emerging Environmental Problems

Developed from Nominal Group Technique - 1981 Workshop

Increases in environmentally caused disease rates
Loss of prime agricultural land
Ocean vs. land disposal
Automation accelerating industrial electrification*
  (pollution from power plants)
Interbasin water transfers to industralizing regions
Accelerated degradation in developing countries
Unpredictable natural disaster effects
Profileration of new chemicals and biological organisms
Environmental effects of another war(s) - limited or
  general
Concentration of industrial toxics around urban areas

Table 2
Examples of Emerging Environmental Problems

Nominal Group Technique for Identifying Topics - 1982 Workshop

Capital budgeting
Economic impact of uniform standards
Mediation methodologies
Alternative management models
Perceived versus calculated risk*
Perception of health and environmental risks in energy
  technologies
Loss of groundwater availability
Water resource utilization: Competing use
Incentive for business involvement

*Topics selected for microassessments (with some modification)

Selection Subsystem

The next logical step after identifying a candidate list of future environmental problems which have a high probability of becoming significant environmental issues. EPA could then carry out further research through mini or microassessments of the selected problems.

The process of selection is also done at the same workshop, mentioned earlier, but at a different session. The process involves the use of criteria in selecting topics from the candidate list of problems. The initial criteria can be classified in three categories: (1) feasibility, (2) desirability, and (3) operationality.

The first step in the workshop is to use structured group techniques to refine the criteria within the three categories. The next step is to apply the criteria to the identified topics. In selecting problems based on such criteria, participants might also be asked to tie problem identification back to specific indicators (quantitative if possible). Such documentation would aid EPA in justifying and substantiating the identification of particular problems to evaluators or outside groups. The outcome of the selection process is a recommended list of environmental problems which could be used in carrying out microassessments.

The final selection of topics for microassessments would be done in-house by EPA, using the recommendations from the workshop. Based on the available funds, EPA could allocate funds for each topic, assuming a range of $5,000 to $30,000 per assessment. EPA might wish to use other in-house criteria in narrowing down the recommended list. After the final decisions on topics have been made, the actual implementation of the microassessment could be carried out.

Scoping Subsystem

The implementation of the microassessment is essentially contained in the scoping subsystem. Scoping involves determining the magnitude or scope of the environmental problem and the appropriate next step for EPA to take. There are four aspects to scoping:

1. to determine whether a potential environmental problem is actually likely to occur

2. to determine whether it is likely to be a significant problem

3. if it is significant, to determine what additional research on the problem should be given a high priority

4. to determine whether addressing the problem is within EPA's mission

The tasks that are involved in scoping include: choice of funding level; selection of experts for microassessments; preparation of preliminary drafts by experts; peer review of the microassessment; interdisciplinary review; and final review by EPA.

From the topics generated at the two workshops, EPA used their internal criteria to select two topics for microassessment.

1. Discrepancies between perceived risk and calculated risk

2. Impact of robotics on utility load curves

Experts were identified at the University of Michigan to carry out the microassessment. Each of the experts was given specific guidelines and asked to prepare a 10-15 page draft report addressing policy issues and impacts. After the report was submitted to project staff and to ensure conformity with the guidelines, it was then provided to two external reviewers--one from the University and one from the private sector. The reviewers were asked to provide their comments at a two-hour meeting in oral form to the experts and project staff. The expert in each case responded to the comments by the reviewers by clarifying issues and by agreeing to include new ideas, delete or change parts of the report, based on the comments provided at the meeting. The revised reports were then reviewed by project staff and sent to EPA.

Both microassessments were related to social management of high technology industries. The first topic, discrepancies between perceived and calculated risk, was more generic and provided suggestions on how the risk management process could be improved. The second topic was directly related to the environmental impact of a high technology industry, robotics, on utility load curves. The purpose of conducting these microassessments was to provide EPA with a preliminary analysis of the problems and issues generated by emerging technologies. EPA may choose to conduct larger scale studies either within EPA or on an inter-agency basis, if such studies are warranted. These studies in turn, may lead to proposals for new programs or legislation to deal with these issues.

CONCLUSIONS

During the conduct of the study, a number of system characteristics which complement and enhance the anticipatory process were defined. EPAS was designed to incorporate as many of these characteristics as possible. Consequently, EPAS is a flexible and cost-effective interdisciplinary process, readily adaptable to changing anticipatory needs and budgetary constraints. It is both comprehensive and participatory, given its sensitivity towards environmental problems emerging from multiple domains, and the involvement of knowledgeable and forward looking people from multiple sectors. Finally, the quick turnaround time on assessing problems enhanced the ability to take timely actions upon emerging problems and issues, rather than react to existing ones.

ACKNOWLEDGEMENTS

Although the research described in this article has been funded wholly or in part by the United States Environmental Protection Agency under assistance agreement CR-8110608-01-0 to the University of Michigan, it has not been subjected to the Agency's peer and administrative review and therefore may not necessarily reflect the views of the Agency and no official endorsement should be inferred.

FOOTNOTE

1. The term "miniassessment" refers to a study that is an order of magnitude smaller than a comprehensive technology assessment. A technology assessment is a form of policy research which identifies policy issues, assesses the impacts of alternative forms of action and presents findings. "Miniassessments" focus on depth or breadth but not both and frequently cost $30,000-$60,000. A "miniassessment" could be used to structure a future comprehensive assessment or it may be used as a pilot or supplementary study associated with some aspect of a comprehensive assessment. A "microassessment" is even smaller than a "miniassessment." It relies heavily on approaches such as brainstorming or structured group techniques. It is useful as a prelude to determine the appropriate scope of a mini or comprehensive assessment. The particular time and cost allotted for mini versus micro assess-

ments will depend on the definition of the problem, its complexity, potential severity of expected impact, the needs of decision makers and the availability of resources. (Arnstein, Susan in Technology Assessment: Creative Futures. Prepared by Mark A. Boroush, Kan Chen and Alexander N. Christakis, North Holland, New York, 1977).

REFERENCES

Helmer, O. 1983. The need for interdisciplinarity. In Looking forward: a guide to future research. New York: Sage Publications.

LaDou, J. 1984. The not-so-clean business of making chips. Technology Review, May-June.

Odell, R. 1983. In a deluge of problems, where are the worst threats? Conservation Foundation Letter, December.

Office of Technology Assessment, U.S. Congress. 1982. Genetic technology: a new frontier. Boulder, Colorado: Westview Press, Inc.

MANAGING HIGH TECHNOLOGY
An Interdisciplinary Perspective
B.W. Mar, W.T. Newell, and B.O. Saxberg (Editors)
Elsevier Science Publishers B.V. (North-Holland), 1985

ISSUES IN TRANSITION FROM SINGLE DISCIPLINES TO MULTI-DISCIPLINARY RESEARCH IN A LARGE FEDERAL AGENCY

Albert H. Rubenstein[1], Eliezer Giesler[2], Stephen L. Rawlins[3], and Lloyd F. Elliott[4]

ABSTRACT

A large federal research-performing agency has identified new and modified national objectives and new areas of opportunity for its research program, which involves over 2,000 in-house scientists. Many of the new areas of emphasis require regrouping of traditional scientific disciplines and mission-oriented specialities. In this paper we examine some of the issues which arise in connection with such a re-orientation to more multidisciplinary research teams, including: organizational impacts, career planning, resource allocation, recruiting, and government-university relations. Examples are given in several problem-oriented research areas.

INTRODUCTION: BRIEF DESCRIPTION OF AGRICULTURAL RESEARCH SERVICE

The Agricultural Research Service (ARS) is the research arm of the U.S. Department of Agriculture (USDA), responding to the needs of interacting industries that produce, process and deliver agricultural products to consumers in the United States and abroad. ARS is a system of Federal science laboraties, with four regions and eight major research centers. The agency has over 2,700 scientists, and its 148 locations span program areas in: Agricultural Research; Human Nutrition Research; Crop Sciences; Livestock and Veterinary Sciences; Post-Harvest Science and Technology; Soil, Water, and Integrated Agricultural Systems.

---

[1]Professor, Industrial Engineering and Management Sciences, Northwestern University, Evanston, IL 60201.

[2]Professor, Department of Business and Management, Northeastern Illinois University, Chicago, IL 60625.

[3]Soil Physicist, National Program Director, Natural Resources, U.S. Department of Agricultural Research Service, Beltsville, MD 20705.

[4]Microbiologist, Land Management and Water Conservation Research, U.S. Department of Agriculture, Western Region, Washington State University, Pullman, Washington 99164.

The ARS is centrally managed by the Administrator, who is located at the USDA Headquarters in Washington, D.C. He is assisted in planning, budgeting and managing by the ARS National Program Staff (NPS), headed by the Deputy Administrator who is located at the Beltsville, Maryland Agricultural Research Center.

Research at ARS is conducted in many centers, laboratories and locations that range in size from 350 scientists at Beltsville, Maryland to some very small stations with less than a dozen scientists. Some locations are designed to address research problems that have a very specific focus. Other facilities, such as Beltsville, have a very wide range of programs of a multidisciplinary nature. Some of ARS's facilities are located on university campuses and State agricultural experiment stations to assure cooperation and interaction with state scientists.

WHY THE SWITCH FROM SINGLE TO MULTIDISCIPLINARY RESEARCH IN ARS AND WHAT FACTORS AFFECT THE SWITCH?

There are a number of factors influencing a major change in research approaches to agricultural problems and a number of factors that are likely to influence the rate and extent of this important change process:

Strategic (Program) Plan for ARS Calls for Identification and Attack on "Important National Problems"

The ARS Strategic (Program) Plan, presented in 1983, and supported by continuing development in implementation and operational plans, structures agricultural research around a series of six general objectives and their constituent Approaches and Approach Elements. All of the entities--Objectives, Approaches, etc., focus on problems involving the agricultural production and distribution process and the related natural resource problems (soil, water) impinging on agriculture. Few of the higher level or intermediate level entities are likely to yield to the theories and techniques of traditional single disciplines or specialties. Even at the lowest or "task" level, knowledge and research techniques from more than one discipline will be required to make substantial progress.

Shift from Disciplinary Pursuit of Knowledge to Problem-Solving Focus

Accompanying the program planning, are shifts in the organizational structure for research in ARS, reflecting the recognized needs for more problem-focused teams vs. groups of people from a single discipline. Although many ARS research efforts have traditionally involved people from more than one discipline, this has often been accomplished on a consultative basis, as compared to the integrated team approach that will be needed to successfully address some of the important national problems (INPs) that are being identified.

Solutions to Many Agricultural Problems Require Knowledge From More Than One Discipline

Although ARS and other agricultural scientists have been very successful in addressing and solving a large number of problems over the past decades via primarily specialized approaches, it is clear that further progress requires, in many areas, that knowledge from diverse fields be applied. In addition, it is clear that new knowledge must be generated, via basic research, in some scientific areas and technologies that have not historically been closely involved with agriculture-motivated problems.

New Research Tools are Becoming Available in Other Disciplines Than Customary Agricultural Sciences

Agricultural scientists have taken advantage of advances in research method in a number of fields, such as biochemistry, mathematical modeling, use of computers and biotechnology. However, advances in sensing, computation, analysis, and other arts are occurring in a wide range of fields which have until now have been outside the general realm of agricultural research. It is necessary now for such peripheral fields to be examined not only for potentially useful techniques, but also for theory that might be adapted and applied.

Identification and Attack on Complex Problems Requires Different Viewpoints

In addition to the substantive content of other fields, it is important that the viewpoints of other kinds of scienists and their intuitive and heuristic approaches be tapped in addressing complex problems. Combining diverse, but potentially relevant, viewpoints can exert considerable leverage on problem recognition and problem formulation, as well as research approaches.

Territories of Traditional Disciplines and Scientific and Technical Specialties Being Invaded by Other Specialties

Whether ARS scientists take a proactive approach or not, incursions into traditional territories of agricultural scientists are occurring and will occur increasingly as scientists in other fields and newly trained agricultural scientists seek for both basic and applied research opportunities. In addition, the interest of commercial companies in developing products for agriculture will support and accelerate such inroads. Competition for recognition and funding make it desirable to combine forces where feasible.

Some Traditional Disciplinary Approaches have Dried Up in Ideas for Attacking Particular Problems

In many areas of agricultural research, useful scientific and practical ideas are still being generated through primarily dedicated specialty groups and approaches. In others, however, the researchers seem to have lost momentum and new, exciting ideas for either research itself or approaches to practical problems appear to have dried up or slowed to a trickle.

There are Limits to the Extent and Rate of Conversion to Multi-Disciplinary Approaches

Despite the urgency and importance of the desired conversion, such a transition will not be quick or easy. Many of the barriers mentioned in Section 3 will slow down the conversion and even prevent it in some areas. In addition, there may be a desired balance between problem-focused multidisciplinary teams and narrowly focused groups specializing in one field or approach. In many instances, both approaches will live side by side, each contributing what it can do best.

There is a need to Inventory and Integrate the Current State of Knowledge in Fields and Problems and to Crosswalk Between the State-of-the-Art of Neighboring Disciplines

Although the traditional state-of-the-art papers and summaries exist in all scientific and technical fields, updated periodically, this will not be enough to support an effective transition. In order to design research teams according to the kinds of knowledge systems and research techniques required for particular problems, it is essential to have a comprehensive inventory of what the knowledge is in a given field and relative to a specific problem. A later section addresses this issue in terms of the emerging fields of Artificial Intelligence, Expert Systems, and Knowledge Engineering.

POTENTIAL BARRIERS TO CHANGE

In the transition from single discipline to multidisciplinary research, ARS may encounter several factors which act as barriers to such change and smooth transition. These barriers may be grouped into four categories:

1. Organizational factors
2. Individual factors
3. Scientific/Technical factors
4. Environmental factors

Organizational Factors

a. Current organizational structure in terms of location and allocation of funds.

b. Management policies and reward structure emphasizing publications in scientific journals.

c. Pressures to maintain a university-type structure in disciplinary units.

d. A program plan with classification of research in approach elements and projects in well-defined areas, many of which are amenable to continued disciplinary approaches.

e. Lack of organizational incentives to engage in multidisciplinary and problem focused research. (Some are now being developed.)

f. Diverging policies, attitudes and interests of the various regions, areas, and facilities of the ARS.

g. Difficulties in starting up research teams and research activities; coordination and gearing up of multidisciplinary teams.

Individual Factors

a. Age distribution of ARS scientists; average age over 45.

b. ARS scientists have indivdual investments in their particular discipline over a lifetime of scientific work.

c. Career paths are defined and encouraged along the scientific ladder, thus putting more emphasis on disciplinary achievements.

d. Creation of individual "turfs" and power centers.

e. Career objectives beyond ARS and Government service tend to lead to university-related research, therefore the desire to maintain high visibility in a discipline.

Scientific/Technical Factors

a. Changes in sciences underlying research areas of ARS and applied areas.

b. Challenges in new developments in individual disciplines, forming subdisciplines of increased narrowness of specialization.

c. Rapid progress of state-of-the-art in many disciplines, absorbing most of the scientist's time in catching up in his/her own discipline.

d. Difficulties in communication among scientists of different disciplines.

Environmental Factors

a. Relations with other USDA and Federal agencies tend to highlight the ARS role as a research agency, with emphasis on fundamental, disciplinary research.

b. Relations with the scientific community, peer pressure, and the scientific external reward system tend to emphasize the disciplinary aspect of ARS research.

POTENTIAL FACILITATORS TO CHANGE

In the transition from a single discipline to a multidisciplinary research framework, other factors may be considered facilitators to the transition. These facilitators may also be grouped into four categories:

1. Organizational factors
2. Individual factors
3. Scientific/Technical factors
4. Environmental factors

Organizational Factors

a. Awareness by ARS management, through elements of the Strategic (Program) Plan, of the cross-cutting activities under ARS responsibility.

b. Central management by ARS Administrator and the National Program Staff facilitates the reassignment of funds and priorities from single to multidisciplinary research.

c. The workshops run by the agency have helped in identifying and prioritizing important problems and research opportunities.

d. The Program Plan assists in identifying and defining important problems for ARS.

Individual Factors

a. Scientists are interested in the challenges of new areas of knowledge.

b. Work with colleagues from other disciplines offers opportunities for learning new skills.

c. Scientists feel a need to contribute to broader national agricultural problems as part of or in addition to pursuing their own scientific activities.

d. Work with a strong, stimulating research team may help rejuvenate a scientist who has hit a plateau in his/her career.

Scientific/Technical Factors

a. Improvements in communication hardware and software tend to facilitate interdiscipline exchanges.

b. Agricultural sciences tend to pose research questions which are increasingly more oriented to multidisciplinary solutions.

c. Inputs from other scientific groups involved in multidisciplinary research.

Environmental Factors

a. Multidisciplinary research helps improve ties and relations with other agencies in USDA, other Federal units, and industry.

b. Criticisms from external agencies contributed to the design of the ARS Program Plan and Objective Number 6, which provides the framework for cross-cutting (multidisciplinary) activities.

c. Emergence of specific national or regional problems (e.g., certain insects of parasites in certain crops) creates a need for research activities on a multidisciplinary level, as demanded by Congress and/or the Administration, which is being pressured by the general public or specific interests affected by the pest damage.

CASE STUDY OF THE NEED FOR CHANGE

Historical

The ARS programs that involve soil microbiology in the solution of important problems have been fragmented and, in many locations, one person efforts. Also, the total number of trained soil microbiologists in the Agency is small. There are a few examples of interdisciplinary cooperation, including: the animal and municipal waste projects, water quality programs, plant nutrient cycling, residue decomposition and management, no-tillage research, and symbiotic nitrogen fixation. Here, cooperation is defined as two or more scientists conducting and publishing research. Generally, the problem should be unsolvable by a single-discipline approach.

More cooperation between soil microbiologists and agricultural engineers, hydrologists, plant pathologists, soil chemists, and soil physicists would be desirable. Because they are few in number and scattered, soil microbiologists have never had a strong disciplinary presence in ARS. As a consequence, there is often a failure to recognize how soil microbiology could be an asset to interdisciplinary efforts.

Research Needs

Erosion and Prediction. There is a pressing national need to predict the consequences of cultural practices on erosion and soil productivity. Thus it is necessary to adopt management practices that maintain or increase soil productivity while protecting the environment. Initial research results indicate the residue management is an essential element in any conservative crop management system. No-till cropping systems that leave all of the crop residue on the soil surface are well-suited to control erosion, but presently they are not always economically acceptable, because they leave new niches for pests and diseases to attack the crop. Also, nutrient, structural, biological and environmental changes occur.

It is clear that the development of acceptable conservation crop production systems will require a coordinated interdisciplinary effort. Physical, chemical and biological processes within the soil are so closely linked that changing any part of the system impacts many other components. The only practical means to predict the ultimate consequences, in terms of crop yield, of any single change in the production system is to link these interacting processes in a predictive systems model.

One result of preliminary analyses of predictive cropping systems models is the discovery that our scientific information is not ideally distributed to make accurate predictions. In some cases, unnecessary data were collected while essential data were not. These analyses are useful in directing the future course of research to fill these knowledge gaps and to eliminate duplication of effort.

One of the knowledge gaps most seriously limiting the development of optimum conservation production systems is the basic relationship of biological processes to soil aggregation and soil resistance to detachment and transport by wind and water. We badly need to know the effect of alternative residue management practices on residues decomposition and the effect of decomposition products on soil structural properties. Equally important is the need to predict the effect of residue management practices on soil organic matter and its relationship to nutrient cycling and other critical soil processes. The development of necessary predictive models and the knowledge bases to suport them will require highly integrated team efforts. Among the essential members of these teams, soil microbiologists are now too often missing.

Plant Vigor and Pest Management. Limited coordinated research occurs between soil microbiologists and plant geneticists, plant pathologists, plant physiologists, and microbial geneticists; yet the payoff in terms of improving plant vigor by managing microorganisms at the soil-root interface is potentially great. As with erosion predition, success will require a highly integrated effort. Evidence suggests that the plant has some genetic control over root surface colonization, and it is likely that there is microbial control over it also. Much of the early work conducted by single disciplines alone has yielded only limited success. Powerful new tools developed by biotechnology research now offer the means to solve some of the more resistant problems. But even with these tools, the complexity of the problems is so great that it is highly unlikely that one or two scientists working alone would possess the basic expertise in the necessary disciplines to solve them. Genetic staility of the organisms has been an extreme problem. It has not been possible to know if the organism studied in the laboratory was genetically the same as the one that came from the root surface. Newly-developed genetic tools provide the mechanism for stabilizing these organisms. Results of these studies will be essential for biocontrol of pests and for managing the colonization of roots by beneficial organisms, or the elimination of deleterious organisms.

Interdisciplinary teams will also be essential for progress in manipulating the chemical itself, associated biological processes, or the soil environment to control herbicide persistence within a range to be effective but not be carried over to the next crop or to the environment. Numerous problems are similar in nature, and require team efforts to solve them.

Although a scientist may recognize the need for expansion of effort beyond the limits of his/her discipline, he/she often resists leaving the security of his/her own niche to become intimately involved in a team effort. He/she may view making a request for help from others as an admission of weakness. Budgets are often tied to individuals rather than to projects, and he/she may resist sharing limited funds with others. The perception that the results of team research will not qualify for publication in a disciplinary journal, that publication will likely be multi-authored, and a perceived decrease in other factors related to peer recognition may be powerful deterents to join an interdisciplinary team.

Effective communication among microbiologists and increased infusion of new ideas, enthusiasm and questions by involving graduate students and post-doctorate researchers, should help increase the awareness for the need to change. These new team members lack the experience to tackle whole interdisciplinary problems, but within a team with mature leadership they can do in-depth research to fill vital knowledge gaps while they develop the overall vision ultimately required. Most importantly, team involvement gives younger members role models to follow. Recognition from team members can help supplant recognition from disciplinary peers, cushioning the negative impact of multi-authorship and nondisciplinary journal publications. Team involvement should allow younger members to make a mark within their peer group before they are required to tackle larger problems that do not fit within a single discipline.

Allocation of money to programs directed toward producing specific technologies or to fill essential knowledge gaps to sovle problems, rather than to individual scientists, can help foster team research.

CASE STUDY OF AN ATTEMPT TO CHANGE

Recognition of Need

In mid-1983 the ARS National Program Staff recognized that research conducted by ARS microbiologists was fragmented and poorly coordinated. Because the efforts of the few microbiologists within the Agency seemed to be sorely needed to solve a number of important problems, a workshop was convened to develop strategies for more effectively involving them in interdisciplinary research programs.

Planning Procedure

Leaders were chosen for each of several problem areas to prepare statement-of-need documents to initiate the workshop discussion. The documents included opinions and ideas solicited from a wide cross section of scientists. A smaller multidisciplinary group of scientists reviewed the document and sent it forward as a basis for definition of the workshop objectives. At all stages, the chairperson was careful to solicit and to use a broad range of ideas and opinions.

Using this information, a multi-disciplinary team was selected to plan the program and agenda. A well-defined planning procedure was developed to serve as a road map to help assure that a product emerged from the workshop that could be used for future program direction. A preliminary meeting was held in conjuction with the American Society of Agronomy national meetings in August of 1983 to hear a number of presentations on the state-of-the-science, highlighting research needs. It is felt that this preliminary meeting was very important to the success of the workshop. This procedure allowed a minimum of presentations at the formal workshop (one half-day) with the majority of the time (two days) being spent on the planning and problem identification process.

The Workshop

Key scientists from ARS, State Experiment Stations, and a few from private industry were invited to the workshop. Also, there was an effort to invite young scientists in ARS who related to the workshop. Although attendance was purposely limited to allow small group interactions, most scientists who expressed strong interest were able to attend. The participants were aware that they needed to produce a finished product at the end of the workshop. Group leaders and recorders were chosen from well-established scientists to guide and to stimulate the discussion in the four work groups. It was felt that these leaders must be carefully selected. They would be in a position to prevent or to moderate "protection of turf" and irrelevant dialogue and to stimulate progress.

Results

A multidisciplinary research plan evolved that enumerated areas of research needs. Approaches to these needs were listed in priority order. Scientist and financial resource needs were established, and lead locations were designated, in some cases.

IMPACT OF TRANSITION ON THE INDIVIDUAL SCIENTIST

When an individual scientist is exposed to a transition from research in a single discipline to a multidisciplinary program, there are two kinds of impacts the scientist may experience. The first is a set of beneficial impacts. The scientist may be able to assess some of the benefits from the transition sometime during the transition period. Some examples may be:

a. Exposure to new ideas, new skills, new techniques and methods

b. Enlargement of the scientist's network of professional relations

c. Increase in the possibilities for career changes and scientific branching, primarily from such well-established disciplines as chemistry and some areas of engineering

Examples of negative impacts which may hinder the scientist's performance are:

a. Increase in the scientist's uncertainty, apprehension and pressure

b. Uncertainty as to the reward system and its implications in the multi-disciplinary scheme; possible effects of a negative reward outcome on the scientist's career and personal life

c. Difficulties with the initial stages of the transition may create unwillingness to continue and may serve as excuses or rationalization of why the scientist should return to single discipline research

The specific impacts on the individual scientist vary according to the stage in the transition. In the stage prior to the transition, there may be an increase in the uncertainty, many unknowns, the desire for a clarification of policies, managerial attitudes and rules and the possible impacts on the individual. It is in this stage that the major effort on the part of management should emphasize the definition and the clarification of policies regarding personnel actions and career paths and options.

In the stages of the actual transition period, most of the factors that impact the scientist are related to the actual management of research teams and the problems and difficulties which tend to arise in the start-up of such a project, taking it up to speed, and the maintenance of its momentum. A counterbalance to the initial problems is the spirit of camaraderie and the feelings of a challenging experience and big hopes which tend to permeate a newly created research team.

In the stages following the initial run of the project, scientists may be able to conduct a preliminary assessment of the impacts the transition had on them. During these stages, management can act to further clarify policies in order to reduce the uncertainty and help the scientist in obtaining a positive evaluation of his/her experience in the transition.

THE ROLE OF COMPUTER NETWORKS, ARTIFICIAL INTELLIGENCE, EXPERT SYSTEMS, AND KNOWLEDGE ENGINEERING

Selecting the Research Problems

When many ARS scientists learned that ARS was refocusing from primarily disciplinary research to a focus on selected important national problems, they wondered how the problems would be chosen. Would this shift change the agency from one primarily concerned with long-term, basic problem-focused research to one concerned with applied research aimed at delivering products for solving short-term problems? The answers to these questions depend on how the problems are selected and who assumes the responsibility for solving them.

Underlying the goal of the Agricultural Research Service Program Plan is the premise that ARS will assume responsibility for delivering the knowledge and technology required for future generations to feed and clothe themselves. To do this, ARS must first decide which problems must be solved to accomplish this goal and then determine what knowledge and technology are needed to solve these problems. In turn, ARS scientists must be able to thoroughly assess existing knowledge in each of the fields related to these problems and to identify the significant gaps in that knowledge.

In the book Megatrends John Naisbitt (1984) points out the magnitude of the problem of assessing what is known. He say, "We are drowning in information, but starving for knowledge." He defines knowledge as information in an accessible form so that it can be used to solve problems. He states that we live in an "information society" where new information is generated at an ever-increasing rate. By 1985, he predicted that scientific information would be doubling in the period of less than two years! There now is so much information to sort through that some scientists find it easier to run a new experiment than to search the literature for an existing answer (Rubenstein, 1974).

Clearly, if ARS is to contribute significantly to the solution, it must effectively target research to fill real knowledge gaps. There is a need for organizing existing information into a form in which it can be brought to bear on important national problems. It may then be possible to develop a rational approach to problem selection and prioritization.

Knowledge Engineering and Expert Systems

Fortunately, the same computer technology which is partially responsible for the vast increase in information flow also has the potential for limiting and routing its flow to productive uses. The purpose of knowledge engineering is to shape knowledge into mental models that lead to systematic principles, better designs, and faster and cheaper solutions of problems.

The principle is that bodies of knowledge can be produced, and put in place for a variety of purposes. Feigenbaum and McCorduck (1984) state in their book The Fifth Generation[3] (p. 55): "This kind of engineering is also an answer to the very legitimate complaint that most of us are overwhelmed by information; our built-in, natural capacity for about four different items we can consciously and simultaneously attend to is at the breaking point. Well-engineered knowledge will subsume the confusion of details, data points, and ever-changing information under orderly, general, and plausible interpretations that allow us to note and forget, or subcontract out to the machine, while we humans use our remaining processing power to attend to more imortant matters."

They cite as an example of knowledge engineering, a networking adventure by Lynn Conway as a means to solve a complex problem in designing and producing very large-scale integrated computer chips. It used the computer network ARPANET (Advanced Research Projects Agency of the Department of Defense), to tie scientists from several locations and from different disciplines into a single team. Feigenbaum and McCorduck point out that Conway's experiment shows that even in delicate enterprises that require the utmost creativity, the old adage that too many cooks spoil the broth simply does not hold. The potential problem of over-seasoning by some zealot was avoided by real-time communication available to everyone. The close linkage and short turnaround provided by the electronic network allowed an idea to be examined, tested, and adopted quickly and effortlessly if it was good--or abandoned quickly if it wasn't. The second problem of too many cooks, that the soup will be too bland as the result of committee compromise, was avoided in Conway's experiment by establishing a set of goals that were accepted by all participants. Those goals had been refined by the same process of massive, speedy trial and error.

An important issue is that even when using 10-year old technology, this network adventure demonstrated the computer's capability to assist in not only what we think, but how we think. Feigenbaum and McCorduck state that despite the grave warnings about how the computer would inevitably dehumanize us, it has not.

Many people think that conducting cooperative research through the use of electronic networks is still futuristic. Computer technology has advanced more rapidly than almost any other technology. Computer capacity per unit of cost continues to double approximately every year. This has resulted in a 1000-fold increase in computer power per unit of cost for the past 10 years. If we are planning for research to be conducted 10 years in the future, we might expect another 1000-fold increase--a million-fold increase in computer power per unit of cost since the development of the technology used by Conway! Certainly, computer technology itself is not likely to be a limiting factor in developing scientific research teams linked together in electronic networks.

The first task to prepare for this future is to catch up to the present. In any field moving as rapidly as computer technology we can expect most noncomputer specialists to be far behind the leading edge. To benefit from the opportunities offered by this technology in putting geographically dispersed interdisciplinary teams together to solve important problems, two things must be accomplished: (1) We need to make up for lost time by bringing our scientists up to an acceptable level of computer literacy. (2) We need to adopt knowledge engineering as an integral part of agricultural science.

One way to encourage scientists to increase their computer competency would be for ARS to establish a computer network including the necessary data bases and personal computers that would be required to implement a modern system such as ARPANET. It may be temping to wait a year to purchase computer hardware because its cost is dropping rapidly. But doing so delays by a year the first step toward developing futher computer literacy in the agency.

Developing capability in knowledge engineering leading to the development of agricultural expert systems will require more than just catching up with the present. It would be a strategy to capitalize on the tremendous technological breakthroughs that the Japanese visualize in their plans to develop 5th generation computers. Their goal is to develop computers for the 1990s and beyond--intelligent computers that will be able to converse with humans in natural language and understand speech and pictures. These systems are being designed specifically to increase the productivity of knowlege into packages and products that will permit farm managers to rise above the confusion of details and data points in the sea of information in which they are now drowning by subcontracting much of the detail to the machine.

The simulation modeling now being conducted by ARS and other scientists will serve as building blocks for expert systems that will run more efficiently on tomorrow's computers. These models will be effective frameworks on which productive multidisciplinary research teams can be assembled. But most important, the will serve as analytical tools to help focus on real knowledge gaps. They represent a concrete product of research that can bring all available information to bear on the solution of those problems.

Only a small percentage of the scientific manpower of ARS needs be devoted to knowledge integration as such. Packaging research parts to solve important problems will probably uncover enough knowledge gaps that need to be filled to keep the remainder of the scientists productively employed. As a federally-funded agency with responsibility to deliver products that solve problems, ARS can leverage its limited resources by encouraging other research performers to help fill critical knowledge gaps. In doing so, private industry might assume responsibility for those with a profit potential, and universities might pick up a number of them that could form graduate student thesis projects. The kinds of knowledge gaps that would tend to be left would be those that require long-term funding commitment, unique facilities, or interdisciplinary teams. They will tend to be the basic, long-term research projects which ARS has stated as its primary research mission. They will tend to be these, however, more by default than by asserting an exclusive right. Rather than becoming an applied research agency, under this scenario, ARS would develop two major areas of expertise: (1) Knowledge integration with the primary objective of packaging research knowledge to solve important national agricultural problems. (2) Long-term, high-risk, basic research projects that are most effectively accomplished by a centrally funded research agency such as ARS. The latter would always be the major function of ARS, but it would look to the interactive function for direction and justification for its research projects. Close cooperation with and integration of university scientists into these projects will continue to be important. Although other research performers may well fill many of the gaps revealed by the systems analyses effort of ARS, ARS would always remain a hungry customer for their results to fit them into the products that customers of ARS Research need to solved important agricultural problems.

SOME INDICATORS OF CHANGE FROM DISCIPLINARY TO MULTIDISCIPLINARY RESEARCH THAT CAN BE USED TO MONITOR PROGRESS AND IMPLEMENTATION OF THE PROGRAM PLAN

Many Research and Development (R&D) organizations have tried to change from primarily disciplinary approaches to those involving multi-disciplinary teams and methodologies. Some have made such transitions successfully and quickly. Others have not. In either case there is a need for a set of criteria or indicators which can serve as guides for evaluating how well the transition is going and signaling for corrective action when it is not going well.

In the case of ARS and its planned reorientation toward more problem-focused research, involving greater use of multi-disciplinary teams, the follwing indicators may be useful in monitoring and assessing the rate and extent of transition.

Funding Distribution

This can indicate whether there have been significant funding shifts from support of single disciplinary projects or areas of research to problem-focused team efforts, involving multiple disciplines.

Sources of Ideas

An increase in the origination and flow of ideas from new sources, addressing long-standing problems, may be an important indicator of the participation of disciplines and specalties other than those which have traditionally owned the problem area.

Fields of Science and Technology Used

A listing of the fields contributing to an attack on a particular problem can indicate the changing makeup of the disciplines brought to bear, including new hybrid disciplines or specialties. In a recent discussion of problems of groundwater contamination, for example, the following specialties are mentioned: hydrology, geology, hydrogeochemistry, microbiology, and knowledge about the nature of wastes that enter the subsurface (*Science*, 22 June 1984, p. 1292, letter from John D. Bredehoeft and Thomas M. Usselman).

Requests for Assistance and Collaboration Across Disciplines

The interaction pattern, particularly that aspect involving initiation of interaction, may be a good indicator of changes in working relations between disciplines.

Communication Patterns in General

Gross communication flow and travel between individuals, groups, and locations, reflecting different specialties, may be useful indicators of change in collaboration patterns.

Emergence of New Hubs and Centers of Excellence

The shift in or sharing of lead scientific or problem-focused roles may indicate the progress of diversification away from tightly held turfs by particular disciplinary groups.

Entry of New Disciplinary Mixes

The emergence and recognition of new groupings of disciplines, more hybrid specialties, and other new combinations may indicate progress in the direction of more multi-disciplinary research.

REFERENCES

Agricultural Research Service. 1982. The mission of the Agricultural Research Service, USDA/ARS,.

Elliott, L.F. and Follett, R.F. 1984. Report of: Work Planning Conference on Soil Microbiological Issues as Related to Conservation Tillage, Seattle, WA, March 20-22, 163 pp.

Feigenbaum, J., and McCorduck. 1984. The fifth generation: artificial intelligence and Japan's computer challenge to the world. Reading, MA: Addison-Wesley.

Naisbitt, J. 1982. Megatrends: ten new directions transforming our lives. New York: Warner Books.

Office of Technology Assessment. 1981. An assessment of the U.S. food Agricultural Research System.

Rubenstein, A.H., Kernaghan, J., and O'Keefe, R. 1974. Search versus experiment. College and Research Libraries.

United States Department of Agriculture/Agricultural Research Service. 1983. Agricultural Research Service program plan, January 1983; and the Implementation plan, February 1983.

Section II :

INTERFACE ISSUES

MANAGING HIGH TECHNOLOGY
An Interdisciplinary Perspective
B.W. Mar, W.T. Newell, and B.O. Saxberg (Editors)

INTERFACE ISSUES - AN INTRODUCTION TO SECTION II

(Overview by the Editors)

INTRODUCTION

Interdisciplinary efforts are encountered in every phase of high technology activities and are not limited to only research and development. The previous conferences in this series focus on interdisciplinary research and development management provided a solid basis to approach the broader issue of other interdisciplinary interfaces. Epton, Payne and Pearson (1983) have urged that the term interdisciplinary be used to identify a group of individuals from different disciplines that belong to the same organizational unit, and that the term multidisciplinary be used to denote a set of individuals from different organizational units that pool their inputs towards a problem solution. Thus the terms multidisciplinary and interdisciplinary are used to describe organizational forms. Epton et al. (1983) further argue that it is necessary to clearly separate the task, the organizational form used to complete the task, and the evaluation process used to determine how well the task has been performed. When a task requires the contributions from several disciplines, the term cross-disciplinary should be used. This distinction was proposed to clarify the situation when a task was observed to consist of subtasks, and some subtasks employed interdisciplinary forms of organization and others employed multi-disciplinary forms. A common observation was that there is a time that working together can help clarify the goals, define the tasks, and plan the integration of the contributions, but once this is done each individual should then address his or her own subtask. This observation can apply to all phases of high technology activities and is not limited to research and development.

NOMENCLATURE

In high technology efforts, the term discipline is not even well defined. When university or research organizations are the subject of study, the definition of discipline is based on academic degree granting units that have a national or worldwide constituency of graduates and practitioners that have common professional organizations, common journals for their publications, and recognized leaders of their paradigms. The discipline is the basic unit in university and research institutions. Many industrial and governmental organizations do not use the academic disciplines to define their basic units. Instead they may adopt a project basis to define units, or many define functions that are crossdisciplinary in nature. In the preceding set of papers many examples were given of project groups such as those developed to produce a spacecraft, a new drug, a new regulation, or a new piece of electronics. Each of these groups have crossdisciplinary tasks which require various contributors with skills or knowledge from at least one of several disciplines. When a matrix form of organization is used as discussed by Reeves (Paper No. 6), functional groups are organized to support project groups. The functional groups provide the disciplinary knowledge essential to support the projects, but the functions may not coincide with the academic definitions of disciplines. Many of the functions defined by high technology efforts have become the basis of new academic disciplines. For example, a materials and processes functional group is common in many high technology organizations. Such a unit provides technology on materials and processes to produce these materials, and contains individuals with backgrounds in disciplines such as chemistry, physics, metallurgy, ceramics, chemical engineering, etc. Is this functional group interdisciplinary or disciplinary? Is this definition important in the management of high technology efforts?

One answer to a working definition of a discipline is to use Kuhn's (1970) paradigm concept. A paradigm is the current accepted way to view and solve a set of problems. Kuhn suggested that in science there is a revolutionary process where failed paradigms are replaced by new ones. With this definition, a discipline would be defined as a group that accepts a given paradigm. Diesing (1982) expressed concern that many of the social sciences (this would include the management sciences) do not have one paradigm, but continue to propose many alternative paradigms. Thus disciplines such as economics have splinter groups each with their own paradigms. Should each splinter group be defined as a discipline? Since Kuhn's definition and model does represent the physical sciences and engineering, the concept of paradigm may be useful in discussing high technology organizations, but caution must be exercised when extending this concept to the management aspects.

The previous conferences in the series have examined, in detail, the interface between disciplines. Given enough time and resources, interdisciplinary groups become highly productive and efficient. The key to success of these groups is the ability to communicate and resolve conflicts. Diesing (1982) suggests that communication is necessary, but not sufficient to the reduction of conflicts between individuals. Defining the differences between paradigms as cognitive barriers, Diesing identifies personal barriers as a more significant problem. Personal barriers are the result of life long experiences, they can be social, cultural, religious, or due to other environmental factors and may inhibit a person to try to communicate. Successful communication requires one party to take the standpoint of the other and learn to think in the categories of the other person. If a neutral language and paradigm could be constructed, it may be possible for both parties to learn the neutral media. If a person's personal barriers prevented the learning of the necessary communication media, then communication may not be possible. An aggressive macho personality may not even entertain the thought of learning another language. Race, religion, economic status, etc., can be personal barriers to communication. These cultural barriers will be explored later.

Effective interdisciplinary groups adopt the language of the leader or create a group language, but either option takes years (Mar, et al., 1974). If the group is stable, then it remains effective. If new members are added, then the language may change due to change in leadership or shift in the majority's preferences for improved communication. The key to interdisciplinary group management is to develop effective communication between group members.

INTERFACES AND CHANGE

The theme of this book is that the real problem of high technology management is the ability to manage change and complexity. In the case of individuals that graduate with only an exposure to one academic discipline, two forms of change will be encountered when they enter the workforce. If the individual is employed in a group where all members have degrees in the same discipline, the individual must learn the culture of the group, and if the group interfaces with other groups or directly with an external customer, the individual must learn customers' preferences. Since all members of the groups belong to the same discipline, there is a common paradigm and language to share, and the changes for the new graduate will be a personal adjustment to organizational norms and the environment in which the organization operates. In the case in which the new graduate joins an interdisciplinary group, an additional barrier must be overcome. The new graduate must resolve how to communicate his or her disciplinary expertise to others in the group, and must learn how to operate in an arena that may contain disciplinary bias, disciplinary intolerance, and disciplinary bigotry. Fortunately, the unit has a culture which will moderate some of the disciplinary differences. If the graduate joins a high technology organization that has a matrix organizational structure, the protection of the common unit is lost, and many more interfaces with other functional and project units must be resolved.

The adjustment to changing environments both within and external to an individual's group is not limited to the new employee. In high technology organizations, the role of a group can change rapidly as new products are developed or new institutional alignments are created to develop the skills, knowledge, and/or resources necessary to compete in the market place. As an individual's group changes, so may the role of that individual. If the group size changes, the individual must interact with more or less people. If the internal interfaces between the group and the rest of the organization changes, then these interfaces must be managed. If the group must interact with subcontractors, new customers, or new regulators then new organizational policies may be needed and implemented. As the number and magnitude of changes to be accommodated increases, so do the management problems.

Brown (1983) has summarized and analyzed four types of interfaces that can change for an individual in any organization. The departmental interface is one between groups of similar stature at any level. In a university the interface between a chemistry and physics department would be such an interface. In a high technology form the interface between the engineering and manufacturing departments may be another. The second type of interface is the level interface, or the interface between two units at different levels. In a university, the interface between a chemistry department and the provost's office would be an example, another example would be the interface between doctors and nurses in a hospital. The third interface is a culture interface where roles and stereotypes based on external cultures may produce misunderstanding, bias, or institutionalized discrimination may disrupt an organization's performance. Brown (1983) does not mention disciplines as a culture, but many of the characteristics of cultural interfaces apply to interdisciplinary issues. The final interface used by Brown is the organizational interface, or the interface between units with different operating goals and strategies. Organizational interfaces are controlled by the external environment--increasing interdependence between competing or complementary organizations, and rapid technological and social change. These four types of interfaces can be summarized as the interface between individuals in the same unit (cultural or internal), the interface between organizations at the same level (departmental), the interface between organizations in the same institution at different levels, and the interface between institutions.

The matrix organization increases the complexity of these interfaces. Functional and project groups are usually placed at the same level in an organization, so each individual must face the departmental type interface identified by Brown (1983). In addition, a person can be a manager of a functional unit yet be assigned to a project or vice versa. Thus the level type interface will not only exist between individuals but be faced by a single individual assigned two different roles. Furthermore, if high technology organizations form coalitions, or employ subcontractors, an individual may face institutional interfaces when members of different firms are assigned to the same project teams. Thus the complexity of interface that can be encountered in high technology efforts are much greater than those encountered in other efforts.

Interdisciplinary teams need to be examined with respect to the various types of interface problems. Many of the issues encountered in the management of interdisciplinary teams may be traced to interface issues common to any organization. If the only unique issue in the management of interdisciplinary groups is the addition of an academic discipline to cultural interface issues, then much of the management of interdisciplinary groups can be resolved with existing management concepts. The example of the new graduate presents many of these existing management interface issues. The new graduates must learn to interact with their counterparts who may have different cultural and institutional backgrounds; they must learn to function under supervision from several people; and they may have to work with people from different units.

Brown's (1983) thesis is that some level of conflict is necessary in these interface organizations to have bargaining and problem solving. Too little conflict results in withdrawal and suppression and too much conflict results in escalation, increasing distrust and antagonism on both sides. The dynamics of these interface groups may also follow a shift in conflict levels. More mature interfaces may tend towards too little conflict because the participants have become too familiar with each other's position, while young interfaces can have too much conflict if the participants cannot find a communication channel. High technology's characteristic of rapid change can be interpreted as a characteristic of rapidly changing organizational interfaces and major interdisciplinary conflicts. Are the issues of organizational conflict resolution, organizational management, and interdisciplinary group management similar?

COMPLEXITY AND CHANGE

Researchers of interdisciplinary processes have been concerned with the size of an organization as a possible factor in determining management needs, but size may be a surrogate for complexity. As the size of an organization increases, the number of interactions between employees need not increase. Management practices usually group employees into small clusters (units, groups, departments, etc.) that have similar talents or duties. The basic size ranges from a few to several dozen, but not to hundreds or thousands. As presented earlier, the traditional organizational structure is a vertical hierarchy or inverted tree. Frequently the expectation is that a person does not interface with another unit at the same level. The path is a series of steps upward until a common manager is reached, and then downward to the unit of concern. This simplistic management concept limits the interfaces to the "level type." In reality informal interpersonal and intergroup relations develop to cope with the awkward formal structure. High technology or other organizations that use a matrix structure introduce all four types of interfaces. As the number of projects and functions increases or decreases in such organizations the complexity of these interfaces will change. Since the ability to resolve complex interface conflicts may be a major task, it becomes important to understand the nature of these complexities as well as the dynamics or change of the interfaces.

Large aerospace organizations represent the extreme of organizational complexity, since they employ matrix organizations and produce high technology products. It is common in such organizations to create new projects requiring hundreds of people in weeks or months. The complexity of interfaces is managed using system engineering, a process that is relatively unknown to other organizations. Lockheed (1983) has described system engineering management as an exhaustive, iterative process where each interface is examined and the important functions are defined and managed using an iterative analytical process that increases in detail and specificity at each subsequent iteration. Practitioners of system engineering are developed on the job, and the practices and methods are seldom recorded for others to learn. The current practice of system engineering is a brute force, exhaustive examination and accounting of all interfaces. Few shortcuts, or simplifications are practiced.

SET II OF PAPERS

Previous conferences have thoroughly investigated the problems of disciplinary interfaces, but have confused these problems with size of the organizations, the cultural and social backgrounds of the participants, and the nature of the environments surrounding the organizations. In the following set of papers the impact of changes in these factors will be explored and an analysis of their relative importance to understanding the management of high technology efforts are presented. Insights from various professional organizations studying these similar issues are a unique characteristic of these contributions. Members of INTERSTUDY represent research administrators and students of large-scale research processes. Members of TIMS/COLRAD represent industrial managers and students of industrial management practices. Members of 4S represent social scientists and humanists studying the social behavior of scientists, and members of AIS are teachers of interdisciplinary approaches to various topics. Each of these groups has a different perspective on the management of an interdisciplinary effort, and the relationship this may have with the management of a high technology activity.

REFERENCES

Brown, L.D. 1983. Managing conflict at organizational interfaces. Reading, MA: Addison-Wesley Publishing Co.

Diesing, P. 1982. Science ideology in the policy sciences. New York: Aldine Publishing Co.

Epton, S.R., Payne, R.L. and Pearson, A.W., eds. 1983. Managing interdisciplinary research. Chichester: John Wiley and Sons.

Kuhn, T. 1970. The structure of scientific revolutions, 2nd ed. Chicago: University of Chicago Press.

Lockheed Missiles and Space Company. 1983. System engineering management guide. Fort Belvoir, Virginia: Defense System Management College.

Mar, B.W. 1974. Problems encountered in multidisciplinary resources and environmental simulation model development. Journal of Environmental Management 2:82-100.

MANAGING HIGH TECHNOLOGY
An Interdisciplinary Perspective
B.W. Mar, W.T. Newell, and B.O. Saxberg (Editors)

INTRODUCTORY REMARKS

Donald R. Baldwin*

Office of the Provost, AH-20
University of Washington
Seattle, Washington 98195

*Mr. Baldwin is Assistant Provost for Research at the University of Washington, Founder of INTERSTUDY, Coordinating Member of the INTERSTUDY International Advisory Board and First Editor of the INTERSTUDY Bulletin.

On behalf of INTERSTUDY, it is a pleasure for me to welcome all of you to the University of Washington and to our great Emerald City, Seattle. We are delighted to have you here and hope that your visit to the U.S. Pacific Northwest proves both enjoyable and memorable. Please let us know if there is something more we can do toward that end.

Turning now to the purpose of our Conference, I suppose we would all agree that "Interdisciplinary Research" is still not a familiar household term. But I submit that awareness and understanding of the term and of the importance of interdisciplinary research to many of the international and societal problems we face today has progressed a long way in the last five years. The First International Conference on Interdisciplinary Research was held in April 1979 in West Germany. About a dozen persons in this room were among the 30 attendees from ten different countries who participated in that Conference. For me, it was a marvelous professional experience at another outstanding location. We met at the International Institute for Scientific Cooperation, housed in an historic castle called Schloss Reisenburg, located about midway between Stutgart and Munich on the banks of the Danube River.

In addition to the information and viewpoints exchanged concerning interdisciplinary research, two concrete decisions came out of that first Conference. One was that there should be a series of working Conferences on interdisciplinary research to be held about every two years.

The other was that there was a need for some definite entity to hold continuing and structured attention of the study of interdisciplinary research. As is evident by now, both of those decisions have been successfully implemented. These were the beginnings of INTERSTUDY founded nearly five years ago.

From the beginning, the objectives of INTERSTUDY have been to: (1) advance the art and science and interdisciplinary research, (2) organize and sponsor meetings, conferences, and symposia to bring together researchers and research administrators for discussion of problems and methodologies related to interdisciplinary research, (3) publish and otherwise disseminate information concerning all aspects of interdisciplinary research, and (4) educate the members as to the status of interdisciplinary research and its application to various local, regional, national, and international goals/needs.

The INTERSTUDY organization has been pretty informal, partly by design and partly because we haven't had enough resources and active participants to do otherwise. Nevertheless, we have succeeded in maintaining a continuing visibility in the scientific community and have earned at least meager recognition as a viable organization. There are several notable accomplishments. A dozen issues of INTERSTUDY Bulletin have been published and distributed during 1981-84. The Second International Conference was cosponsored with the Manchester Business School in England in 1981. A two-year planning effort has finally succeeded in bringing us together for this Third Conference.

Further, a Proceedings of the First Conference and a book based substantially on the Second Conference have been published. We certainly expect a publication from this Third Conference, too. In addition, about half a dozen workshops on various aspects of interdisciplinary research have been carried out in the U.S. and England by persons operating at least in part under the INTERSTUDY banner. Some of the articles fostered by INTERSTUDY have been published in various refereed journals including the Journal of the Society of Research Administrators. The more optimistic among us continue to talk about a Fourth Conference and I suspect it will come to pass.

Clearly, INTERSTUDY has enjoyed some success and is beginning to show some new signs of maturity. But it is still fairly fragile and requires attentive leadership. All is not a "bed of roses." The modest voluntary subscription fees for the bulletin have never been sufficient to cover more than about half of the actual out-of-pocket costs for printing and mailing. Although I prefer not to be specific about the sources, I have continued to find funds to assure continuation of the Bulletin because I fear that without it, the INTERSTUDY organization would soon disappear. But the University of Washington, Brian Mar, and I have contributed to all we can, so I anticipate that our final session at the close of this Conference will include discussion of INTERSTUDY's future. I hope commitments will be forthcoming that indeed assure a future. Certainly, I pledge my continued assistance to the organization but necessarily at a lower level of effort than has been possible during the last five years.

Thanks to Brian Mar and his Program Planning Committee, and the contribution all of you will make, an outstanding Conference will unfold here during the next three days. It is both a pleasure and a privilege for me to be part of it.

MANAGING HIGH TECHNOLOGY
An Interdisciplinary Perspective
B.W. Mar, W.T. Newell, and B.O. Saxberg (Editors)

SOCIETY FOR SOCIAL STUDIES OF SCIENCE (4S) PERSPECTIVES ON INTERDISCIPLINARY RESEARCH

Nicholas C. Mullins[1] and Daryl E. Chubin[2]

[1]Dept. of Sociology, Virginia Polytechnic Institute and State University, Blacksburg, VA 24061

[2]Georgia Institute of Technology
Atlanta, Georgia 30332

The Society for Social Studies of Science was founded in 1975, and will hold its Ninth Annual meeting this year in Ghent, Belgium. We are an interdisciplinary society, founded by sociologists, historians, philosophers, policy researchers, and others who study science. We meet jointly with the Society for the History of Technology (SHOT), History of Science Society, and the Philosophy of Science Society about every four years, including in Washington, D.C. in 1986. These joint meetings increase the interdisciplinary range and depth of our interests.

When we were asked what did we have to say about interdisciplinary studies, our first thought was that it is like Moliere's gentleman speaking prose, we have been doing it all along. When one asks about research being done by interdisciplinary teams we cite, for example, Table 1.

Table 1
Recent Social Studies of Science

Disciplinary Category	Names	Works
philosopher-sociologist	Latour-Woolgar	LABORATORY LIFE (Sage Books)
astronomer-sociologist	Edge-Mulkay	ASTRONOMY TRANSFORMED (Wiley)
Information Scientist-historian	Griffith-Small	series of papers in *Social Studies of Science*

We can also cite work on interdisciplinary subjects such as the discovery and growth of reverse transcriptese (which each of us has done independently and with divergent results, Mullins et al., 1977; Chubin and Studer, 1979).

This work in the biological sciences combined with work in the physical sciences leads us to observe that discipline and problem areas in the biological sciences do not seem to be as clearly defined as those in the physical sciences.

However, this is not the important part of what we have to offer to this meeting. 4S members bring a variety of perspectives to questions of interdisciplinary research and research management.

For example, at the Ghent meeting there is a session on cross-disciplinary interaction which Brian Mar has been asked to chair, and one on evaluating priorities. However, this name-dropping approach does not get at how questions are tackled.

Disciplines and interdisciplinary are question-begging terms. On one level, physics is certainly a discipline, but exobotany probably isn't. There are programs in NSF called developmental biology and cell biology. These are just definitional questions to some, but to many social scientists, they represent a class of phenomena involving the taxonomy of the scientific culture and its representations in social situations.

One study (Gieryn, 1983) points out that not only are the boundaries between disciplinary areas elements in an ongoing political struggle between competing groups and interests, but the boundary between science and engineering, science and law, science and religion, etc., are all boundaries that are continually defended and modified in a process. These processes are more likely to interest us than the specifics of the disciplinary boundaries or what is interdisciplinary at the moment.

Management of scientific units, the support and dependence of science and scientists on support, and the representation of the scientific ideas in various media are other phenomena that interest 4S members.

Robert Kohler's (1979) study of the differences in the development of a new interdisciplinary area, biochemistry, in Europe and America in response to quite different institutional arrangements points directly to the analysis of interdisciplinary studies, and why they are a problem that would interest us.

One of us is presently completing, with others in attendance at this meeting, Alan Porter and Fred Rossini, a source book on interdisciplinarity that will be published next year by Lomond Publications (Chubin et al., 1985). The disciplines involved in this one collaboration reflect a range of interests reflected at this meeting--solid state physics, philosophy of science, chemical engineering, psychology, and sociology.

Works such as this begin to address important questions. For example:

1. Where do interdisciplinary researchers come from?

   which institutions?
   which fields?
   which major professional influences?
   which phases of a researcher's career?

2. Which research areas feature interdisciplinary work?

   which pairs of areas are connected?
   which areas do not connect?
   which areas are intellectually promiscuous?

3. What factor or combination of factors appears to cause and/or support interdisciplinary work?

   intellectual curiosity?
   social problem urgency?
   emphasis of research sponsors?
   local constraints or rewards of employing organizations?

4. What are the differences between settings that have practiced interdisciplinary work (under different rubrics) as the routine mode of operation and those that have not?

Tentative answers to some of these questions have already been offered in the literature; others will be aired in papers to follow. To answer these questions, social scientists would point out, we need to realize that interdisciplinary research is more than an organizational form, if is also an attempt to surmount the divisions of knowledge. Most of the mechanisms of certification of knowledge and knowers as sound or competent are bound to bureaucratic divisions in universities, separate and specialized journals, specific credentials of degrees and license, peer panels of certified experts in recognized fields, bureaucratic divisions of sponsoring organizations into programs, and the personal and corporate interests of groups of scientists and engineers in maintaining these divisions.

Interdisciplinary research is often undertaken because competence in one field is not enough to solve problems and that scholarly journals do not always capture disciplined knowledge. As Gieryn points out we have been socialized to stake out and defend an intellectual territory for self interest and preservation (not necessarily to push forward the frontier). Interdisciplinary work cuts across those lines. We need to examine its successes and failures in both intellectual and social terms.

Interdisciplinary research challenges all that is comfortable. 4S is one professional society which recognized this reality. Indeed, it was founded in the recognition that social studies and not a single discipline or mix of them, would advance knowledge about science and technology. 4S thus has provided and will continue to provide a forum based on the object of study, not on approaches, theories, or methods alone. The commingling of disciplinary specialists at our annual meetings occasions new relationships granted legitimacy and substance between meetings. 4S helps to create interdisciplinary collaboration for studies of science. The Society for Social Studies of Science welcomes you to what it has been nurturing for nearly a decade--a window to the social processes which sustain the production of interdisciplinary research.

REFERENCES

Chubin, D.E., Rossini, F.A., Porter, A.L. and Connolly, T. 1985. Interdisciplinarity: theory and method of problem-focused research and development. Mt. Airy, MD: Lamond.

Chubin, D.E. and Studer, K.E. 1979. Knowledge and structures of scientific growth: measurement of a caner problem domain. Scientometrics 1(Jan):171-193.

Gieryn, T.F. 1983. Boundary-work and the demarcation of science from non-science: strains and interests in professional ideologies of scientists. American Sociological Review 48:781-795.

Kohler, R.E. 1979. Medical reform and biomedical science: biochemistry - a case study. In M.J. Vogel and C.E. Rosenberg, eds., pp. 27-66, The therapeutic revolution: essays in the social history of American medicine. Philadelphia: University of Pennsylvania Press.

Mullins, N.C., Hargens, L.L., Hecht, P.K. and Kick, E.L. 1977. The group structure of coditation clusters: a comparative study. American Sociological Review 42(Aug):552-562.

MANAGING HIGH TECHNOLOGY
An Interdisciplinary Perspective
B.W. Mar, W.T. Newell, and B.O. Saxberg (Editors)
Elsevier Science Publishers B.V. (North-Holland), 1985

THE NATIONAL SCIENCE FOUNDATION AND HIGH-TECH INTERDISCIPLINARY RESEARCH

Richard R. Goulet

Office of Interdisciplinary Research
National Science Foundation
Washington, D.C. 20550

The world's most scientifically and technically advanced nations recognize the tremendous value and potential of interdisciplinary research (IDR), since IDR may be the single most important organizational contribution to technological advancement. Today, an international awareness is evolving of the importance of successful collaboration across scientific disciplines on increasingly complex scientific and societal problems. Yet, despite some spectacular successes such as the discovery of the structure of DNA and the landing of men on the moon, many collaborative efforts between disciplines are difficult to nurture and often yield limited successes.

Nonetheless, interdisciplinary research is critical to addressing the complex problems of restoring the vigor of innovation, productivity, and economic growth. Not an end in itself, interdisciplinary research should be facilitated when required to address critical scientific, technological, and societal problems through cooperative inquiry among the sciences and engineering. Though strenuous and high-risk, interdisciplinary research can provide a richness of innovations in both product and disciplinary realignments.

Recognizing the risks and holding that the vitality in science is usually at the interface of disciplines, the National Science Foundation (NSF) has assumed an active role through leveraging funds to promote the processes of cross-disciplinary communication and research. The NSF can serve as a catalyst of potential enrichment among the disciplines, and can provide leverage for promoting research through federal, state, local, and industrial arrangements.

The Foundation's essential mission is the health and vitality of science, and the health of science depends upon the interaction of disparate disciplines to yield new insights and principles of science. The Foundation is assuming an important role in supporting research efforts that follow the continuum from university-based efforts to opportunities for state and local application as well as industrial participation. The Foundation can provide leverage for interdisciplinary research through an array of alternative mechanisms for academic association with industry that facilitate the free flow of information. It can play a catalytic role by bringing together disciplines with common interests in scientific and technological problems, but with little common research experience. Perhaps most important, it can heighten the vitality in research through support for open, fundamental research that will underlie further improvements in an evolving interdisciplinary area of scientific or economic importance.

The Foundation provides only a thin slice of the total resources made available by the Federal Government to universities and colleges for research and education. But, members of these institutions can do more to enhance the creative spirit in both science and engineering. They can give our technical community the tools it needs, and build a bank of knowledge and experience on which others may draw. The thin slice provided by the Foundation is a critical element in the innovation chain that leads from the library and the laboratory out into the marketplace.

The National Science Foundation is organized to support research ranging from cellular biology to the design of buildings that can withstand earthquakes. The Foundation's structure reflects the fact that it has statutory mandates to support basic research in all the sciences and engineering, and also to investigate the application of scientific and engineering knowledge in the service of national needs. This requires a structure that is flexible enough to accommodate creativity in both the search for new knowledge and the seizing of exciting and useful technical opportunities.

To reinforce such an organizational perspective, in November of 1981, the Office of Interdisciplinary Research was established. The need for this Office became clear when an examination of the distinctions between basic and applied research revealed that some of the lines drawn between the sciences and engineering had begun to fade. Although engineering was born as a practical and intuitive art, today it embraces a range of knowledge from the highly theoretical to the very practical. Engineering disciplines embrace interdisciplinary efforts in addressing scientific and societal problems. Engineering builds linkages with physical, social, and life sciences on one side, and with industrial practices on the other. Here, we can see the vitality as well as the totality of research, an exciting tapestry representing the process of scientific inquiry, with practical applications stimulating and reinforcing research, and vice versa.

The Office of Interdisciplinary Research (OIR) has undertaken many activities to meet its ongoing objective of facilitating and coordinating NSF support of interdisciplinary research since its establishment in 1981. Its accomplishments have included organizing and coordinating team activities in the areas of Biotechnology, Robotics and Automated Manufacturing, Soils Research, and Science and Technology to Aid the Handicapped, directed toward coherent support of research by several NSF programs. Activities involve joint funding, joint announcements, and a cross-cut of support. The Biotechnology committee, for example, comprises 18 programs in four directorates (Engineering; Biological, Behavioral, and Social Sciences; Astronomical, Atmospheric, Earth, and Ocean Sciences; and Scientific, Technological, and International Affairs) and uses informal means as well as workshops and conferences to involve other agencies and industry.

The OIR encourages industry-university coupling through joint funding of industry-university research and centers, and through the support of workshops involving participation and contributions from the academic, industrial, and government sectors. Strong examples of such cooperative efforts are shown in the Robotics and Biotechnology area. At the University of Rhode Island, for example industry is supporting the major portion of a four-year effort to establish a Robotics Center focusing on integrated robot systems to develop a data base for industrial applications. Research thrusts are in vision systems for robot control and integration of robots into production systems, and in articulated robot hands with sensors. Industrial support began at 75 percent for the first of the four-year effort, and will increase to total support in two more years.

Another area of OIR activity is the support for workshops and conferences to highlight emerging and complex areas of interdisciplinary research, determine research needs, and encourage cross-disciplinary communication. Recent areas of investigation have included Magnetic Information Technology (MINT) and the social impacts of the microelectronics revolution.

Magnetic Information Technology (MINT) is a critical growth industry for the United States since all sectors of the economy, both public and private, need to store and retrieve ever increasing volumes of information. Annual sales of magnetic devices to meet these needs are growing at about 35 percent per year.

The NSF role in MINT is to encourage research in a wide range of fields that address the needs expressed by industry for fundamental knowledge underpinning further technological advances in MINT. NSF also is in a unique position to encourage interdisciplinary collaboration in a wide range of needed research identified by industry in areas such as information theory, signal processing, tribology, materials, and magnetic transducers.

Another NSF conference entitled, Microelectronics in Transition: Industrial Transformation and Social Change, held in May 1983 at the University of California, Santa Cruz, examined the various social impacts of the microelectronic revolution and of the industry that it has produced. Conference participants were drawn from local and federal governments, academia, industry, labor, and the affected community; and topics included the effects of microelectronics on industrial structure, on labor organization and the workplace, as well as on urban development and community formation.

The OIR also supports state-of-the-art review papers from recognized experts in emerging interdisciplinary research areas that are ready for critical analysis and synthesis, in order to identify long-term research needs and thrusts. Papers supported by OIR have included diverse topics such as Adhesive Bonding, Hydroponics, Software Psychology, Combustion Research, CAD/CAM, Microbial-Enhanced Oil Recovery (MEOR), and Fluidization. The review on "Software Psychology," for example, was written by Ruven Brooks of the ITT Advanced Technology Center, Shelton, Connecticut, with collaborators from Yale University. It addresses those aspects of man-computer interaction that are particularly concerned with the way properties of human behavior interact with software structures and systems, and with the effect different structures and systems have on that behavior.

In contrast to other areas of psychology, software psychology emphasizes developing models and theories of behavior that are useful specifically for describing human interaction with software. The paper emphasizes the cognitive factors in software development and use, and reviews selected achievements in the area of software psychology, and then analyzes the methodologies used to obtain those results.

Nine papers currently are available from OIR as well as from NTIS in hard copy and microfiche.

The OIR also is supporting research on IDR methodologies at several levels of analysis in order to improve interdisciplinary research processes. In addition to the effort supported through the Georgia Institute of Technology to refine the conceptual framework for further study and improvement of IDR processes to be discussed by Alan Porter and Fred Rossini at this conference, OIR supported a project to test a technological innovation matrix approach to R&D evaluation and priority-development. Another example is the application of a magnitude-estimation scale for polling authorities in order to establish priorities. This study used a series of proposed remedial policies and research topics identified in an NSF report on Flood Hazard Mitigation.

An OIR call-for-proposals issued in December of 1983 was designed to encourage research in interdisciplinary fields of inquiry. Projects are expected to produce proof-of-concepts or prototypes of the desired instruments. The effort is aimed at improving the quality and broadening the scope of interdisciplinary research at the proposing institutions, and includes four major areas of support: Biotechnology and Bioprocessing Engineering, Biotechnological Information Acquisition Systems, Lightwave Technology, and Hazardous Waste Management. Sixty proposals were received in response to the call-for-proposals, and we expect to announce nine awards totalling about $800,000 by September of this year. The program is scheduled to continue and an announcement should be made by November.

NSF's new high-tech interdisciplinary initiative known as the Engineering Research Centers (ERC) will be administered by the OIR. Last year, a National Science Board panel recommended the establishment of 100 or so ERCs over the next decade, and by next July, we will be selecting the first of a series of engineering research centers based at universities. Currently, $10 million dollars have been apportioned for this effort, and the figure is expected to expand greatly over the next few years.

The ERC program is seen as an overall effort to raise support sharply for engineering manpower, research and equipment in key fields of international technological competition, besides raising consciousness for the need to see modern engineering in terms of cross-disciplinary systems efforts. While individual centers will differ from one another, all will provide for working relations between students and faculty on the one hand, and industry engineers and scientists on the other. In addition, they will emphasize the synthesis of engineering knowledge and seek to integrate different disciplines in order to bring together their requisite knowledge, methodologies, and tools to solve issues important to practitioners. The Center programs must also be designed to contribute to the increased effectiveness of all levels of engineering education.

These, then, are the broad functions and responsibilities of the OIR directed toward facilitating IDR by the NSF. These responsibilities are based on the principle that virtually all research is potentially applicable. We know that interdisciplinary research has been the source from which most modern technologies have arisen. One need go back no further than the discovery of DNA to see a result of basic research that very quickly has led to a technology of enormous potential social and economic importance through interdisciplinary research efforts.

Interdisciplinary research must be viewed as an essential process for improving the linkages among the sciences directed toward the solution of scientific and technological problems. Never an end in itself, this process should be facilitated when required to addresss critical problems. Though often a high-risk and arduous process, interdisciplinary research may yield enormous rewards as well as produce innovative and needed realignment of disciplines (e.g., earth sciences and materials sciences), making it increasingly clear that the vitality in science is usually at the intersection of disciplines. Perhaps, then, the isolation of disciplines, following the generation of fundamental knowledge, is intellectually and economically wasteful.

MANAGING HIGH TECHNOLOGY
An Interdisciplinary Perspective
B.W. Mar, W.T. Newell, and B.O. Saxberg (Editors)

INTERDISCIPLINARITY WITHOUT DISCIPLINES? A VIEW FROM THE SOCIOLOGY OF SCIENCE

Arie Rip
University of Amsterdam
The Netherlands

Interdisciplinarity and interdisciplinary research tend to be diffuse issues, but it is possible to bring a focus to the discussion by choosing a particular theme. In this conference, it was management of high tech, and it did the job well. What it also did, at least for me, was to bring out the tension between practitioners, who try to solve their problems as well as possible and do not bother to check whether there is interdisciplinarity or not, and analysts, scholars who study processes of interdisciplinary research and try to come up with general conclusions.

Schematically, one might distinguish three categories of experience and study, ranging from primarily action-oriented to primarily knowledge-oriented (Figure 1):

(1)	PRACTICAL PROBLEMS & EXPERIENCE OF PRACTITIONERS
(2)	MANAGEMENT STUDIES, DESIGN OF TEAMWORK GUIDELINES FOR INTERDISCIPLINARY RESEARCH
(3)	SCHOLARLY STUDIES OF: - group processes, communication - innovation in organizations, technology - scientific processes, epistemology

Figure 1
Knowledge Relevant to interdisciplinary Research

The contributions to the INTERSTUDY conference can be allocated to these categories. Papers on practical problems and experience of practitioners were much more numerous (and strong) than is usual in academic conferences (thanks to the choice of focal theme and the efforts of the organizers). Papers in the management and design category were often directed to very specific or location questions, but did have the potential of being embedded in an analytic framework. The third category, which might be the specific contribution that INTERSTUDY has to make, was represented quite well for the first two subjects, group processes and communication, and innovation, organizations, technology. Sociology and epistemology of science were nearly absent, although some papers drew upon these fields.

I am not just making a plea for my own field (although I will do so in a moment). Too often, interdisciplinarity or interdisciplinary research is used as an unproblematic label, while it may cover very different things. For instance, in Europe a conference on interdisciplinarity would draw a heavy contingent of philosophers and some sociologists of science. While in the U.S., as I have been told, it is the engineers that are interested in interdisciplinarity. Clearly, the concepts have to be specified. Previous meetings of INTERSTUDY appear to have done this; see especially the definitions given by Epton and Pearson (1984) in their introduction to the Proceedings of the Second Conference. But problems remain: the focus upon tasks to be completed implies that only part of the connotations of interdisciplinarity is kept (ideals of unified science, or at least a new level of doing science, are not taken into account). And more importantly, the concept of discipline remains an unanalyzed resource or one that is adapted to whatever particular end. For example, Epton, Payne and Pearson (1984, p. 76) do not want to be limited by philosophical concerns: "It is important that we use the widest possible definition of discipline. This must be extended to include all groups of people with common interests and affiliations whose opinions, expertise and cooperation are required at a particular stage of the project."

If discipline remains an unanalyzed resource, one misses an opportunity to increase our understanding of interdisciplinarity--or perhaps misunderstand it. For example, it is not implausible to claim that disciplinarity is the product of university education, and in that sense an artifact that has little to do with scientific research. If this claim is valid, the implication is that a call for (more or better) interdisciplinarity has little to say about scientific research, but flows from a dissatisfaction with present arrangements in universities. The dissatisfaction may well be justified, but it does not touch the substance of the issue of interdisciplinarity.

From the occurrence of calls for interdisciplinarity, one can conclude that there is a demand, be it unspecific. Interdisciplinarity is too often a motherhood issue, and so the (potential or actual) supply is never described very clearly. As soon as examples are offered, it is equated (or seen in line with) problem-solving--the terms were used interchangeably during the conference! Or problems of communication and collaboration in research are presented as instances of barriers to interdisciplinary research. With such looseness of usage, the study of interdisciplinarity may easily become baroque, and will have great difficulties to get out of the wallowing phase. Phrased in a different way: the dependent variable in our studies is insufficiently specified (it may also be redefined) so that it is of little use to analyze independent variables, and comparison of cases and cumulation of insights are hindered. For this reason, it is important that attempts at further articulation and conceptualization continue, e.g., Epton, Payne and Pearson

(1984), or the STRAP framework developed by Chubin, Porter and Rossini (see e.g., Porter and Rossini, 1984). In addition, systematic description of multi/interdisciplinarity as encountered must receive more attention. This may well start from simple operationalizations, e.g., as in Gerpott and Domsch (1984). I have tried my hand at such descriptions by tracing disciplinary components in articles appearing in a biotechnology core journal; for references and further experiments and mapping the cognitive structure of the field (see Rip and Courtial, 1984).

Let me now introduce the view from the sociology of science (Mullins and Chubin, 1984) and attempt to analyze discipline. In fact, I shall also come up with my redefinition of the issues of interdisciplinarity. Even if you do not accept the redefinition, I hope to have shown that a contribution from the sociology of science is necessary.

To understand the dynamics of scientific development and the role of disciplines, it is useful to introduce a conceptual scheme in which scientific fields are distinguished from research practices (Figure 2).

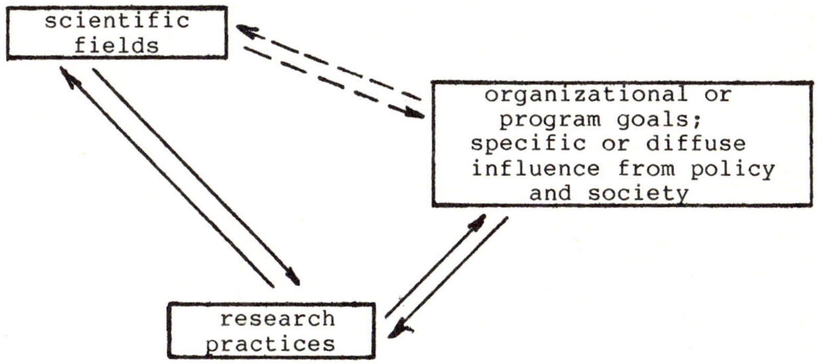

Figure 2
Dynamics of Scientific Development

The conceptual scheme is an abstract one: it does not talk about laboratories, groups, organizations but about the activities and products. These activities and products have a cognitive, as well as a social dimension; both are important when one wants to understand the dynamics of science. The scheme is based upon studies of the development of scientific specialties and research areas, an important subject in recent sociology of science. Its relevance may be appreciated by considering how it fits points made during the conference. For example, when Martha Russell notes that one should look at people, disciplines and administrative units, she is referring to the three components of the scheme. And when Harvey Gold says that the issue is how research goals (in research practices, A. Rip) can be reached while maintaining disciplines as a source of expertise, he wants to maintain the triangular arrangement in good health.

The different components of the scheme have received differential attention in the sociology of science. The 1960s and 1970s saw many studies at the level of scientific fields, e.g., reward system, stratification, communication, development of specialties. In the 1970s there was an upsurge of detailed, sometimes ethographic studies of research practices: Laboratory Life was the title of an influential book by Latour and Woolgar (1979). More recently, the influence of external goals, contexts of research and science policy programming on scientific fields and on research practices is being examined. Most of the arrows in the scheme can now be explicated; one of the outcomes is that it is impossible to generalize quickly because of differences in dynamics across fields and sectors.

What do the results of sociology of science imply for the concept of discipline? Note first that scientific field is a more general concept: it encompasses all levels and interactions in science that are nonlocal (in contrast to the localized research practices). Part of the work in research practices is to transform the local production of knowledge items into more universally valid (or at least accepted) knowledge claims, e.g., in a scientific paper. (One might, if one wishes, also use the concepts "context of discovery" and "context of justification," drawn from an older, philosophical tradition.) The example serves to make clear that the difference local/nonlocal is not one between individual and collective.

There are different levels of nonlocality, which can be arranged from smaller to larger units. But the hierarchy is only partial: the processes on the different levels are different in kind.

research area/ problem area	frame for ongoing work, network of knowledge, problems, skills, people
specialty	paradigm, cognitive style, ideal of explanation
discipline	educational, organizational and librarian category
branch of science or scholarship	idem; possibly also differences in object, method, dynamic (e.g., science vs. humanities)
science as a whole	focus of public policy & attitudes of public; institution in Western societies; component of Western civilization ("progress of science")

Figure 3
Level of Increasing Non-Locality

From this specification of levels of nonlocality, one can conclude immediately that there will be different kinds of interfield issues: interbranch is C.P. Snow's well-known denouncement of the Two Cultures. Interspecialty are many issues of conflicting problem-definitions and styles, of communication barriers and of struggles for intellectual turf. Interresearch area are problems of transfer of knowledge and skills, possible merging of frameworks, etc. Interdiscipline, finally, are issues that fall between institutional boundaries, especially in academe. Barriers to collaboration may thus have different origins, and different approaches are required to overcome theory.

There are some further implications. In the management of interdisciplinary research, now taken as the label for research practices where skills, styles or paradigms (and thus people) linked to different disciplines or specialties are necessary for task performance, there are different issues and constraints. One set is related to group processes, communication, etc. (general management problems), a further set derives from the organization or program within which the research is conducted, while the third set is related to scientific fields. The influence of the relation with scientific fields varies in strength. In some fields, e.g., theoretical physics and some areas in experimental physics, research practices are seen, by the physicists, as "settlements," in some locale, of their field. Thus, they try to mold environmental constraints to the perceived requirements of the field ("autonomy of science"). Collaboration with other fields is part of the environment of their "settlement," and thus acceptable only if it serves physics. Scientific fields can have less independent existence than physics. In some cases, they are just epiphenomena of training programs (e.g., management science), or labels for important objects of study (women's studies, black studies, studies of interdisciplinary research) that may or may not evolve into clusters of research areas or even specialties. In other cases, there is more intellectual coherence (e.g., in psychology or chemistry), but it is based on an earlier historical phase, and maintained in university education and organization, while the research areas and specialties within those broad classifications branch out in all directions, and the dynamics of disciplinary development are thus very different from those in physics. (This analysis of different relationships between fields and research practices replaces the not very productive discussion during the conference if disciplines exist (where physics was pitted against psychology.)

The conceptual scheme of fields, research practices and programs and goals is also useful in analyzing mission-oriented research (in industry and government institutions, also in some university centers or programs) in contrast with nonmission-oriented research (the traditional task of universities). Mission-oriented organizations are interested in research practices and their (hopefully relevant) outputs. Solving the problem at hand is an overriding concern, and if that requires information from several scientific fields or collaboration between people with different backgrounds, then that's what must be managed. Mission-oriented organizations are only interested in scientific fields as information retrieval classifications and as indicators of recruiting categories. They have no stake in them, other than making sure that their service function is in order. Interdisciplinarity is not at issue.

Universities, on the other hand, are not only organized along disciplinary lines, but house researchers, especially in the graduate schools, who do have a stake in scientific fields. Their careers depend, to a quite large extent, on their status within the field, and so they tend to consider their department or research group as a settlement of a scientific field. (In practice, academics often participate in more than one field, which complicates the picture further.) In line with my argument, the orientation toward scientific fields is less visible and important in those universities (or parts thereof) with a strong mission-orientation (e.g., some of the land grant colleges in the U.S., some Scandinavian universities with a regional orientation).

Territorial struggle, or "turf," appears to be a human universal. But it takes different shapes. In industry, turf is organizational and cognitive matters (knowledge and skills, etc.) are viewed for their instrumental value, not as potential hegemonies. In universities, however, turf is both organizational/ institutional and cognitive. This is why universities are such a difficult place for proponents of interdisciplinarity, in spite of the fact that there is a lot of interresearch area and interspecialty work going on within academic departments (Porter and Rossini, 1984).

In analyzing discipline, I have presented the concept as one element within a more comprehensive framework drawn from the sociology of science. In doing so, I have come quite close to saying that interdisciplinarity is not a useful concept, which should not be taken as implying that there are no problems. I share the dissatisfaction with current arrangements, especially in universities, and I realize that there are barriers to communication, collaboration and problem solving when several scientific fields are involved. In concluding this paper, I shall list four points that approach interdisciplinarity from a more positive perspective.

Bricolage

The world of science is full of metaphorical argument, conceptual shifts, knowledge transfer between areas, serendipity. One does not need sociology of science to realize this. One could call this ad hoc interdisciplinarity or bricolage. "Bricolage," a term used by Levy-Strauss, has been adapted to describe that aspect of science where the researcher scouts for whatever he can use and is prepared to make do with what he finds. So if there is dissatisfaction, it should not focus on a presumed absence of interdisciplinarity, but on the challenge of sustaining or improving existing interdisciplinarity--provided one can specify, at least provisionally, what can be gained this way.

Redefinition

To do justice to the dissatisfaction with current arrangements, without falling into the trap of a too narrow focus on discipline, the problem can be redefined in a more comprehensive way:

> There are structures and processes at the level of scientific fields, which hinder (or interfere with) research practices in realizing preferred goals (mission-oriented or academic). (Or are only perceived to do so, e.g., by dissatisfied or failed innovators who blame the system.)

The advantage of such a redefinition can be illustrated with an example. Epton, Payne and Pearson (1984) say that the monodisciplinarity of scientific journals is a barrier to interdisciplinary research. But scientific journals are seldom monodisciplinary (even if they have disciplinary names): the variety in the articles they carry is very large. They may form a barrier to publication, but that does not depend on disciplinarity or interdisciplinarity: many papers are first rejected and have to be rewritten, or will be published in another journal. If Epton and Pearson had not forced themselves to talk in terms of disciplinary and interdisciplinarity, they would have been able to describe the barrier and to specify which goals, in which research practices, were actually obstructed by it.

Surplus Value

The preceding redefinition focuses, as it were, on the negative side of the concern about interdisciplinarity: the existence of barriers to realize goals. But to bring out the specific and positive side of the concern, one has to say something about the research product or the ways to go about achieving this product. Interdisciplinarity then appears as a way to do research, combining several scientific fields, so as to create a surplus value (if necessary, in spite of the hindering structures and processes of these fields). The surplus value may be problem solving (especially of complex problems, says the ideology of interdisciplinarity), innovation, including cultural innovation (think of recent trends toward having biosocial approaches, and integration or unification of science.

This point has been made before, but it remains important to stress that the surplus value need not always be better management for problem solving. There may also be intellectual gains to be had. In addition, the way I have phrased this point allows me to use it for the issue of science policy measures and interdisciplinarity.

Policy for Interdisciplinarity

There are many science policy measures that have implications for the possibilities of interdisciplinary research. For instance, budget cuts will make university departments less willing to support ventures that seem marginal to their concerns. But interdisciplinarity is itself a policy issue, e.g., in stimulating biotechnology, there is a problem of interaction between the biological sciences and the physical and engineering sciences, and sometimes special measures have to be taken to overcome this barrier. Such science policy problems occur more often in Europe than in the U.S., because government patronage of science in Europe extends to a concern for the development of whole fields and the establishment of review committees and planning and programming agencies or committees. In contrast with the situation of mission-orientation organizations, strategic planning is now much more difficult. There is no mission, with which different contributions can be linked. If there is management of R&D, it is indirect management at most. The science policy agency now does have a stake in scientific fields at least nationally: it must identify structures and processes at work there, that may hinder the achievement of the policy goals. And these goals always contain a specification of a surplus value to be had from interaction between scientific fields--in the case of biotechnology, more systematic and successful innovation in production or services with the help of biological agents.

For me, the problem of interdisciplinarity at the level of science policy making is as much of a challenge, or more, as the management of interdisciplinary research because there is now a direct link to the nature and dynamics of scientific fields. The sociology of science is my "turf," and my priorities are derived from the cognitive framework it provides. Perhaps I am not (yet) truly an interdisciplinarian.

REFERENCES

Epton, S.R., Payne, R.L., and Pearson, A.W. 1984. The management of cross-disciplinary research. R&D Management 14(2):69-79.

Gerpott, T. and Domsch, M. 1984. Cross-disciplinary structures of industrial research and development (R&D) units. Third International Conference on Interdisciplinary Research, Seattle, WA, August 1-3.

Latour, B. and Woolgar, W. 1979. Laboratory life. The social construction of scientific facts. Beverly Hills and London: Safe.

Mullins, N.C. and Chubin, D.E. 1984. Society for social studies of science (4S): perspectives on interdisciplinary research, Third International Conference on Interdisciplinary Research, Seattle, WA, August 1-3.

Porter, A.L. and Rossini, F.A. 1984. Interdisciplinary research redefined: multi-skill, problem-focused research in the STRAP framework. R&D Management 14(2):109.

Rip, A. and Courtial, J.P. 1984. Co-word maps of biotechnology: an example of cognitive scientometrics. Scientometrics 6(6):381-400.

MANAGING HIGH TECHNOLOGY
An Interdisciplinary Perspective
B.W. Mar, W.T. Newell, and B.O. Saxberg (Editors)

INTERDISCIPLINARY/INTERDEPARTMENTAL CONTRASTS IN INDUSTRY

Brian W. Mar
Department of Civil Engineering, FX-10

William T. Newell and Borje O. Saxberg
Management and Organization, DJ-10

University of Washington
Seattle, WA 98195

ABSTRACT

Given the nature of complex scientific and social problems, their solution requires an integrated interdisciplinary team effort. The compartmentalization of the university into discipline departments has proven a serious obstacle to interdisciplinary study and research. Industry addresses problems in science and technology through interdisciplinary task forces. A definite goal orientation related to conditions of employment makes the scientist/engineer subject to the expectations of bureaucratic authority and of creative and innovative contribution to the work at hand. The success of the resulting system of interface conflict is predicated on common interest and compatible behavior in a managed system.

INTRODUCTION

The rapid pace of scientific and technological development, and the great number of unresolved social issues have contributed to a sense of generally increased complexity in the tasks which face engineers and scientists. The solutions to these problems are worked on either in the university or other nonprofit setting, or in industry. In either case, success hinges on the creation of interdisciplinary team effort.

ACADEMIC ENVIRONMENT

The university exists as a sheltered environment for academic and scientific study. Problems do not come conveniently defined in terms of a university's particular departmental structure. Thus the need has existed for addressing these problems through interdisciplinary research. In the past, this work has progressed and at times given rise to new departments in recognition of the viability of a new interdisciplinary research area, e.g., bioengineering, chemometrics, molecular biology. With a quickened pace, particularly in calling for applied research to solve societal problems, the university's effort to respond through interdisciplinary research has met with problems (Mar et al., 1981; Epton et al., 1981; Mar, 1984).

As the authors have reported elsewhere (Mar et al., 1976; Saxberg et al., 1981), the university's organizational system discourages forays away from the departmental base. This is intimately connected with the academic tenure system whereby faculty on tenure career paths are evaluated by their colleagues who are superior in rank.

Inevitably there will be a specialist view applied to the evaluation in judging whether an individuals's work is central to the department's discipline and has received appropriate professional recognition within the profession's refereed journals or presentations at professional meetings and symposia. The suspicion has been voiced that (Mar et al., 1981) interdisciplinary research is an escape from rigorous research requirements in the specialist area, given the suspect evaluation process for interdisciplinary research projects. Similarly, interdisciplinary programs of study have labored under handicaps--also raising the question of the credibility of an interdisciplinary background for demands of expertise representative of a particular discipline.

Given that interdisciplinary applied research sponsored by outside funding agencies frequently concerns issues and problems, young faculty members would by inclination and academic background be tempted to become involved but dissuaded from doing so given the impact that such an involvement might have on their academic career. The question could be raised whether the university is the place for interdisciplinary research under these circumstances--or under what conditions might interdisciplinary research be carried out successfully in the university setting (Newell et al., 1980).

Earlier research on interdisciplinary management (Mar et al., 1981) studied several hundred groups in over twenty universities in the United States. These studies reported serious problems of organization and many failures due to the inability of individuals from different disciplines to create an acceptable organizational framework. The strong departmental structure that exists in universities was hypothesized and shown to be a major factor in this inability to organize for interdisciplinary work. Cheng and McKinley (1983) examined data from an international sample (N=288) of academic research units and concluded that there are three characteristics of usable research; practical relevance, applicability of findings, and specificity. More importantly, they found that scientific groups with highly developed paradigms can be influenced by bureaucratic control, but groups with less developed paradigms will not be influenced by bureaucratic control and may respond negatively to such control. Since most interdisciplinary groups by definition lack a highly developed paradigm, they would not respond to bureaucratic authority.

Cohen, March, and Olsen (1972) developed a simulation model of the university based on the assumption that university problem solving groups are organized anarchies that are described by four parameters that can be treated independently: (1) choices looking for problems, (2) issues and feelings looking for forums, (3) solutions looking for questions, and (4) decision makers looking for work. This model assumes that energy is available in the form of decision makers and solutions, and that energy is the basis for resolving issues and making choices. The model was used to illustrate the response of different kinds of universities to a reduction of organizational slack. Knorr-Centina (1982) has provided a critique of economic models and science describing the decision making of groups of scientists as reflections of bargaining and exchange. The Cohen, March, and Olsen model would explain the failure of interdisciplinary groups to respond to bureaucratic control since the problems they address are ill-defined and require large amounts of energy to resolve. The problem in a participant's own field would require less energy to solve.

INDUSTRIAL ENVIRONMENT

The other question that then looms has to do with how industry deals with interdisciplinary research. Are the problems the same? The scientists and engineers in their past backgrounds very likely pursued discipline oriented study programs. They expected that in their employment they would find use for their hard-won knowledge in projects in their specialty area. This would appear to be a relevant question in the high technology area where many leading edge projects are housed within industry laboratories.

Our preliminary investigation indicates that the management problems of high technology may be more complex than conventional technological management since no one person nor discipline has the knowledge or experience to guide the work. There is also a higher degree of professional risk involved for the participants in these efforts, since the outcomes are unpredictable. In order to assemble a team to participate in these efforts, individuals from departments or units having the appropriate skills and knowledge, must be identified and assigned.

The management of a high technology task force is complex, since individuals and their home units may have agendas that are perceived to be more important than those of the task force. Lambright and Teich (1981) point out that industrial research laboratories have more in common with government laboratories than with universities. They cite several studies from the 1960's that revealed tensions caused by scientists' desires for autonomy as against the pressures for integrated team work in the industrial laboratories (cf. Saxberg et al., 1968). However, the literature has little empirical evidence to offer on the impact of different organizational forms on the effectiveness of innovation. Furthermore, the management tasks are often assigned to scientists and engineers who have not been formally exposed to management concepts and may have greater continuing involvement and commitment to research and development than to administrative work.

A critical issue in the management of high technology or any other innovative effort is the need to understand and control the conflicts between the individual's relationships to their home department, their current assignment, and their personal goals.

Many groups in the industrial setting are interdisciplinary in nature but the individuals have a single supervisor who evaluates their performance. The goals of the group involve team work to produce a project or a service. Individuals are hired to provide a service to the group and will be released if their performance is unacceptable. Such an environment eliminates many of the problems that frustrate interdisciplinary team building in the environment. Furthermore, there usually is a pool of individuals willing and able to be hired into the group.

Interdisciplinary problems as observed in the university setting do not appear to be common in the industry setting. However, industry has the need to find an effective mechanism to assemble interdisciplinary research teams from line departments for large projects or programs. While industry does not face the problem of interdisciplinary team building, the problem of cooperation remains.

MODELS OF INTERDISCIPLINARY ORGANIZATION

It may be appropriate to review at this point the organizational models which have been proposed for dealing with organization design for innovation and creativity. Galbraith (1973; Galbraith, 1977) regards the organizational design problems as one of addressing uncertainty, which he defines as the difference between available and required information. Thus the organization structure reflects an attempt to improve information. In particular, after traditional organizational planning has been exhausted, there is a need to consider further decreasing the ability to process information through slack resources or semiautonomous work groups. If further complexity so requires, the organization's capacity for information processing can be further increased by introducing management information systems and lateral relations, among which the matrix organization represents the most complex variant.

Davis and Lawrence (1977) have elaborated further on the matrix design, involving multiple command with the functional matrix managers in charge of how tasks will be carried out and who will be assigned to do them, while project managers have responsibility for definition of work to be done and the schedule according to which it must progress toward completion. This leaves both with the challenge of resolving conflicts and cooperating with each other in mutually advantageous ways. The ability of the matrix organization to provide flexibility and innovation may be offset by the interpersonal conflicts brought about by diffusion in authority and responsibility.

Minzberg (1979) proposes his model of adhocracy for the organization design which is characterized by flexibility and fluidity, and thus able to provide creative and innovative solutions to problems in a dynamic environment (Emery et al., 1965). The complexity of the organization design corresponds to the dynamic nature of the environment--thus the demands of requisite variety must be fulfilled (Weick, 1979). However, the adhocracy--a temporary organization as suggested by Minzberg's choice of label--is continually under pressure to move toward some form of bureaucratic organization with established routines.

McCann and Galbraith (1981) have argued that because matrix organization is a complex and recent response to the dual needs of specialization and coordination, its effectiveness is difficult to appraise and rigorous research is lacking. Gunn (1982) has related organizational complexity to the disciplinary institutional barriers that arise from hierarchical bureaucratic structures in manufacturing organizations where the changes tend to be incremental and conventional. In his review of the matrix organization, Knight (1976) identified the principal problems such a structure engenders as conflict, stress, balance of power, and increased costs of communication and management. Golden and Thorndyke (1980) have edited a series of papers in a Rand study of team building in the military, which focuses on individual perception and behavior in group work and their impact on cooperation.

While the application of models from general administrative science to the specific problems of interdisciplinary organization is lacking, the model proposed by Brown (1983) provides a comprehensive framework potentially for such an effort. It is a conflict model that postulates a meeting ground where units come face to face as parties represented by individuals who may then represent groups of similar or different status, or departments, or as officials of different organizations. Brown's model is a system theory approach to the analysis of organizational conflict. The assumptions include: "continued interaction between interdependent social units produces interfaces that are social units themselves. Continued contact encourages the development of boundaries and shared expectations that regulate the interactions of interface participants--the interface becomes an organizational entity." The model considers four different types of interface conflict, the conflict between groups of equal power, the conflict between groups that have unequal power, the conflict between participants who have different cultural backgrounds, and the conflict between individuals who represent different organizations. Since too much conflict will result in unproductive exchanges of aggression and too little conflict results in suppression or denial of potential confrontation the model proposes the need for a third party to maintain an optimal level of productive conflict.

Diesing (1982) has characterized disciplines as separate cultures and has suggested that the goals and values of scientists in each discipline may be stronger than cultural values. The cultural interface suggested by Brown may require consideration of both societal and discipline defined cultures. The four different types of conflict have been reported in studies on interdisciplinary research (Mar et al., 1981; Mar et al., 1983). Whether the Cohen et al. (1972) model of organized anarchy or the Brown (1983) model of critical conflict level represent the organizational issues faced by high technology interdisciplinary groups coping with these four types of conflicts, remains to be resolved. When the interface group adopts a formal structure such as the matrix organization, then bureaucratic authority may regulate the decision making.

Efforts to understand the control of complex organizations involves two major topics. The first is to define an optimal control strategy for complex systems, which are well understood, and the second and more difficult is to develop an understanding of systems where individual decision making may be based on subjective as well as objective information. Titli and Singh (1981) have edited proceedings discussing alternative control concepts for hard systems (physical, well understood)--decentralized control appears to be as effective as central control. Sage (1981) has described the application of the concepts of systems engineering including estimation theory and optimal information filters to organizational design.

Progress is also reported on the understanding and simplifying of complex systems. Geoffrion at UCLA has been funded by the Decision and Management Science (DMS) program at NSF to define aggregation, and his work may provide insights into simplifying complex organizational behavior. MacKinnon and Wearing (1980) raise an interesting issue on the need for experts to manage innovative or unique efforts.

They observed that novices managed as well as experts when new situations were presented. Leung (1983) has approached the management of conflict resolution from a subjective view and is attempting to apply the fuzzy set methods of Zadeh (1965) to quantify the subjective perceptions of conflicts. Traditional conflict resolu tion methods such as those summarized by Rapoport (1974) and Nijkamp (1978) may also provide alternative models for reducing conflicts in these innovative efforts.

CONCLUSIONS

In summary, the literature on organizational structure assumes that the tasks can be defined and the integration of these tasks is understood. When the tasks are ill-defined and the potential participants are recruited from complex existing organizations with strong bureaucratic procedures, then management must recognize the possibility of conflicts between individuals and groups in complex settings such as high technology interdisciplinary/interdepartmental groups. The existing literature provides clues on the aggregation and understanding of complexity, the effectiveness of matrix versus project management, and the recognition of subjective aspects of decision making. A beginning has been attempted by the authors in empirical applied research in the high technology area to verify the nature of cooperative and conflict systems. This will provide an opportunity to verify the models we have described above and thus move more firmly on the successful management of high technology as an interdisciplinary/interdepartmental challenge.

REFERENCES

Brown, L.D. 1983. Managing conflict at organizational interfaces. Reading, Mass: Addison-Wesley.

Cheng, J.L. and McKinley, W. 1983. Towards an integration of organization research and practice: a contingency study of bureaucratic control and performance in scientific settings. Administrative Science Quarterly 28:85-100.

Cohen, M.D., March, J.G. and Olsen, J.P. 1972. Garbage can model of organizational choice. Administrative Science Quarterly 17:1-25.

Davis, S.M. and Lawrence, P.R. 1977. Matrix. Reading, Mass: Addison-Wesley.

Diesing, P. 1982. Science and ideology in policy science. New York: Aldine.

Emery, F.E. and Trist, E.L. 1965. The causal texture of organizational environments. Human Relations 18:21-31.

Epton, S.R., Payne, R.L. and Pearson, A.W., eds. 1981. Managing interdisciplinary research. In Second international conference on interdisciplinary research, Manchester Business School, United Kingdom.

Galbraith, J.R. 1977. Designing complex organizations. Reading, Mass.: Addison-Wesley.

Galbraith, J.R. 1977. Organization design. Reading, Mass.: Addison-Wesley.

Geoffrion, A.M. 1983. Decision management service award 1982-1983. Washington, D.C.: National Science Foundation.

Golden, S.E. and Thorndyke, P.W. 1980. Improving team performance. Proceedings of the Rand Team Performance Workshop R-2606. Santa Monica: Rand Corporation.

Gunn, T.G. 1982. The mechanization of design and manufacturing. Scientific American 247(3):114-130.

Knight, K. 1976. Matrix organization: a review. The Journal of management Studies 7:111-130.

Knorr-Centina, K.D. 1982. Scientific communities or transepistemic arenas of research? A critique of quasi-economic models of science. Social Studies of Science 12:101-130.

Lambright, W.H. and Albert, T.H. 1981. The organizational context of scientific research. In Handbook of organizational design, vol. 2, P.C. Nystrom and W.H. Starbuck, eds., pp. 305-319. New York: Oxford University Press.

Leung, Y.A. 1983. A value based approach to conflict resolution involving multiple objectives and multiple decision making units. In International and regional conflicts: analytic approaches, eds. W. Isard and Y. Naao. Cambridge: Ballinger.

MacKinnon, A.J. and Wearing, A.J. 1980. Complexity and decision making. Behavioral Science 15:285.

Mar, B.W., Newell, W.T. and Saxberg, B.O. 1976. Interdisciplinary research in the university setting. Environmental Science and Technology 10:650-653.

Mar, B.W. 1984. Sampling design for aquatic ecological monitoring, Phase II report, methods development, vols. 1-2. Palo Alto: Electric Power Research Institute.

Mar, B.W., Newell, W.T., Saxberg, B.O. and Adams, S.A. 1983. Applied interdisciplinary/interdepartmental management? Interstudy 4:3.

Mar, B.W., Newell, W.T., Saxberg, B.O. and Adams, S.A. 1981. Applied interdisciplinary research in the university: managing the university's role in solving society's problems, final report to NSF NM 44380. Seattle: Management Improvement Program, University of Washington.

McCann, J. and Galbraith, J. 1981. Interdepartmental relations. In Handbook of organizational design, vol. 1, eds. P.C. Nystrom and W.H. Starbuck, pp. 60-84. New York: Oxford University Press.

Minzberg, H. 1979. The structuring of organizations. Englewood Cliffs, New Jersey: Prentice-Hall.

Newell, W.T., Saxberg, B., Mar, B.W. and Adams, S. 1980. Guidelines for applied interdisciplinary research in the university: how to manage the university role in solving society's problems. Seattle: Research Management Program Project Report No. 11, University of Washington.

Nijkamp, P. 1978. Compromise choices in spatial interaction and regional planning models. Amsterdam: North-Holland.

Rapoport, A., ed. 1974. Game theory as a theory of conflict resolution. Dordrecht: D. Reidel.

Sage, A.P. 1981. Designs for optimal information filters. In handbook of organizational design, vol. 2, eds. P.C. Nystrom and W.H. Starbuck, pp. 104-121. New York: Oxford University press.

Saxberg, B.O., Newell, W.T. and Mar, B.W. 1981. Interdisciplinary research--a dilemna for university central administration. Journal of the Society of Research Administrators 13:25-43.

Saxberg, B.O. and Slocum, J.W. 1968. The management of scientific manpower. Management Science 14:B475-B489.

Titli, A. and Singh, M.G. 1981. Large scale systems theory and applications. In Proceedings of the IFAC symposium. Oxford: Pergamon press.

Weick, K. 1979. The social psychology of organizing. Reading: Addison-Wesley.

Zadeh, L.A. 1965. Fuzzy sets. Information and Control 8:338-353.

MANAGING HIGH TECHNOLOGY
An Interdisciplinary Perspective
B.W. Mar, W.T. Newell, and B.O. Saxberg (Editors)

INTERDISCIPLINARY RESEARCH WITHOUT DISCIPLINES

Frederick A. Rossini and Alan L. Porter

Technology Policy and Assessment Center
Office of Interdisciplinary Programs
Georgia Institute of Technology
Atlanta, Georgia 33032

INTRODUCTION: CONCEPTUAL FRAMEWORKS AND EMPIRICAL INVESTIGATIONS

The purpose of this paper is to present the current status of our work on the problem of conceptualizing interdisciplinary research. In the process we will discuss some of the major issues presently facing us, as well as the directions we see as most fruitful in our ongoing efforts at conceptual development.

We have been developing conceptual frameworks for understanding interdisciplinary research since 1976 when we began the study of National Science Foundation sponsored technology assessments (Rossini et al., 1981). Guiding this effort has been the view, especially common in the physical sciences, that there is an interplay between theoretical or conceptual frameworks and empirical observations. Frameworks serve to focus and determine the empirical observations to be made, while empirical observations provide information useful in revising the frameworks. This ongoing feedback relationship produces a spiral effect of development through conceptual refinement and empirical advances. These successive changes are not necessarily continuous. Indeed, as Kuhn (1970) has pointed out, scientific knowledge often makes its most important advances through discontinuities from the existing state of knowledge. Paradigm change also provides new focal points for empirical activity. Thus changes in knowledge consist of framework shifts as well as progressive developments within preexisting frameworks.

THE STRAP FRAMEWORK

Prior to 1982 our work was based on a loose conceptual framework that incorporated the conventional wisdom about interdisciplinary research. The conventional wisdom was based on experiential writings about interdisciplinary research. These were mostly written by academics about their experience in working on one or more projects (see for example, Petrie, 1976; Burdge and Opryszek, 1983). The experiential focus on what seemed to be important at the time produced a disconnected set of emphases. Our study of technology assessment projects began by trying to organize some of this wisdom into factors affecting integration into interdisciplinary research (Rossini, 1977). This preliminary agglomeration of putative linkages was eventually refined into a causal model (Rossini et al., 1981). This model dealt with the role of factors internal to a project in integrating the output of that project.

At that time there was not a great deal of empirical work that went beyond individual case studies. A notable exception was the work of the University of Washington group (see for example, Saxberg et al., 1981; Birnbaum, 1977; Birnbaum, 1981). The interdisciplinary work with which we dealt was policy related and applied rather than basic research. We began to realize that basic research could be interdisciplinary through an investigation of Anatomy Departments (Rossini et al., 1984) which were in the process of transforming their research from gross anatomy to interdisciplinary neurosciences. In fact 45 percent of the faculty and 75 percent of the postdocs of eight anatomy departments studied were from disciplines other than anatomy. Moreover the concentration of the literature on academic interdisciplinary work and the problems attendant on it masked the great concentration of interdisciplinary work in nonacademic contexts. We had encountered such research in the technology assessment study where nonacademic organizations had been found to present a different, but equally real, set of barriers to interdisciplinary research.

The final factor that led to the establishment of a new framework was the realization that "discipline," the bedrock concept on which the conventional notion of interdisciplinary research was based, was almost impossible to reasonably operationalize--either intellectually or organizationally. The upshot was the creation of the STRAP framework described by Porter and Rossini (1984).

In sum, the STRAP framework offered a set of variables that described the static structure of an interdisciplinary research project. The principal departure was in the substitution of "intellectual skills" for "disciplines." Intellectual skills were divided into substantive knowledge and techniques, the S and T of STRAP. These were considered as being either at the expert level (at which original knowledge could be generated) or the journey man level (at which common knowledge was applied. Range, R, referred to the scope of the S and T involved in the project, whether within a clearly defined research area, within the same grand category of knowledge (such as physical sciences, life sciences, social sciences or engineering), or across more than one of these grand categories. A was an administrative variable dealing with the relationships among the organizational units to which the project team members belonged. P was a personnel variable dealing with the number of team members and the permanence of their relationships. These variables constituted a static structural description of a particular research project.

EXPERIENCE IN OPERATIONALIZING THE STRAP FRAMEWORK

We developed the STRAP framework in the early phases of our study of 40 National Science Foundation research projects from five programs (Porter et al., 1984). In order to test the major features of the framework, it was necessary to operationalize its elements. In the case of intellectual skills, the issue was to communicate the notion to researchers so that they could relate to it and specify the substantive areas and techniques required for the project, as well as their needed depth. After some initial faltering, we were able to communicate with the principal investigators (PIs) about "skills." We found our analysis of the skills required for the projects drawn from the review of proposals proved to be reasonably accurate when reviewed and embellished by the PIs.

A number of interested findings resulted. One was that there were few if any substantive areas used at the journeyman (or technician level), a finding that was not true of techniques. Substantive areas at a journeyman level of knowledge appeared as background to the main thrust of the project. This background knowledge was often tacit to the extent that PIs did not allude to it explicitly in describing their projects. Typically any substantive area significantly included in the project was seen as involving frontier knowledge development. Techniques, such as basic statistical analyses, were more readily acknowledged to be at a low level. Another finding was that in some instances substantive areas included techniques. In some cases where these were not broken out independently, there did not appear to be techniques involved in the project. Thus the level of disaggregation of intellectual skills proved important in analyzing the project. This latter issue needs further clarification since agreement on skills required appeared quite good. As suggested by Torsten Gerpott (private communication), it would be useful to catalog skills to compare them inductively with Institute of Scientific Information subject categories, UNESCO subcategories, and other classification schemes to see if any uniformity in skill categories emerges.

Range seemed relatively easy to operationalize. However, the projects appeared in many cases to divide dramatically--some lying within a well-defined research area, others spreading across grand categories of knowledge. A possible explanation of this bimodal split is the following: research areas may congeal to include skills that are nominally related to different grand categories of knowledge, such as mechanical engineering and physiology in some research areas of biomedical engineering. This forces the analyst to characterize the range either as a well-defined research area or as a spread across grand categories of knowledge.

Administrative arrangements appeared to be significant from our earlier work on technology assessments (Rossini et al., 1981). We had found that having a supportive organizational environment improved the conduct of research and that increasing organizational distance within the project group was a negative factor. Organizational distance was not a significant factor in research performance. Another aspect of administrative arrangements loomed as potentially important. In our former work interdisciplinary research in academic departments appeared to be fraught with peril, in accordance with the common wisdom. In this study, on the other hand, academic PIs reported in a number of cases that their work was facilitated by their departments, and essentially no problems were reported from the academic units involved. This was a great surprise. Yet, on reflecting on our experience with the Anatomy Departments, it should not have been. The basic research areas covered in the current study were areas with demonstrable history of openness to a range of techniques and substantive areas. Consider geophysical sciences. A colleague of ours, P. Kelly, recently noted that Georgia Tech's School of Geophysical Sciences, had three of the four elements, earth, air, and water, within its purview. We responded that it could acquire the fourth, fire, by hiring a volcanologist. The history of neurosciences, both at NSF and in the academic and research worlds was related to us by Dr. James Brown of the Foundation. Like the Anatomy Departments, many diverse intellectual skills were incorporated into this area. Archeology also has a long tradition of viewing a site from a number of different perspectives, including culture, fauna and flora, and climate.

The apparent difference between the supportive nature of the academic units involved with our present sample and those in which some of the technology assessments of the earlier study took place was striking. To deal with this difference, we introduced the concept of "open" and "closed" organizations into this context. We will discuss this in the next section.

An analysis of personnel patterns on the 40 NSF projects reinforced our newly won perception that multiskilled research could be performed by individuals. In addition it indicated that for team research longer-term arrangements were somewhat preferable to ad hoc teams.

In sum we were able to begin to operationalize the STRAP framework. Major challenges remain in refining measures of intellectual skills and in exploring the nature of open and closed organizations. We now turn to this latter task.

OPEN AND CLOSED ORGANIZATIONS

Perhaps the most helpful way to begin to consider open and closed organizations is to view them as systems with different kinds of boundaries (Von Bertalanffy, 1968). A closed system has an impermeable boundary, while an open system has a boundary that is penetrable from both sides by the sort of elements that are contained in it. Clearly there are not perfectly open or closed systems among the organizations we are considering. However, we could characterize a relatively open research organization as one into which new substantive areas and techniques could enter with low barriers and from which substantive areas and techniques could disappear under the same conditions. Likewise, but less immediate for our purposes, the organization is willing to change its approaches to managing research. A closed research organization is one that defines itself as involved in certain substantive areas using a well-defined set of techniques. It does not question its accustomed approach to managing research.

In fostering multiskilled research, the open organization will support innovations, whether from within or without, that lead to solution of the research problems of interest. The closed organization will stop at predefined boundaries, either tackling only part of the problem, turning it over to another organization across the boundary, or labeling it unimportant by definition since it lies outside the organization's purview. "Innovation by invasion" can transform the open organization by the infusion of new intellectual capital. The closed organization, when it changes, accepts its lead from among its high status members. Outsiders lack the necessary credentials to be taken seriously.

We have not yet attempted to operationalize these concepts for multiskilled research. Clearly we developed these ideas from academic organizations, yet they can and should be applied to any research organization. In fact, our experiences over the years with contract research organizations indicate to us that these include both open and closed organizations.

A list of proposed marks of open and closed research organizations follows. These marks are such that they can be empirically observed and recorded. Open organizations have a steady flow of professional personnel through the organization. These persons hold degrees from the different areas. Their publications and citations to their work appear in a broad range of literature. Internal organization is flexible, with groups being formed for projects or programs, and then dissolved when reason for being disappears. While identifiable research programs exist, these are evolving both in scope and content in response to the development of new knowledge and techniques, from whatever source in the program.

Closed organizations have a stable complement of personnel with suitable credentials. The main concern in new research personnel, whether for long- or short-term appointments, is that they fit a clearly defined "slot." They hold degrees in the same area. Their publications and citations are narrowly concentrated in the literature. Internal organization is more set with emphasis being given to subunits organized along the lines of disciplines or subdisciplines. These units have continuing structure throughout a succession of research projects and programs. The development of knowledge is in accord with the bounds of the unit's charter. Evolution of knowledge and techniques is in line with developments as accepted by disciplinary leaders. There is a clear delineation between problems belonging within or without the unit.

These descriptions obviously refer to extreme types of organizations. No organization is purely open or purely closed, and identificaton as such refers to a dominant orientation. Further empirical work should determine whether open or closed organizations can be readily identified by these or other marks.

In addition empirical work should investigate causal factors that make some organizations of one type (e.g., academic departments) open and others closed. From the perspective of research policy, the ability to identify and muster needed intellectual skills within the context of an open organization is desirable to facilitate multiskilled research, and the study of the openness of organizations should prove practically useful.

CONCLUSIONS: CURRENT CONCEPTUAL STATUS AND FUTURE DIRECTIONS

The conceptual work summarized herein represents a qualitative transition in our understanding from interdisciplinary research to multiskill research. By focusing on multiskill research within an open organization, we escape at once the traps of discipline, team, and academe. The widespread applicability of the concept of multiskill research becomes apparent. At the same time a structural framework is advanced for its static description at the project level. This should be further articulated and explored, especially in the areas of intellectual skills and open and closed organizations. The analysis of project dynamics has not yet moved beyond the earlier work of Rossini et al. (1981) and Birnbaum (1977; 1981). Areas such as research project management, intra-project communication, and integration need to be further explored within the newer structural context.

Issues in research policy and research program development have been addressed in a recent project report (Porter et al., 1984). At the level of the research funder, organizational issues, such as openness, can be influenced by policies and incentives.

Further developments should iterate between conceptual refinement and empirical observation. Then as our knowledge of the research process increases, we can better manage it to be more effective in meeting human needs.

REFERENCES

Birnbaum, P.H. 1977. Assessment of alternative management forms in academic interdisciplinary research projects. Management Science 24:272-284.

Birnbaum, P.H. 1981. Integration and specialization in academic research. Academy of Management Journal 24:487-503.

Burdge, R.J. and Opryszek, P. 1983. On mixing apples and oranges: the sociologist does impact assessment with biologists and economists. In Integrated Impact Assessment, eds. F.A. Rossini and A.L. Porter, pp. 107-117. Boulder, CO: Westview Press.

Kuhn, T.S. 1970. The structure of scientific revolutions. Chicago: University of Chicago Press.

Petrie, H.D. 1976. Do you see what I see? The epistemology of interdisciplinary inquiry. Educational Researcher 5:9-15.

Porter, A.L. and Rossini, F.A. 1984. Interdisciplinary research redefined: multiskill problem-focused research in the STRAP framework. R&D Management 14:105-111.

Porter, A.L., Rossini, F.A. and Chubin, D.E. 1984. Interdisciplinary research (problem-focused, multi-skilled research) - National Science Foundation experiences, final report. National Science Foundation Grant OIR-8209893. Atlanta: Georgia Institute of Technology.

Rossini, F.A. 1977. How can we put it together? A first model of technology assessment integration. In Retrospective technology assessment - 1976, J.A. Tarr, ed., pp. 57-76. San Francisco: The San Francisco Press.

Rossini, F.A., Porter, A.L., Kelly, P. and Chubin, D.E. 1981. Interdisciplinary integration within technology assessments. Knowledge 2:503-528.

Rossini, F.A., Porter, A.L., Chubin, D.W. and Connally, T. 1984. Crossdisciplinarity in the biomedical sciences: a preliminary analysis of Anatomy. In Managing interdisciplinary research, S.R. Epton, R.L. Payne, and A.W. Pearson, eds., pp. 176-184. Chichester, United Kingdom: John Wiley.

Saxberg, B.O., Newell, W.T. and Mar, B.W. 1981. Interdisciplinary research - a dilemma for central university administration. Journal of the Society of Research Administrators 13:25-43.

Von Bertalanffy, L. 1968. General system theory. New York: Braziller.

MANAGING HIGH TECHNOLOGY
An Interdisciplinary Perspective
B.W. Mar, W.T. Newell, and B.O. Saxberg (Editors)

CONTEXTUAL ISSUES IN MANAGING CROSS-DISCIPLINARY RESEARCH

S.R. Epton[1], R.L. Payne[2] and A.W. Pearson[1]

[1]Manchester Business School

[2]Department of Psychology
University of Sheffield
Sheffield S10 2TN United Kingdom

In the Manchester Conference on Interdisciplinary Research held in 1983 we made two major points and from them drew a number of implications for the management of interdisciplinary research groups. This paper has just appeared in R&D Management (Epton et al., 1984). the first point was that the generic term for this area of activity should be "cross-disciplinary" to distinguish between interdisciplinary research that was genuinely cooperative and collaborative, and possibly successful, and interdisciplinary research which is multidisciplinary or an unsuccessful attempt at interdisciplinary relationships. We are not holding our breath waiting for this usage to become universally acceptable. In the present paper, however, we would like to reiterate some of the points we made about the differing contexts in which interdisciplinary research takes place. We see a potential danger in ascribing to a general concept such as interdisciplinary research global similarities which may not reflect the true complexities of the real world. These contextual features of cross-disciplinary research are important for the management of research, and for research on its management.

In Epton et al. (1984) we focused on differences in the organizational contexts in which interdisciplinary research activities have been carried out. We will present this material to illustrate some of the ways in which research groups will differ just because they are located in different organizational contexts. In the second half of the paper we will describe a framework for looking at the more micro aspects of context enabling us to examine the role demands of managers leading teams of researchers engaged in interdisciplinary activity. This will include examples of some empirical studies using the framework outlined.

CROSS-DISCIPLINARITY AND ORGANIZATIONAL FORMS

An organization and the social environment in which it exists live in a state of mutual dependence. What the organization can do for the environment depends on what the environment supplies it with: i.e., resources of money, technology, information, machines, materials, and people; if the environment restricts their supply then the achievement of the organization will be constrained.

But the achievement is also constrained by the internal workings of the organization itself, by the way its members relate to one another, by the perception each member has of the possibilities for self-fulfillment, by the way rewards are distributed among them, by the skill of the leadership in controlling, coordinating, and motivating.

How far these two forms of constraint, external and internal, operate in cross-disciplinary research depends on a number of factors, in which institutional setting and size are the most important. Therefore, in the rest of this chapter we look at the problem of how to make such research most effective in the face of the constraints and within this two-dimensional framework. The following section contains a description of this framework.

A TAXONOMY OF ORGANIZATIONAL FORMS

Group Size

A major determinant of the complexity and degree of structuring of activities in task-oriented groups is the number of people they employ. Social psychology recognizes that the nature of social relationships changes when a group becomes too large for a member to know most of the others well and to identify with the total group that makes up the organization. Sociologists have traditionally distinguished between primary groups and secondary groups. How large a goup can be and remain primary is impossible to say, but we shall assume that in an industrial context a group is unlikely to be primary if it is difficult to regulate people's behavior without formal rules and standard procedures. In contrast, behavior in primary groups is regulated informally by implicit agreement among members as to who is entitled to exert authority and have power over the others.

Institutional Setting

The second element in our taxonomy likely to influence the structure of a cross-disciplinary group is its location relative to particular social organizations. We shall use the following simple classification of such reference organizations:

University-based research units
Industry or government based units
Inter-institutional units

The last group is those units that are set up to serve several social institutions.

Table 1 summarizes the elements of our taxonomy and provides examples of each subdivision. We distinguish between primary and secondary groups (small vs. large) and the three institutional locations. Together these generate six potential settings for the execution of cross-disciplinary problem solving which can be expected to call into being different organizational forms. In Table 1 we have reduced these six settings to five because we doubt that interinstitutional groups will ever grow very large.

This classification is perhaps rough-and-ready, but we believe it to be adequate to make the point, as we show below, that cross-disciplinary research in different institutional settings faces different kinds of difficulties.

Table 1
Issues Arising from Doing Interdisciplinary Research in Different Organizational Contexts

	Organizational Context				
KEY ISSUES	SMALL, UNIVERSITY BASED	LARGE, UNIVERSITY BASED (ORU)	SMALL, INDUSTRY BASED	LARGE, INDUSTRY BASED	INTER-ORGANIZATIONAL PROBLEM SOLVING SYSTEM
Number of Disciplines Involved	Small	Moderate	Moderate	Moderate-large	Large
Typical Problems	Pure	Pure-->Applied	Applied-->Pure	Applied and pure -->Applied	Applied
Funding	Short-term & uncertain	Relatively certain	By contract & short-term	Relatively certain	Relatively certain but limited duration
Source of Problems to Be Solved	Found by team	Found by team (largely)	Given by environment	Found by team and given by environment	Given by environment
Structure of Organization	Flat	Hierarchical (selective decentralization)	Flat	Hierarchical (selective decentralization)	Flat
Main Source of Power/Influence	Expertise	Role in Organization (may be expert based)	Expertise and ownership of problem	Role in Organization (may be expert based)	Role in system (experts are advisers)
Leader's Involvement	Part-time	Full-time	Part-time	Full-time	Part-time, full-time and temporary
Member's Career Prospects	Uncertain	Relatively certain/predictable	Uncertain	Relatively certain/ predictable	Almost irrelevant
Key Leadership Roles	Outside liaison	Outside liaison and coordination--management of interdependence	Outside liaison--resource acquisition	Management of inter-interdependence--motivation and performance	Coordination communication
Dominant Values	Scientific/ team based	Discipline based	Problem-solving/ entrepreneurial	Managerial	Problem-focused: pragmatic
Work Contracted Out	Unlikely	A little	Quite a lot	Some	Quite a lot
Communication Problems	Few potentially: mainly scientific liaisons if funds	Internal and external	External and scientific liaison if funds low	Internal	Internal due to physical separation and variety of disciplines
Access to Scientific & Technical Resources	Good	Good	Not so good	Good if affordable	Good if affordable

We can now return to our main theme which is to examine how the constraints referred to in the introduction to this section apply to the five settings of interdisciplinary problem solving groupings.

SMALL, UNIVERSITY-BASED GROUPS

Financial Resources

These groups typically consist of university teachers who have found a problem which crosses their disciplinary boundaries and who then find some resources, usually from outside the university, to enable them to work on the problem. Thus, the project leaders are often part-time leaders. They may have some full-time researchers working for them and some full or part-time technicians. Technical facilities are often obtained from the host institution when it has capacity to spare.

Restriction of financial resources, and uncertainty about their continuation are major environmental constraints upon staffing these small problem-solving groups. It is also possible, though the evidence is scarce, that the poor career prospects offered by such employment induce high staff turnover and discourage the very best people from accepting them. This may be particularly exacerbated when the problem is cross-disciplinary, for there are clear career risks for people who stray from the well-defined career paths of the established disciplines.

The limited financial resources also cause other problems. Successful research/innovation depends on having good sources of information outside the group. The information-providing activities take up much time in scanning and contacting the environment. The contacting is best effected by personal contacts at conferences or by direct visits to other institutions. This involves much traveling and extensive traveling requires extensive funds. They are frequently not adequate, since a small group of researchers may need to do almost as much traveling as a much larger group, but the travel grant may be calculated as a proportion of the total grant. When the research problem requires inputs of information from several disciplines these difficulties increase in severity.

On the other hand, one of the advantages of locating a cross-disciplinary problem solving activity in a university is that a wide range of expertise and services is readily at hand to help speed up the identification of relevant sources of knowledge and skills.

Even here the question of adequate resourcing is important. Pineau and Levy-Leboyer (1984) and others have shown that small size is quite strongly correlated with lower levels of performance or effectiveness. The removal of this constraint is easy, but only if the resources are available.

The removal of the resource barrier usually depends on the generosity of funding bodies, but in times of economic recession they tend to strive even harder to get value for money. Outside pressure to meet deadlines in an activity as uncertain as cross-disciplinary problem solving can lead to poorer research, or to the failure to apply that careful polishing which can transform the quality of the finished product. This is particularly true when that product is a written publication rather than a patent. Where pressure is excessive, its effects may be negative since quantity is less valued than quality by university-based researchers (French and Caplan, 1973).

Cultural Factors

Differences in values and training create major problems of cooperation and coordination for people from different disciplines. These within-system problems, however, are possibly least troublesome in small university groups. This is because there is very little compulsion in the university for people to work on projects they do not like. This does not mean that such conflicts are absent, but it does mean there is more often than not a prior commitment to accepting and trying to work them through.

The process is facilitated by two factors. Firstly, in universities expertise rather than formal authority is the dominant basis of social power (French and Raven, 1959). Secondly the groups are small, and this in itself tends to reduce the variety of disciplines and the values associated with them. It is well known that participation increases in small groups (French and Caplan, 1973), and participation reduces strain and conflict; but there is no guarantee that small size will lead to either of these desirable outcomes.

Though some advantages have been identified for this sort of research group we might do well to heed the following comment about its suitability for solving cross-disciplinary problems:

> A university is well structured for small-scale basic research involving one or two members of the faculty working closely with a handful of graduate students. Such groups usually represent a single academic discipline. A university relates to large-scale, applied missions only with great difficulty. Organized research sub-units such as centres, institutes, and programs abound on practically every campus. Although many claim to be interdisciplinary most are closely tied to, or dominated by, one academic department (Lambright and Teich, 1981, p. 313).

LARGE UNIVERSITY-BASED GROUPS

As implied in Table 1, these large groups are usually located in institutes and research centers attached to universities. Such institutes are frequently involved in monodisciplinary or multidisciplinary projects rather than interdisciplinary ones.

Handcombe's paper (1984) described an organization that falls into this category. It employes 27 full-time staff, but also involves 100-150 members of the staff from the rest of the University of Salford. Many of its contracts are cross-disciplinary and the Center itself provides the management expertise to ensure that the necessary coordination and cooperation takes place effectively.

The Institute of Research established at Gallaudet College and described by Trybus (1984) is another example of a large university-based group. Trybus indicates that research actions are largely at the discretion of the researchers themselves, but the functions of budget planning, expenditure of funds, equipment acquisition and other administrative actions are centrally controlled. It is likely that this high level of research autonomy may be eroded in time as research teams become directed to solving problems defined relative to the institute's general goals. All organizational designs are compromises, however, and the economies of scale, security of tenure, and resources which institutes provide allow the tackling of larger problems and the assembly of the variety of expertise needed to deal with them. These are obviously to be considered as reasonable trade-offs with the freedom of choice such as is enjoyed by smaller, more independent groups.

ORGANIZED RESEARCH UNITS

The establishment of an organized research unit or ORU as Teich (1979) described them, by a college or university is not a common event. Teich points out that many such ORUs have been established to meet the needs of specific clients such as government departments. Some of these have become *de facto* independent of the universities that first gave birth to them. Most remain part of the university but with different degrees of attachment to and involvement in its affairs.

Friction can develop between the university and the ORU. Administrative and financial control can become an issue as well as factors to do with the image of the two institutions. Many ORUs are much more problem applied/centered because of the nature of the contracts they undertake. Universities have traditionally favored more fundamental research and promoted values of independent thought and action. The university environment is then a major constraint on those ORUs which have become large enough to develop power and influence in their own right. The institutes can offer financial and other inducements to the better graduate students so that they may compete directly with the host institution. Very large ORUs may even be able to offer careers and conditions that overcome the problems faced by those in small ORUs where researchers often have unfavorable terms and conditions when compared with the teaching faculty.

Teich was sufficiently impressed by the evidence indicating tension between universities and larger ORUs to suggest that universities may be well advised to resolve some of the difficulties by setting up what Brooks (1977) has called buffer institutions. "These intitutions are sufficiently distinct from their university bases that their staffs do not see themselves as second-class citizens in comparison to regular faculty members, yet they are close enough to contribute to the vitality of the university's research environment. Since part of these institutions may be considered government laboratories, their senior staffs may even have the job security of civil servants." (Teich, 1979). In the United Kingdom the Medical Research Council has many units which have this sort of relationship with the university.

While this provides a structural answer to dealing with relationships between institutions good relationships still depend on having leaders who are skilled at and dedicated to maintaining good communications with other important organizations in the proximal environment.

ROLE OF THE SPONSOR

For many ORUs the most important organization in the environment is the sponsor. This relationship appears particularly delicate when the projects are genuinely interdisciplinary.

Hattery (1979) feels that this particular linkage is so crucial to the development of cross-disciplinary research that it should be the object of concerted research itself. He claims that monitoring of cross-disiciplinary projects by sponsors had led to much unhappiness, but the monitors who monitor least are not always the most successful monitors. The tension arises from the sponsor's need to ensure they get quality and get it on time, and the desire of educated and articular professional scientists to control their own destinies. Since little empirical knowledge exists about this relationship, it is clear one cannot offer firm prescriptions as to how to solve it.

Both Cutler (1979) and Hattery (1979) point out that cross-disciplinary research requires more time and money in the earliest stages of the research process than does conventional research. Several models of how successful negotiations can be achieved exist and this experience may well be useful to the personnel who are involved in the management of this crucial boundary (Carlisle and Leary, 1981). Miles (1980) reviews the literature on boundary roles and suggests some of the qualities important for being successful in them. Both sponsors and large-scale ORUs might well need to train specialists for handling this interface, and Miles' review may help in the successful selection of candidates.

INTERNAL STRUCTURES

The strong desires for autonomy and independence expressed and expected by scientists are probably the main determinant of the internal structures that develop in larger cross-disciplinary research groups. Mintzberg (1979) argues that organizational structure arises from two main forces. One is the need to achieve coordination among the major parts of organizations, and the other is to deal with the conflicts that arise within the organization itself. Mintzberg describes five organizational parts and five mechanisms for achieving coordination among the parts.

The five parts are the strategic apex (top management), the middle line (middle management), the operating core (the scientists), the support staff (personnel administration, secretaries, etc.) and the technostructure (corporate planning, O&M, etc.). The five coordinating mechanisms are direct supervision, standardization of work processes, standardization of worker skills, standardization of outputs, and mutual adjustment. Mutual adjustment achieves coordination by the simple process of constant and continuing information communication.

Mintzberg would claim that research scientists working on uncertain projects would prefer to achieve coordination through mutual adjustment. His assumption is that any particular structure has arisen from a resolution of conflicts among the five parts of the organization. Thus, if the strategic apex wins the conflict then it can probably use direct supervision to achieve good coordination. If the operating core wins then the dominant coordinating mechanisms will be mutual adjustment. In ORUs the operating core are scientists and they tend to win the conflict.

Organizations with this pattern of relationship are described by Mintzberg as adhocracies. Adhocracies are characterized by much horizontal specialization, high levels of professional training, little formalization of procedure, and organic rather than mechanistic relationships. A considerable amount of autonomy is given to lower levels in the organization, but the autonomy is given selectively. The main control exercised from the top in adhocracies relates to project management. The other roles of those in the strategic apex are to do with external liaison and negotiation. Even policy formation is spread throughout the organization in the sense that what gets done is the policy. Innovation requires flexible and fast responses, and the lack of reliance on standardization combined with decentralization enables the adhocracy to achieve this response.

Managers in adhocracies are not, however, engaged in the direct supervision so often identified with the management process. Their time is spent on liaison and negotiation with other teams, outside advisers, sponsors, etc. Even as managers they are part-time for they are themselves experts and are very likely to be involved in carrying out their own research: they are operators as well as members of the middle line.

The barriers to achieving successful cross-disciplinary research in large, university-based institutes may be diagnosed as ones which inhibit the workings of the adhocracy. People who have been socialized into university systems tend to have been socialized into particular disciplines and socialized out of the idea that management is a professional skill worthy of the same efforts as their science. To them their supervisory role is a matter of directing research, assisting their student assistants to solve day-to-day problems, and monitoring the quality of the output. The reward system encourages this narrow view of management. The main problem then is one of recruiting good scientists who enjoy the uncertainty and ambiguity of working in an adhocracy and who are committed to sustaining it as an organizational form. Professional scientists are pulled towards creating what Mintzberg calls a professional bureaucracy (universities adopt this form of structure) and this tendency must be strongly resisted, particularly in cross-disciplinary research situations where such standardization based on skill or training is inimical to the interdisciplinary spirit so necessary to achieve successful integration among disciplines. It is perhaps clearer now why it is difficult to develop adhocracies within a university setting where the basic structure and values are inherently those of the professional bureaucracy.

SMALL, INDUSTRIAL/GOVERNMENT CROSS-DISCIPLINARY GROUPS

Small, relatively independent groups solving complex cross-disciplinary problems are most likely to be funded from industrial sources where the industry wishes to set up its own research and technical advisory bodies, or by technical entrepreneurs who offer problem solving skills through privately-owned consultancies. In either case, they are small organizations that have highly trained professional and technical personnel with secretaries as the main form of support staff. Being both small and highly professionalized means that they have a strong pull towards operating under mutual adjustment. Mintzberg (1979) distinguishes between operating and administrative adhocracies (see below) and these organizations have all the characteristics of the operating adhocracy. Small size, physical proximity to each other and common professional values enable such organizations to effectively operate on mutually understood expectations and continual negotiations. No one in the operating adhocracy monopolizes the power to innovate or to act to meet the fluctuating demands of an enquiring environment.

It may be that the strength of such a system is the source of its ultimate weakness. The flexibility and innovativeness of these consultancy groups often goes hand in hand with a reluctance to supply services that meet a customer's or sponsor's standard of reliability.

In any case if innovators dominate a group as they tend to do in the consultancy groups then the use of standard procedures and formal systems will be discouraged. On the other hand if one attempts to apply the remedy of bringing professional organizers into the system then the organization is set on the road to becoming what Mintzberg calls a professional adhocracy. Here standard, reliable, and efficient services are the preferred output.

But when this happens, the group may no longer be as effective in attacking the sorts of complex problems that interdisciplinary work involves.

Thus one of the barriers to effective performance in such groups is the rational attraction of specializing roles and services. This can become particularly acute when the organization has successfully solved a particular kind of problem more than once, and when the market for solving the problem is large. This situation offers a tempting niche to occupy, where income is increased, and uncertainty decreased. Unfortunately the fun, excitement, challenge that motivates such scientific entrepreneurs also disappears, and the skills and qualities that make for efficient professional bureaucracies are very different from those demanded by effective (they can't be efficient) operating adhocracies.

Another external constraint frequently encountered is lack of financial resources. This may sometimes lead to lack of physical resources too: the right piece of equipment cannot be bought or even hired at a price that is affordable. Again, the dilemma is apparent--people who are creative and innovative risk-takers do not tend to make good accountants. Such organizations need to recognize the need for first-class services in the financial area. On the face of it, this problem is easily overcome by hiring a good finance man, but he too will need to have the commitment to interdisciplinary working and risk-taking if he is to successfully mesh with the dominant values of the operating adhocracy. His role will not only be financial control but will contain a strong element of liaison and persuasian with sponsors, banks, and scientific equipment suppliers.

Manpower may be another scarce resource. Small organizations need to keep down operating expenses and avoid long-term commitments such as pension rights. This often puts them into the situation of having to make do with the skills they have rather than employ an expensive professional. The challenge to these high innovators who set up such organizations is perhaps too readily accepted, but the combination of novelty, difficulty and pressure linked to attractive intrinsic as well as extrinsic rewards make them vulnerable to self-imposed role overload.

In systems where the existence of slack resource is at a minimum, it is vital that the right people are chosen. Scientific and professional expertise need to be balanced by commitment to collaboration and the openness of mind that requires acceptance of the time that interpersonal relationships take up and the ability to tolerate ambiguity and frustration.

LARGE, INDUSTRY-BASED GROUPS

Large, industry-based groups have two origins. Firstly, they come into being when governments invest in large research institutions parts of which become involved in cross-disciplinary research which is carried out in an interdisciplinary form, although some of the research is also carried out in a multidisciplinary form. Secondly, they arise when industrial organizations or whole industrial consortia set up more or less autonomous research organizations. Again, only parts of these may operate in a genuinely interdisciplinary mode, but these parts may be quite large in themselves, creating secondary rather than primary groups. Even if the groups are small their location in the larger organization affects their context and their form. Mintzberg calls these large and (normally) older organizations Administrative Adhocracies.

Administrative adhocracies differ from the operating adhocracies in that instead of serving the social environment, they tend to serve themselves. They set up projects because they, or their parent organization, wish to know the answer to a problem. The administrative adhocracy makes a sharp distinction between its administrative function (strategic apex and middle line) and the operating core. The operating core is separate from the administrative function. This is achieved in three main ways according to Mintzberg. A separate company is formed to actually carry out the work once the adhocracy has decided the basic direction the research should take. A second way is to contract out the development and production elements in a problem: again this is done once the basic research and strategy have been decided. This was how NASA tended to function (Sayles and Chandler, 1971). The third solution to separating the operating core from the administration is to automate the production process and thus leave the administrative adhocracy to concentrate on the problems of the day after tomorrow. Some chemical and pharmaceutical companies have adopted this sort of structure.

INTERDEPENDENCE AND IDEOLOGY

Whatever variation of these three forms the organization takes one of the essential problems of the administrative adhocracy is the management of interdependence. This may be within the main part of the organization itself where there are large numbers of line and support staff with no clear bureaucratic divisions to signal who is responsible for what, and few stable policies to guide decision-making. This uncertainty exacerbates the problems of distributing resources and rewards both to departments and to individuals. The dual-career ladder is a symptom of these difficulties, for it is in administrative adhocracies that technical and support staff can be seen to make as important a contribution as the so-called line personnel. These conventional separations into line and staff do not sit that well in this kind of organization in practice, but greater size almost inevitably leads to increased segmentation and variety in organizational roles and titles (Pugh et al., 1968).

McCann and Galbraith (1981) describe some of the problems that face organizations which of necessity must handle high levels of interdependency, this being the main problem of the administrative adhocracy:

> Differing perceptions of interdependence, lack of agreement about strategies, and conflict are promoted when
>
> a. asymmetries in status and access to information exist
>
> b. different people and departments use different performance criteria and reward structures
>
> c. some people and departments are pursuing short-run objectives and others are pursuing long-term objectives
>
> d. job dissatisfaction and task ambiguities stimulate unilateral action to change roles and to reduce ambiguities
>
> e. some shared resources are scarce
>
> f. different training and knowledge bases create semantic barriers and perceptual diversity.

These authors also point out that whether interdependence proves beneficial to the organization depends on creating shared appreciations of the value and necessity of making interdependence work. Administrative adhocracies clearly face most of the problems identified by McCann and Galbraith and the successful ones must, therefore, be effective in creating this shared appreciation or ideology.

Starbuck (1982) and others have recently raised the issue of organizations as ideological systems. Its appearance as an issue reflects a growing concern in organizational theory to move from understanding how organizations work only by examining their context and their adaptability to that context (Gerwin 1981). The cited paper does not contain detailed prescriptions about how to create corporate ideologies, and indeed there is some debate as to whether they can be rationally and deliberately created at all (Brunsson, 1982). The role of heroic myth in creating ideologies and of ritual in sustaining them is much discussed by these authors, but myth-making in modern work organizations is a little understood process. Several of the papers in the special issue stress, however, the effectiveness of ideologies in dealing with the irrationality that pervades organizational life in such a way that it facilities the return of a new rationality.

It would seem clear that the role of ideologies in the large administrative adhocracies is important in generating commitment and a dominant element of their ideology must encompass a belief in the virtue or value of the process of problem solving itself, rather than in solving the problems of particular disciplines. In the present context this line of thinking suggests that physicists, engineers, biologists, administrators, and so on need to see themselves as part of a problem solving community or clan (Ouchi, 1980) rather than as the hired hands their financial contracts imply. An ideology of pragmatic problem solving may be more conducive to this than an ideology concerned with being scientific.

It must not be forgotten, however, that it is the good financial rewards and resources possessed by large organizations that attract the best scientists, and the most interesting problems, in the first place. As we have tried to show, the challenge is to create the intellectual and emotional bonds that synthesize the talents of these gifted scientists within the constraints of instrumental relationships, difficult goals and ambiguous roles. We are suggesting that in large adhocracies in particular, ideology may supply a powerful lever for generating and sustaining motivation and commitment. Managers may do well to put more effort into creating an ideology suited to the dynamics of the adhocracy.

Large size brings both benefits and constraints as we have seen already. The administrative adhocracy is particularly susceptible to the pressures to standardize and centralize control. Yet its existence was generated to innovate for itself and its industry, so it must fight this pressure if it is to remain viable. This may mean the creation of new organizations or the continual change of key persons so as to provide renewal. "Change for change's sake" may not be a bad slogan for an ambitious adhocracy.

This high rate of change and interpersonal communication is highly inefficient and produces very uneven workloads but individuals must be selected and socialized so that they can tolerate such inequities, and then work like mad when they need to. As Mintzberg says, in adhocracies "the games are played without rules."

INTER-INSTITUTIONAL GROUPS

Size takes on a different meaning for inter-institutional groups for they are not really large in the sense that some other types of organizations are large. Their members are often representatives of very large constituencies such as the general public or the government, but the inter-institutional systems themselves are unlikely to grow really big. Since they involve people from a range of organizations or institutions, however, they may well have the characteristics of secondary rather than primary groups for a majority of the members of the network. The effect of size in these groups is therefore different from that in large organizations where its main effect is to produce very different structures and relationships. In the case of inter-institutional groups increasing size is just as likely to increase the problems of coordination and communication, but the problems will surface as differences of degree rather than kind. We shall not, therefore, make the size distinction in this section.

Inter-institutional groups are the least researched in the literature of interdisciplinary research. Both the first and the second conferences received papers about them: Bents and Horsmann (1979) in the former and Sharp (1984) and Etzold (1984) in the latter. All three examples are about problems in the public domain. The spread of disciplines they involve is much wider than many of the interdisciplinary scientific projects represented in the other papers (e.g., the biomedical teams described by Moser and Levy-Leboyer, 1984). Another distinctive feature is that they start out as temporary groups. In the terminology of Aldrich and Whetten (1981) they are action-sets: "groups or organizations that have formed a temporary alliance for a limited purpose." They may last several years, as appears to be the case in the Etzold example related in the paper to the second conference; but when the problem is solved, their dissolution should occur.

Cooperation requires communication and that is a major problem for inter-institutional groups. Not only is there the problem of physical separation across towns and even countries, there is a wider than average professional gap between the scientists and technologists on the one hand and the politicians and administrators on the other. Sharp's (1984) group relied heavily on computers to handle the vast amounts of data and yet provide timely access to summaries of it that could be used as guides by non-specialists. Bents and Horsemann (1979) designed a structure where a small, permanent group collected, collated, and summarized the data requested by the policy group. This permanent group organized and then transmitted the information from the experts to the policy group. Both experts and the permanent group were satisfied since the experts dealt with knowledgeable colleagues directly, and the cross-disciplinary synthesis were achieved by the small, highly motivated group.

The problem of motivation and commitment is considerable in such groups, for there may be little reward for some of the members of the action-set. Bents and Horsemann make some useful comments about how the system they developed encouraged motivation:

> It furthered motivation if the study work was also acknowledged in the home institution of the participant, and if the person or group was made fully responsible for its work output.
>
> Motivation and flexibility were enhanced because some of the work was budgeted and paid for as private, additional work, not belonging to the home institution.
>
> The time/money budget should contain some unassigned reserves. These can be used as creativity packages and/or as a buffer against additional or unanticipated work packages.
>
> The project should not be fully allocated to persons at the beginning. Their interest in follow-up packages also motivates an efficient preparation of the already assigned ones.

Despite their success in managing this project Bents and Horsmann still acknowledged the "delicate balance between organizational activities and the international motivation of study workers. Even internally highly motivated workers need some amount of organizational 'coercion' to reach the study objectives."

It is possible that coercion is a little strong here, but other studies of work motivation have shown it is important for people to know what they are supposed to be doing and how it fits into the general scheme of things so that they can develop some personal identification with the task (Campbell and Pritchard, 1976). These conditions are obviously not easy to achieve when the team may include people who never meet each other. In such cases the carrots need to be particularly attractive since in such action-sets there are often no big sticks: inter-institutional groups are not usually organized hierarchically. They are loosely coupled with each part having no obvious power over other members of the action-set. In many cases it does not matter to particular individuals if they leave the action-set halfway through the project. Many individuals or groups may be giving their services voluntarily away. As Cutler (1979) observed at the first conference, cross-disciplinary problem solving costs a lot more than the mere sum of the parts, and sponsors have to budget for that.

Keeping the action-set to a manageable size in very complex problem solving systems may well involve making use of experts but directing their attention strictly to matters inside the confines of their own expertise, that is, building a relationship of the kind Gold and Gold (1984) define as contractual. This avoids the possibility of their being drawn into the main network permanently where too much variety might overload the communication system and lead to chaos. Johnssen (1982) describes a failure in a municipal authority (a clear example of a cross-disciplinary group of the kind under consideration) which arose because a subcommittee's lay members had not been given the time to absorb the information provided by the authority's officials. This prevent them from undergoing the cognitive tuning necessary to them before they could be committed to action flowing from the information.

Nothing was therefore done. Cognitive tuning takes time especially when nonexperts are exposed to the disagreements which so often occur between experts. Reduction of complexity must be a major concern in dealing with transactions between lay persons and experts.

The above analysis reveals how organizational context produces different organizational forms and some of the issues this raises for research managers and research scientists working in them. We now wish to present a conceptual framework for looking at job contexts and use it to ask how these different organizational forms might produce very different jobs for individuals charged with managing interdisciplinary research within them.

THE DEMANDS, SUPPORTS, CONSTRAINTS FRAMEWORK

This framework developed out of research on occupational stress. Research on work stress has followed two main lines of explanation. One was that stress was caused by having tasks which were too difficult or too difficult to do in the time available. The other was that stress was caused by not having control over what happened at work. In 1979 Karasek showed that stress is better predicted by looking at both these variables together. The most stressful jobs are those which are highly demanding and simultaneously low in control. They happen to be blue-collar jobs rather than white-collar jobs on the whole, but the conditions can clearly exist in professional and managerial jobs too. Another strand in the recent literature on stress has been that the effects of environmental stress can be buffered by having good social support available from bosses, colleagues, friends, family, etc. (House, 1982). Payne (1979) has taken these concepts and put them together to suggest that the degree of environmental stress can be described by the balance between three aspects of the environment: the degree to which the environment places demands upon the person, the degree to which the environment constrains the person in meeting those demands (takes control away from him/her), and the degree to which the environnment supports the person in meeting its demands. Figure 1 illustrates how the three concepts can be used to describe six different environments which have the effect of producing (in most people) different psychological outcomes or emotional states.

ENVIRONMENTAL STATE			PSYCHOLOGICAL STATES	
Demands	Supports	Constraints		
L	L	H	S = Submitted	Deterioration
L	L	L	T = Torpid	
L	H	L	R = Relaxed	Development
H	H	L	E = Energized	
H	L	L	S = Struggling	Destruction
H	L	H	S = Smashed	

H = High
L = Low

Figure 1
Framework for Relating Environment States to Psychological States

In Figure 1 demands, supports and constraints are simply assumed to have high or low values and the various combinations of them then lead to different psychological states. The framework enables us to see how managerial jobs may be highly demanding, but not really stressful, because they are often combined with low constraint and high support.

In collaboration with Alan Pearson and Roy Payne, Jabri is using this framework to study the climate and effectiveness of individuals and groups in R&D (Payne et al., 1984). He has studied over 150 scientists in a variety of R&D organizations and created questionnaire measures of demands, supports and constraints. One interesting result from the demands questionnaire is the important distinction between demands which are felt to be negative (attributional demands) and those that are actually regarded as positive demands (attractional demands). In designing the demand measure each subject was asked to rate the strength of demand scores and attractiveness scores showed some demands to be strongly correlated, some strongly negative correlated and many demands were neither liked or disliked. Surprising examples of demands which were not consistently liked or disliked are: work to tight deadlines; undertaking more work than I have the time to do well. This is an important finding since in the stress literature demands are usually assumed to be disliked and to lead to stress. Factor analysis of the attractional and attributional demands items showed both to consist of several factors showing just how complex the microenvironments of R&D workers are. Table 2 shows the means and SDs for the items in the first factor of the attributional demands items. We have named the factor project team demands. The scale used to assess strength of demand appears below, and the mean scores on the items in Table 2 indicate that for most R&D managers the attributional demands are not perceived as existing to a very great extent. This rating of relatively low demand is consistent with data on the support-constraints measures too where R&D managers report much lower levels of constraint than do managers from manufacturing and commercial organizations. The other factors of attributional demands are: demands from higher authority, workload demands, unclear goals, and bureaucratic demands.

Table 2
Attributional Demands: Project Team Demands

		Mean	SD
1.	Work with people who do not find it easy to operate together as a team	2.06	1.11
2.	Work with people with whom I find it difficult to communicate	1.88	0.95
3.	Undertake work which might damage my own career prospects	1.51	0.85
4.	Manage projects which are not given sufficiently high priority	2.12	1.03

N = 15
Factor 1: accounts for 26% of variance

The attractional demands produce three factors and items from the third factor are presented in Table 3 as an illustration. It has been labeled technical/scientific demand.

Table 3
Attractional Demands: Technical/Scientific

	Mean	SD
1. Work in interdisciplinary teams	3.37	1.32
2. Work on projects which are scientifically exciting	3.18	1.21
3. Work with people who are intellectually demanding	3.33	1.03

N = 155
Factor 3: accounts for 13% of variance

The other two factors concern liaison/communication and figurehead demands. Table 3 shows that working in interdisciplinary teams is positively regarded and with a mean score of over 3 indicates that it happens reasonably regularly in industrial R&D. The size of the standard deviations on all these items indicates sufficient spread of scores to distinguish between the contexts of individuals and groups.

DEMANDS RESPONSE SCALE

Score

1 = To a very low extent
2 = To a low extent
3 = To some extent
4 = To a high extent
5 = To a very high extent

The items we have constructed to measure environmental supports and constraints have also produced some interesting dimensions.

As already indicated, R&D managers do not rate any aspects of their environment as constraining, but seven items were rated as constraining and they separated into two factors. The first accounted for 33 percent of the variance and concerned internal rivalry between departments while the second reflected the bureaucratic system as a hindrance rather than a support in getting the research done. Table 4 is taken from Dill et al. (1982) and shows how two of the factors from the support items differentiate between high and low performing units. The ratings of performance are from the same individuals who completed the Demands, Supports, Constraints questionnaire. These data show that even the lower performing units are rated as supportive since support is measured on a five-point scale and their mean scores are between 3.5 and 4.1, but the better rated units are seen as significantly more supportive, backing up our earlier claim that these scales can discriminate between different environments.

Table 4
Significant Items of Support for the High and Low Performing Groups

	HI (N=20)	LO (N=40)	SIGN.
<u>Factor I: Job Stimulation</u>			
1. Degree to which job stimulates and interests	4.45	4.02	.027
2. Amount of variety (opportunity to do a number of different things in job)	4.50	4.07	.047
3. Degree to which job provides opportunity to use knowledge, skills and abilities that you value.	4.35	4.02	.084
4. Variety of jobs on which you work	4.15	3.61	.009
<u>Factor IV: Success/Quality</u>			
5. Current level of success in area of activity	4.20	3.44	.001
6. Degree to which you are encouraged to meet the standards of best scientific practice	4.28	3.62	.004
<u>Other Distinguishing Items:</u>			
7. Degree of authority to decide what kind of equipment to buy and when to buy it	4.00	3.38	0.38
8. Degree of authority over who works on the project	4.00	3.59	0.64
9. Degree to which you have authority over decisions within the project	4.25	3.84	.024
10. Level of skills and knowledge of subordinates	4.15	3.75	.027
11. Morale of people in the organization	3.60	3.15	.046
12. Adaptability of people within your organization	3.70	3.35	.099

These results show the R&D manager to have low negative demands, high attractional demands, good support and low constraints. In terms of Figure 1 they are in relaxed and energized states, being relatively stress-free.

CONCLUSIONS

If it can be taken or demonstrated that the micro and the macro environments of IDR groups differ in significant ways then it can be assumed that they present different managerial problems. It follows from this that the sorts of people who will successfully manage these situations will themselves differ from each other.

The analysis of contexts ought to help reveal the sorts of people who are most likely to succeed in the different contexts. The literature already contains relevant examples. Stankiewicz (1979) has shown that the success of a research group depends on the size of the group and the experience of the leader. Dewhirst et al. (1978) found that the relationship between information accessibility and performance or satisfaction differed for research as opposed to development teams. Both examples illustrate the influence of context on managerial problems and the behaviors managers need to adopt to cope with them. The prediction of stress symptoms has been significantly improved when a person-environment fit model has been used to design the measures and guide the analysis of the data (French et al., 1982).

If researchers in the IDR field can bring together measurement of the context and measurement of personality, ability and past experience then both research and the practice of management can be improved. The second part of this paper is a modest step towards designing measures specific to IDR contexts. The next step is to relate these to individual and team performance and test for the moderating influence of personal characteristics as implied in the person-environment fit model.

REFERENCES

Aldrich, H. and Whetten, D.A. 1981. Organization-sets, action-sets and networks; making the most of simplicity. In Handbook of organizational design, vol. 1: Adapting organizations to their environment, eds. P.C. Nystrom and W.H. Starbuck. New York: Oxford University Press.

Bents, D. and Horsmann, W. 1979. Determination of the research of computer-based technologies for urban traffic systems. In Interdisciplinary research groups: their management and organization, First Int. Conference, eds. R. Barth and R. Steck. I.R.G.I.P.

Brooks, H. 1977. Office of technology assessment invites comments on health of science. F.A.S. public interest report, 30(1):6.

Brunsson, N. 1982. The irrationality of actions and action rationality: decisions, ideologies and organizational actions. Journal of Management Studies 19(1):29-44.

Campbell, J.P. and Pritchard, R.D. 1975. Motivation theory in industrial and organizational psychology. In Handbook of industrial and organizational psychology, ed. M.D. Dunnette. Chicago: Rand McNally.

Carlisle, J. and M. Leary. 1981. Negotiating groups. In Groups at work, eds. R. Payne and C.L. Cooper. London: J. Wiley.

Cutler, R.S. 1979. A policy perspective on interdisciplinary research in U.S. universities. In Interdisciplinary research groups: their management and organization, First Int. Conference, eds. R. Barth and R. Steck. I.R.G.I.P.

Dewhirst, H.D., Arvey, R.D. and Brown, E.M. 1978. Satisfaction and performance in research and development tasks as related to information accessibility. I.E.E.E. 25(3):58-63.

Dill, D., Pearson, A.W. and Jabri, M. 1984. The organizational climate of team leaders in R&D: demands, supports and constraints. Working paper, R&D Management Unit, Manchester Business School.

Epton, S.R. Payne, R.L. and Pearson, A.W. 1984. The management of cross-disciplinary research. R&D Management 14(2):69-79.

Etzold, D.J. 1981. Management, planning and organization in initiating, supporting and coordinating an ongoing fresh water diversion project. In Managing Interdisciplinary Research, 2nd International Conference, eds. S.R. Epton, R.L. Payne and A.W. Pearson, 1984. London: J. Wiley.

French, J.R.P., Jr. and Raven, B. 1959. The bases of social power. In Studies of social power, ed. D. Cartwright. University of Michigan: I.S.R.

French, J.R.P., Jr. and Caplan, R.D. 1973. Organizational stress and individual strain. In The failure of success, ed. a.J. Marrow. New York: Amacon.

French, J.R.P., Jr., Caplan, R.D. and Van Harrison, R. 1982. The mechanisms of job stress and strain. New York: J. Wiley.

Gold, H.J., and Gold, S.E. 1981. Some elements of a model to improve productivity of interdisciplinary groups. In Managing Interdisciplinary Research, 2nd International Conference, eds. S.R. Epton, R.L. Payne and A.W. Pearson, 1984. London: J. Wiley.

Gerwin, D. 1981. Relationships between structure and technology. In Handbook of organizational design, vol. 2, Remodeling organizations and their environment, eds. P.C. Nystrom and W.H. Starbuck. New York: Oxford University Press.

Handscombe R.D. 1984. University based technology transfer. In Managing Interdisciplinary Research, 2nd International Conference, eds. S.R. Epton, R.L. Payne and A.W. Pearson, 1984. London: J. Wiley.

Hattery, L.H. 1979. An assessment of the present state of research on I.D.R. Management. In Interdisciplinary research groups: their management and organization, First Int. Conference, eds. R. Barth and R. Steck. I.R.G.I.P.

House, J.S. 1981. Work, stress and social support. Reading, MA: Addison-Wesley.

Jonssen, S. 1982. Cognitive tuning in municipal problem solving. Journal Management Studies 19(1):63-74.

Karasek, R.J. 1979. Job demands, job decision latitude and mental strain: implications for job redesign. Administrative Science Quarterly 24:285-308.

Lambright, W.H. and Teich, A.H. 1981. The organizational context of scientific research. In The Handbook of Organizational Design, vol. 2, Remodeling organizations and their environment, eds. P.C. Nystrom and W.H. Starbuck. New York: Oxford University Press.

Levy-Leboyer, C. and Pineau, C. 1981. Managerial and organizational determinants of efficiency in bio-medical research teams. In Managing Interdisciplinary Research, 2nd International Conference, eds. S.R. Epton, R.L. Payne and A.W. Pearson, 1984. London: J. Wiley.

McCann, J. and Galbraith, J.R. 1981. Interdependent relations. In The Handbook of Organizational Design, vol. 2, Remodeling organizations and their environment, eds. P.C. Nystrom and W.H. Starbuck. New York: Oxford University Press.

Miles, R.H. 1980. Boundary roles. In Current concerns in occupational stress, eds. C.L. Cooper and R.L. Payne. London: J. Wiley.

Mintzberg, H. 1979. The structuring of organizations. Englewood-Cliffs, NJ: Prentice-Hall.

Moser, G. and Levy-Leboyer, C. 1984. Relations between deciders and research workers on environmental problems: the case of French research on noise. In Managing Interdisciplinary Research, 2nd International Conference, eds. S.R. Epton, R.L. Payne and A.W. Pearson, 1984. London: J. Wiley.

Ouchi, W. 1980. Markets, bureaucracies and clans. Administrative Science Quarterly 25(1):129-141.

Payne, R.L. 1981. Stress in task-focused groups. Small Group Behavior 12(3):253-268.

Payne, R.L., Pearson, A.W., and Jabri, M. 1984. Attractional v. attritional demands: a reconceptualization of job demands. Working paper, R&D Management Unit, Manchester Business School.

Sayles, L.R. and Chandler, M.K. 1971. Managing large systems. New York: Harper and Row.

Sharp, J.M. 1981. A method for peer group appraisal and interpretation of data developed in interdisciplinary research programs. In Managing Interdisciplinary Research, 2nd International Conference, eds. S.R. Epton, R.L. Payne and A.W. Pearson, 1984. London: J. Wiley.

Stankiewicz, R. 1979. The effects of leadership on the relationship between the size of research groups and their scientific performance. R&D Management 9:207-212.

Starbuck, W.H. 1982. Organizations as ideological systems. Journal Management Studies: Special issue 19:1.

Teich, A.H. 1979. Trends in the organization of academic research: the role of U.R.U.'s and full-time researchers. In Interdisciplinary research groups: their management and organization, First Int. Conference, eds. R. Barth and R. Steck. I.R.G.I.P.

Trybus, R.J. 1984. Development of a special purpose, interdisciplinary research institute in an academic setting: a case study. In Managing Interdisciplinary Research, 2nd International Conference, eds. S.R. Epton, R.L. Payne and A.W. Pearson, 1984. London: J. Wiley.

MANAGING HIGH TECHNOLOGY
An Interdisciplinary Perspective
B.W. Mar, W.T. Newell, and B.O. Saxberg (Editors)

CROSS-DISCIPLINARY STRUCTURES RESEARCH AND DEVELOPMENT (R&D) UNITS IN WEST GERMANY

Torsten J. Gerpott and Michel Domsch

Institut für Personalwesen und Arbeitswissenschaft I.P.A.
Postfach 70 08 22
D-2000 Hamburg 70
West Germany

ABSTRACT

It is widely assumed that industrial R&D requires the collaboration of R&D professionals of different scientific disciplines, i.e., that industrial R&D units have cross-disciplinary (CD) structures. However, empirical data on CD structures are rare at best. Based on a discussion of past research on cross-disciplinarity (CDT) this paper conceptualizes CDT as a dimension of an R&D unit's organizational complexity. Two measures of CD structures are developed to assess the relationships between CD structures, R&D unit size and industry in a comparative study of 210 West German R&D units. One measure (number of different disciplines employed in a unit) has affinities with differentiation measures of prior organization research (e.g., the Aston approach), whereas the second distributional measure of disciplinary diversification has been rarely used. It is found that the two measures are differentially related to size and industry variables. The paper concludes that future research and managerial practices need to give greater attention to different aspects of CD structures (complexity). Industry thresholds and barriers for disciplinary diversification are identified and their implications for integration mechanisms in R&D are briefly outlined.

Awareness of the importance and difficulty of obtaining successful collaboration across increasingly specialized scientific disciplines on ever more complex research problems has widely grown. For example, the books edited by Barth and Steck (1979) and Epton et al. (1983) are solely devoted to problems that are cross-disciplinary (CD). Many authors have defined frameworks to classify various types of relationships for studying CD research (e.g., MacDonald, 1983; Wilpert, 1979; Blaschke, 1976, pp. 146-152). A few investigators have studied groups in field settings (e.g., Birnbaum, 1979; Birnbaum, 1981). Investigators have focused on academic CD research in the U.S. mostly using a case study approach (Porter and Rossini, 1984; Porter and Chubin, 1983; Birnbaum, 1982). Furthermore, Epton et al. (1984) noted that empirical studies on CD research are beset with methodological problems resulting from ambiguities in the definition of CDT.

One purpose of the present paper is to provide descriptive information on the prevalence of CD research in West German industrial R&D

units. A second purpose of this study is to extend prior comparative organizational research to the study of cross-disciplinarityy CDT in order to conceptualize CDT in a less ambiguous manner. CDT is conceptualized as characteristic of an R&D unit's organizational structure and is measured by two different variables focusing on the number of different disciplines employed in the unit and on the distribution of its R&D professionals among various scientific disciplines. These measures were used in a comparative study of 210 West German Industrial R&D units. A more complete description of the present research may be found in Gerpott and Domsch (1984).

THEORETICAL BACKGROUND

Distinctions of various types of modes of collaboration (e.g., inter-, multi-, cross-, pluri-, and trans-disciplinarity) are widespread (e.g., Wilpert, 1979, p. 172). Practitioners (e.g., Loveday, 1984, p. 95) as well as organizational researchers (e.g., Darvas and Haraszthy, 1979, p. 36) have indicated that these differentiations are overly subtle. However, a clear definition of cross-disciplinarity should not be left solely to the reader's imagination in order to remove ambiguities from the meanings associated with terms describing collaboration between such professionals. What is needed instead is a nonevaluative and objective definition and measurement of collaboration between disciplines which may be used across a large number of different organizational units.

To guide future research on CDT, Epton et al. (1984) have developed a definitional framework for various terms used in the literature to connote relationships between scientific disciplines. They distinguish the following forms of collaboration: "1. The 'pure' multidisciplinary form in which the portions of the task are carried out by organizationally separate units each of which includes practitioners of only the disciplines. ...2. The pure interdisciplinary form in which the elements of the tasks are carried out within a single organizational unit consisting of the practitioners of the disciplines necessary for the completion of the task (Epton et al., 1984, p. 70). Further, there seems to be some agreement that one main dimension of organizational structure is complexity, that is the degree of differentiation that exists within an organization (Robbins, 1983, p. 45; Dalton et al., 1980, p. 57; Ford and Slocum, 1977, p. 562). Note specifically that role specialization was operationally measured in prior studies by counting the number of discrete roles (position, tasks, cf. Kieser, 1973, p. 79) that were identified in an organizational unit based on a standard role catalog (see Pugh et al., 1976, p. 59, 65-67 for a more detailed description of this measure). Reflecting the work of Epton et al., this research uses the term CD in a nonevaluative manner to describe an organizational unit in which professionals of an indefinite number of disciplines collaborate. As a corollary, CTD is viewed as a component making up organization complexity, that is as a subdimension of organization structure. In sum, the present conceptualization of CDT is based on the assumption that "meaningful comparisons across a large number of organizational units can be made only when there is a common standard for comparison..." (Pugh, 1981a, p. 136). In our case, standards for comparison are the disciplinary qualifications of professionals composing the organization units sampled in a field study.

Research Questions

The empirical part of the research reported here set out to develop simple measures of CD structures of industrial R&D units in order to address the following exploratory research question:

> Q1: To what extent is cross-disciplinary collaboration prevalent in West German industrial R&D units?

In addition, based on general theoretical arguments establishing the relationship between organizational (subunit) size and organizational (subunit) structures (Blau and Schoenherr, 1971, pp. 300-329) the present research investigates the following question:

> Q2: Do measures of CDT deviating more or less strongly from the type of measures employed in the Aston program show a strong positive relationship with R&D unit size?

A number of researchers have investigated and explained industry differences in the size of an average R&D unit in a particular industry (e.g., Bergmann, 1972; Kay, 1979; Brockhoff, 1980; Freeman, 1982; Freeman, 1974; Poensgen and Hort, 1983). Addressing the question of whether the size-complexity relationship is contingent on other variables Child (1976) and Bresser (1984) found no interaction between organization size and industry in predicting role specialization. In contrast, Kieser (1973, p. 130) stated that the relationship between size and complexity as measured by role specialization may be contingent on industry. Given these contradictory conclusions the final questions addressed by this research are as follows:

> Q3: Do measures of CDT show a relationship with R&D unit size while controlling for industry effects?

> Q4: Do relationships between CDT and R&D unit size vary across industries, i.e., does industry moderate the relationship between CDT and R&D unit size?

METHOD

Sample and Data Collection

Data reported in this paper were collected as part of a larger study of personnel management practices in R&D units of industrial companies in the Federal Republic of Germany (FRG). A list of all companies with at least 1,000 employees which had collaborated with the FRG Ministry of Science and Technology in 1980 was obtained. These companies were used as a sample for the present research. Given the wide size range of the responding companies (61 of the responding firms employed no more than 1,999 employees and 34 employed at least 30,000 employees), there is no prima facie reason to expect any systematic bias in the results except that the findings cannot be generalized to small firms employing less than 1,000 employees. A structured questionnaire was used to gather data from the sampled industrial R&D units. Note that in this study questionnaire recipients serve as informants rather than as respondents since they are asked to report facts and not express personal attitudes (Payne and Pugh, 1976, p. 1128; Pugh and Hickson, 1976b, p. 6).

In late 1981, 393 questionnaires were directly mailed to higher level managers of the R&D units. A total of 222 recipients returned usable questionnaires. Of the informants 37 percent were R&D directors, 26 percent were R&D department heads and 37 percent held other higher level management positions in R&D. Because twelve of the responding R&D units had missing data on focal variables of the present research, they were not included in any of the analyses outlined in this paper, resulting in an overall N of 210 R&D units.

Measures

Cross-Disciplinary Structures. In this research CDT was measured by the formal disciplinary qualifications of R&D professionals using the following two measures: (1) Number of Disciplines (NOD): R&D executives were asked to report the number of R&D professionals employed in their unit for each of the following disciplines: engineering, mathematics, physics, chemistry, pharmacy, biology and medicine. Scientific disciplines were then scored on a 1 or 0 basis depending on whether there was at least one R&D professional in each of the different disciplines. The NOD score for an R&D unit was obtained by adding up the binary discipline scores. It is evident that the NOD measure is similar to horizontal differentiation measures of prior organization research (e.g., Blau and Schoenherr, 1971, p. 63; Pugh et al., 1976, p. 49, 65-67; Mansfield and Zeffane, 1983, p. 53; Bresser, 1984, p. 126). (2) Distribution of Disciplines (DOD). As shown by Kieser (1973) there is a need to include an additional measure recording the distribution of a unit's R&D staff throughout different scientific disciplines. Such a measure was derived from sociological and organizational literature (Gibbs and Martin, 1962; Rushing, 1967; Hoffmann, 1980) and, in principle, it is based on the following formula:

$$DOD_I = 1 - \frac{\sum_{k=1}^{7} X_k^2}{\left(\sum_{k=1}^{7} X_k\right)^2} \qquad (1)$$

where DOD_I is the distribution of a unit's R&D professionals among scientific disciplines and X_k is the number of an R&D unit's professionals trained in discipline k (e.g., mathematicians). DOD_I equals zero when the R&D unit's professionals all belong to the same discipline and it achieves its maximum value as the R&D professionals employed in a unit are evenly dispersed throughout the seven disciplinary categories used in this research. The following formula was developed for DOD in order to ensure that the measure reaches a maximum value of 1 when R&D professionals were evenly dispersed throughout the presently used seven disciplinary categories.

$$DOD_{II} = \frac{DOD_I}{1-1/7} \qquad (2)$$

where DOD_I is the distribution of disciplines as defined in formula

(1). The measure defined by formula (2) is referred to in the rest of this paper when the abbreviation DOD or the terms "distribution of disciplines" or "disciplinary differentiation" are used.

R&D Unit Size. In keeping with the tradition of previous research (Robbins, 1983, pp. 107-114) size was principally measured by the total number of R&D professionals employed in a unit as reported by the informant. However, in light of prior research results (e.g., Child, 1976; Child and Kieser, 1979) the natural logarithm of R&D unit size was computed to test for the possibility of an asymptotic relationship of organization size with characteristics of organization structure. The mean of the logarithmized size measure was 3.3 (SD = 1.3; N = 207).

Industry. Informants were asked to indicate the industry in which the company to which an R&D unit belonged operated by checking against 24 industry categories included in the questionnaire. Of the sampled R&D units 24.4 percent were in the mechanical engineering industry, 12.9 percent were in the chemical industry (pharmaceutical companies were also assigned to this industry category) and the remaining 45.0 percent were widely dispersed among various other industries.

ANALYSIS AND RESULTS

To address the first research question, Tables 1 and 2 present descriptive results on the CD structures of the sampled R&D units. The data reveal that CD collaboration in West German R&D units can be described as being modest, corroborating prior findings of Darvas and Haraszthy (1979). More than one out of four sampled units were monodisciplinary.

Research question 2 has addressed the topic of whether an R&D unit's size considerably shapes a unit's disciplinary differentiation and diversification. The correlation between a unit's size and its disciplinary differentiation (NOD) is .60 (N=203; p. < 001) and the size-disciplinary diversification (DOD) correlation is .24 (N=203; p < .001). The correlation of .60 between size and NOD is close to previously demonstrated correlations or size with measures such as "role specialization" (Pugh et al., 1976) of "division of labour" (Blau and Schoenherr, 1971). This study supports the conclusion of former research "...that increasing size promotes structural differentiation, but at a decreasing rate" (Robbins, 1983, p. 108; Hickson and McMillan, 1981, p. 1983; Child, 1976, p. 51; Blau and Schoenherr, 1971, pp. 62-72). The NOD measure is significantly more positively related to size than the DOD measure. Hence, this research highlights the necessity to include measures of horizontal complexity in future organizational research which go beyond the mere counting of the structural parts found in an organization.

To address the final two research questions a number of regressions were run for each of the measures of CDT. Firstly, regressions were computed in which industry variables were entered first and then R&D unit size was added to test for size effects on each measure of CDT while controlling for industry effects (research question 3). Secondly, regressions were run in which the order of variable inclusion was reversed to test for industry effects on each measure of CDT while controlling for size effects. Finally, regressions of

Table 1

Distribution of 210 West German Industrial R&D Units Across Five Value Intervals of the Number of Disciplines (NOD) Measure

Number of Disciplines*	Absolute Frequency	Relative Frequency (Percent)
1	59	28.1
2	51	24.3
3	47	22.4
4-5	48	22.8
6-7	5	2.4

* Mean = 2.60; Standard Deviation = 1.41

Table 2

Distribution of 210 West German Industrial Units Across Five Value Intervals of the Distribution of Disciplines (DOD) Measure

Distribution of Disciplines*	Absolute Frequency	Relative Frequency (Percent)
0	59	28.1
$0 < DOD \leq .25$	34	16.2
$.25 < DOD \leq .50$	46	21.9
$.50 < DOD \leq .70$	50	23.8
$DOD > .70$	21	10.0

* Mean = .32; Standard Deviation = .27

both measures of CDT were performed which included the above independent variables (main effects) plus two interaction terms as predictor variables. Columns 1a and 2a of Table 3 include the standardized regression coefficients of the full regression model using industry dummies, size plus interaction terms as predictors of CDT and R&D units and employing mechanical engineering industry as a reference category. Both regressions were highly significant and accounted for 59 percent of the variance of the NOD measure and 40 percent of the variance of the DOD measure indicating that the CD structures of an R&D unit are considerably shaped by the industry to which the unit belongs and by its size.

To test the differences between the intercepts and slopes of the two industries represented by the dummy variable included in columns 1a and 2a, two additional regressions were run in which the reference category was changed to another industry (electronics, see columns 1b and 2b of Table 3). It was found that intercepts differ significantly between chemical and electronics industries for both regressions. Further the regression slope of the chemical industry dif-

Table 3
Beta Weights for the Regressions[a] on Number of Disciplines (NOD) and Distribution of Disciplines (DOD) with Industry, Log of R&D Unit Size, and Interaction Terms as Independent Variables

Independent Variables	Dependent Variable: NOD (1a) Full Equation Beta Weights	Dependent Variable: NOD (1b) Full Equation Beta Weights	Dependent Variable: DOD (2a) Full Equation Beta Weights	Dependent Variable: DOD (2b) Full Equation Beta Weights
1. Chemical Industry (C)[b]	-.04***	-3.7***	.42***	-0.6*
2. Electronics Industry (E)[c]	.23***	-	.35***	-
3. Mechanical Engineering Industry (M)[c]	-	-.49***	-	-7.0***
4. Log of R&D unit size (US)	.32***	.29***	.05	.02
5. C x US	.59*	.65*	.21	.28
6. E x US	.08	-	.11	-
7. M x US	-	.19	-	.23
R^2	.59		.40	
Overall F	30.57		14.26	
df	5,107		5,107	
p	.001		.001	

[a]Columns 1a and 2a include full regression models with main and interaction effects using mechanical engineering industry as the reference category. Reference group for the full regression models of columns 1b and 2b is the electronics industry. Variables 1, 2 and 3 are 0.1 dummy variables reflecting industry. In testing the significance of the Beta Weights the standard regression method is used for the industry variables (simultaneous inclusion), then the increment in R^2 due to US is tested and finally the interaction terms are added and tested in single steps (hierarchical method). R&D units not belonging to one of the three industries included in the Table are excluded from the analyses.

*$p \leq .025$; ***$p \leq .001$

[b]Including pharmaceutical companies.

[c]Excluding EDP hardware products.

fered significantly from the slope of the electronics industry.

In sum, it was found that industry had a significant effect on both measures of CD structures. R&D unit size, however, was related to the disciplinary differentiation (NOD) but not to the disciplinary diversification of an R&D unit. Figure 1 depicts these findings on the relationship between R&D unit size and each of the two measures of CDT for different industries. The regression lines plotted in Figure 1 are computed only from those raw regression coefficients which were found to be statistically significant (cf. Table 3).

Table 4 reports the unique variance in the measures of CDT explained by R&D unit size (industry) when controlling for industry (size) effects. Both R&D unit size and industry explained a significant unique portion of variance in the disciplinary differentiation (NOD) measure. The interaction terms explained a significant portion of variance only when disciplinary differentiation (NOD) was the dependent variable. R&D unit size accounted only for an insignificant two percent of the variance in disciplinar diversification (DOD) when industry effects were controlled. In other words, in a given industry R&D professionals in large units are not more evently dispersed among disciplines than those working in small units. In contrast, a significant portion of variance in the DOD measure is explained by the fact that the sampled units were in different industries even when the effect of size was controlled.

Table 4
Main and Interaction Effects of Industry (1) and Log of R&D Unit Size (US) with Two Measures of Cross-Disciplinarity as Dependent Variables

	Dependent Variable: Number of Disciplines					Dependent Variable: Distribution of Disciplines				
Regression Variables	R^{2**}	R^2	F^*	df	p	R^{2**}	R^2	F^*	df	p
US	.41					.11				
		.15	19.85	2,107	.001		.29	25.50	2,107	.001
US,1	.56					.40				
1	.36					.38				
		.20	53.74	1,107	.001		.02	2.60	1,107	>0.5
1,US	.56					.40				
1,US	.56					.40				
		.03	3.25	2,107	.05		.00	.22	2,107	>0.5
1,US,1 x US	.59					.40				

*The F-ratio is computed and tested according to Cohen and Cohen (1975), p. 141-142 (eq. 4.4.5).

**All R^2's are significant at $p \leq .001$ (two-tailed t-test).

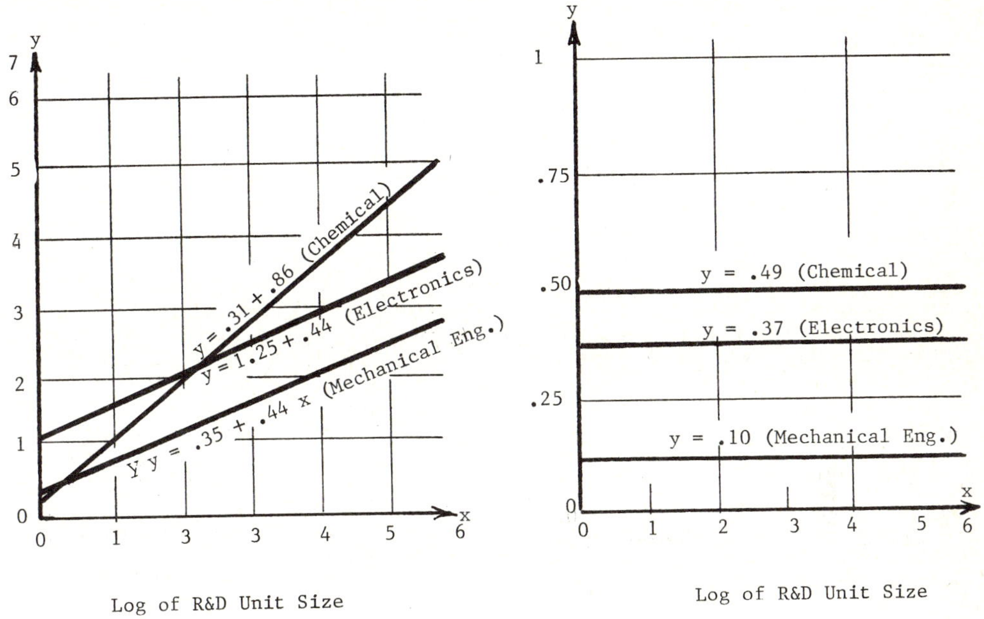

[a]Only those regression coefficients which are found to be statistically significant (see Table 3) are used to compute intersect and slope values for the cross-disciplinarity unit size relationship.

Figure 1
Significant Linear Regression Relationships Between Log of R&D Unit Size With Two Measures of Cross-Disciplinarity as Moderated by Industry[a]

DISCUSSION

Our descriptive results indicate that the sampled R&D units differ considerably in the degree to which they use CD structures to complete their R&D tasks. At the level of future research, perhaps the most noteworthy conclusion resulting from our findings is that greater attention should be given to different aspects of CD structures of R&D units or, in a more general sense, to different aspects of complexity of organization structures. Although the two measures of CD structures employed in this study (NOD, DOD) are not independent, they are differentially related to an R&D unit's size. Whereas prior research has argued that "...size...has pervasive effects on...organizational characteristics" (Blau and Schoenherr, 1971, p. 56) our findings lead to a somewhat different view of the role of an organization subunit's size in shaping organizational complexity. It was found that regardless of industry, increasing R&D unit size fosters disciplinary complexity at a declining rate when employing a measure of disciplinary differentiation (NOD) which is very similar to prior research measures. However, in a particular industry no relationship between size and complexity was discovered when using a measure of disciplinary diversification (DOD) which was not included in most of the prior organization research. These results suggest that larger R&D units employ a smaller number of R&D professionals trained in other disciplines than the majority of their staff as reflected by the positive correlation between size and the NOD measure while controlling for industry effects. At the same time, however, the distribution of R&D staff among the core disciplines of a unit remains basically unaffected by size in a particular industry.

Recalling the shape of the regression lines graphed in Figure 1 and based on further analysis not reported here (Gerpott and Domsch, 1984) it may be concluded that this research has provided some support for the existence of a threshold for the distribution of R&D professionals among scientific disciplines in chemical and electronics R&D units and for the existence of a barrier for the disciplinary diversification in mechanical engineering R&D units. Given that only very few authors have provided firm evidence for their views of the uniqueness of CD structures in different industries (for a notable exception see Mansfield et al., 1971) the present investigation has provided clear empirical support to these arguments along at least two subdimensions of organizational complexity of R&D units.

An obvious limitation of this research has been its reliance on non-evaluative measures of CDT which are based on a somewhat arbitrary distinction of seven categories of scientific disciplines. Secondly, our cross-sectional research design cannot establish causality. Although we have tried to avoid the impression that size or industry causes certain CD structures, the logic behind our associational statements and the statistical techniques used imply that size and industry cause certain CD structures. Finally, the methodology adapted in this study can be characterized as "a comparative approach to the study of organizations" (Blau and Schoenherr, 1971, p. 7) which aims at "developing universal principles of organizational life" (Evered and Louis, 1981, p. 390; Pugh, 1981a). In recent years this approach has been subjected to a number of fundamental criticisms (e.g., Morey and Luthans, 1984; Kieser and Kubicek, 1983; Starbuck, 1981; Zey-Ferell and Aiken, 1981) to which rejoinders have been formulated by advocates of this approach (e.g., Pugh 1981b; Pugh and Hickson, 1976a; Donaldson, 1982). Therefore the general shortcomings of this approach also hold true for this study. To overcome these difficulties, studies linking process and comparative methods in investigating CD structures in R&D are needed.

REFERENCES

Barth, R.T. and Steck, R., eds. 1979. Interdisciplinary research groups: their management and organization. Proceedings of the first international conference on interdisciplinary research groups held at Schloss Reisenburg, 22-28 April 1979. Vancouver-Hannover: IRGIP.

Bergmann, U.H. 1972. Unternehmensgrösse and technischer Fortschritt. Koln: Kölner Universitäts-Verlag.

Birnbaum, P.H. 1979. Research team composition and performance. In: Interdisciplinary research group: their management and organization, eds. R.T. Barth and R. Steck, pp. 207-225. Vancouver-Hannover: IRGIP.

Birnbaum, P.H. 1981. Integration and specialization in academic research. Academy of Management Journal 24:487-503.

Birnbaum, P.H. 1982. The organization and management on interdisciplinary research - a progress report. Journal of the Society of Research Administrators 13/Winter:11-23.

Blaschke, D. 1976. Probleme interdisziplinärer Forschung. Wiesbaden: Franz Steiner Verlag.

Blau, P.M. and Schoenherr, R.A. 1971. The structure of organizations. New York-London: Basic Books.

Bresser, R.K. 1984. Structural dimensions of university departments and their context: the case of West Germany. Organization Studies 5:119-146.

Child, John. 1976. Predicting and understanding organization structure. In: Organization structure: extensions and replications - the Aston programme II, eds. D.S. Pugh and C.R. Hickson, pp. 45-64. Westmead: Saxon House.

Child, J. and Kieser, A. 1979. Organization and managerial roles in British and West Germany companies: an examination of the culture-free thesis. In: Organizations alike and unlike, eds. C.J. Lammers and D.J. Hickson, pp. 251-271. London: Routledge and Kegan Paul.

Cohen, J. and Cohen, P. 1975. Applied multiple regression/correlation analysis for the behavioral sciences. Hillsdale: Lawrence Erlbaum.

Dalton, D.R., Rodor, W.D., Spendolin, M.J., Fielding, G.J., and Porter, L.W. 1980. Organizational structure and performance: a critical review. Academy of Management Review 5:49-64.

Darvas, G. and Haraszthy, A. 1979. Interdisciplinary orientation of research teams in six European countries. In: Interdisciplinary research groups: their management and organization, eds. R.T. Barth and R. Steck, pp. 34-48. Vancouver-Hannover IRGIP.

Donaldson, L. 1982. Comments on contingency and choice organization theory. Organization Studies 3:65-72.

Epton, S.R., Payne, R.L. and Pearson, A.W. 1984. The management of cross-disciplinary research. R&D Management 14:69-79.

Epton, S.R., Payne, R.L. and Pearson, A.W., eds. 1983. Managing interdisciplinary research. Chichester (U.K.): Wiley.

Evered, R. and Louis, M.R. 1981. Alternative perspectives in the organizational sciences: "inquiry from the inside" and "inquiry from the outside." Academy of Management Review 6:385-395.

Ford, J.D. and Slocum, J.W. 1977. Size, technology, environment and the structure of organizations. Academy of Management Review 2:561-575.

Freeman, C. 1974. The economics of industrial innovation. Middlesex: Penguin Books.

Freeman, C. 1982. The economics of industrial innovation. 2nd ed. London: Frances Printer.

Gerpott, T.J. and Domsch, M. 1984. Cross disciplinary structures of industrial research and development units. Working paper available by writing to T.J. Gerpott, Institut für Personalwesen, HSBW, Postfach 700822, D-2000 Hamburg 70, FRG.

Gibbs, J.P. and Martin, W.T. 1962. Urbanization, technology, and the division of labor: international patterns. American Sociological Review 27:667-677.

Hickson, D.J. and McMillan, C.J. 1981. Concluding comments on countries, context and culture. In: Organization and nation, eds. D.J. Hickson and C.J. McMillan, pp. 187-196. Aldershot: Gower.

Hoffmann, F. 1980. Führungsorganisation - Band I; Stand der Forschung und Konzeption. Tübingen: J.C.B. Mohr.

Kay, N.M. 1979. The innovating firm - a behavioural theory of corporate R&D. London-Basingstoke: MacMillan.

Kieser, A. 1973. Einflussgrössen der Unternehmensorganisation. Köln: Habilitationschrift der Wirtschafts - und Sozialwissenschaftlichen Fakultät der Universität zu Köln.

Kieser, A. and Kubicek, H. 1983. Organisation. 2nd ed. Berlin-New York: Walter de Gruyter.

Loveday, D.E. 1984. Factors affecting the management of interdisciplinary research in the pharmaceutical industry. R&D Management 14:93-103.

MacDonald, W.R. 1983. The many meanings of interdisciplinary research. Paper presented at a Working Conference on the Management of Interdisciplinary Research. Manchester (U.K.), August 3-5, 1983.

Mansfield, E., Rapoport, J. Schnee, J., Wagner, S. and Hamburger, M. 1971. Research and innovation in the modern corporation. New York: Norton.

Mansfield, R. and Zeffane, R. 1983. Organization structure and national contingencies. Aldershot: Gower.

Morey, N.C. and Luthans, F. 1984. An emic perspective and ethnoscience methods for organizational research. Academy of Management Review 9:27-36.

Payne, R.L. and Pugh, D.S. 1976. Organizational structure and climate. In: Handbook of industrial and organizational psychology, ed. M.D. Dunnette, pp. 1125-1176. Chicago: Rand McNally.

Poensgen, O.H. and Hort, H. 1983. F&E-Aufwand, Firmensituation und Firmenerfolg. Zeitschrift für betriebwirtschaftliche Forschung 35:73-93.

Porter, A.L. and Chubin, D.E. 1983. An indicator of cross-disciplinary research. Paper presented at a Working Conference on the Management of Interdisciplinary Research. Manchester (U.K.), August 3-5, 1983.

Porter, A.L. and Rossini, F.A. 1984. Interdisciplinary research redefined: Multi-skill, problem-focused research in the STRAP framework. R&D Management 14:105-111.

Pugh, D.S. 1981a. The Aston program perspective. In: Perspectives on organization design and behavior, eds. A.H. van de Ven and W.F. Joyce, pp. 135-166. New York: Wiley.

Pugh, D.S. 1981b. Rejoinder to Starbuck. In: Perspectives on organization design and behavior, eds. A.H. van de Ven and W.F. Joyce, pp. 199-203. New York: Wiley.

Pugh, D.S. and Hickson, D.J. 1976a. Concluding remarks. In: Organizational structure in its context - the Aston programme I, eds. D.S. Pugh and D.J. Hickson, pp. 179-186. Westmead: Saxon House.

Pugh, D.S. and Hickson, D.J. 1976b. The Aston approach. In: Organizational structure in its context - the Aston programme I, eds. D.S. Pugh and D.J. Hickson, pp. 1-16. Westmead: Saxon House.

Pugh, D.S. and Hickson, D.J., eds. 1976c. Organizational structure in its context - the Aston programme I. Westmead: Saxon House.

Pugh, D.S., Hickson, D.J., Hinings, C.R. and Turner, C. 1976. Dimensions of organization structure. In: Organizational structure in its context - the Aston programme I, eds. D.S. Pugh and D.J. Hickson, pp. 1-16. Westmead: Saxon House.

Pugh, D.S. and Hickson, C.R., eds. 1976. Organization structure: extensions and replications - the Aston programme II. Westmead: Saxon House.

Robbins, S.P. 1983. Organization theory - the structure and design of organization. Englewood Cliffs, NJ: Prentice-Hall.

Rushing, W.A. 1967. The effects of industry size and division of labor on administration. Administrative Science Quarterly 12:273-295.

Starbuck, W.H. 1981. A trip to view the elephants and rattlesnakes in the garden of Aston. In: Perspectives on organizational design and behavior, eds. A.H. van de Ven and W.F. Joyce, pp. 167-198. New York: Wiley.

Zey-Ferrell, M. and Aiken, M. 1981. Introduction to critiques of dominant perspectives. In: Complex organizations: critical perspectives, eds. M. Zey-Ferrell and M. Aiken, pp. 1-21. Glenview: Scott, Foresman and Company.

MANAGING HIGH TECHNOLOGY
An Interdisciplinary Perspective
B.W. Mar, W.T. Newell, and B.O. Saxberg (Editors)

ADMINISTERING INTERDISCIPLINARY COLLABORATION

Martha Garrett Russell

Associate Director
Microelectronic and Information Sciences
Institute of Technology
University of Minnesota

ABSTRACT

The personnel, disciplinary, and administrative aspects of research provide a three-dimensional framework for identifying and understanding the nature of collaboration and interdisciplinary activity. This paper discusses the issues which are specific to each dimension: funding, recognition, promotion, equipment and facilities, leadership, teamwork, and conflict. It also points out how the services of research administrators can be more effectively and efficiently targeted to their points of greatest impact by considering the dimension-specific concerns of these issues.

The best research administration is that which enables scientists who ask good questions and who identify appropriate ways to look for answers to obtain the resources they need. This paper explores the administrative issues which have become critical in administering interdisciplinary collaboration in academic research. This paper presents a three dimensional framework for identifying and understanding administrative challenges in terms of the nature of collaboration and interdisciplinary activity. It points out how the efforts of research administrators can be more effectively and efficiently targed to their points of greatest impact by considering the dimension-specific concerns of each of these issues.

The role of interdisciplinary research in the solution of complex problems confronting society today has been widely acknowledged (Birnbaum, 1981a; Cotterrell, 1979; Gusdorf, 1977; Sinaceur, 1977). Thomas Kuhn defines a discipline in terms of its paradigms--its parameters for recognizing, asking and answering questions (Kuhn, 1970). Rustom Roy, on the other hand, defines a discipline as a body of knowledge that is taught by 25 or more departments across the country (Roy, 1977). By definition, interdisciplinary efforts have been recognized for their potential to deal with multifaceted problems and to generate proposed solutions which satisfy multiple criteria. However, the challenges in bringing about scientific collaboration at universities have also been repeatedly lamented.

Collaboration in science and technology has been strongly influenced by two forces. First, the sponsorship of research by agencies external to the research institution has made accountability more complex. Secondly, the development of subdisciplinary specializations has redefined the size and nature of scientific peer groups. Both of these forces have affected how scientists have developed their professional identities and interdependencies. Their collaboration with each other presents challenges to research administrators--challenges regarding funding, recognition, promotion, equipment and facilities, leadership team work and conflict.

Both scientists and research administrators have been encouraged by successful collaborative ventures and frustrated by others which did not meet expectations. Some studies have attempted to identify the characteristics of successful interdisciplinary teams (Birnbaum, 1981; Rossini and Porter, 1981). And several studies have discussed the obstacles to such collaboration (Ikenberry and Freedman, 1972; Saxberg, Newell, and Mar, 1977). But given the interest in interdisciplinary collaboration as a means to solve problems through research, it is helpful to identify the adminsitrative dimensions and strategies which facilitate this collaboration.

DIMENSIONS

Interdisciplinary activity is often a collaborative activity involving the efforts of two more scientists. Equally important, such research activity frequently involves joint efforts of scientists from two or more departments. In general the term interdisciplinary refers to a specific plan, approach or set of efforts which blends components of two or more administrative units within a university or among a set of colleagues drawn from a variety of institutional settings. These dimensions then, when integrated, form a three-dimensional framework for understanding collaboration and interdisciplinary activity: the people, the disciplines and the administrative units.

The People

The investigators are primary in importance, and disciplinary activity usually assumes more than one investigator. Both professional and educational accountabilities in academic research are oriented toward the individual faculty member. Criteria for judging merit and distributing awards to academicians reflect the ability of a particular faculty member to make substantial academic contributions to a particular discipline (Lewis, 1979)--both through expanding the knowledge base and also through attracting support for research. In addition, faculty are rewarded for professional development within their specific disciplines and for educating students and granting degrees. These credentials are awarded through departments and programs comprised of disciplinary components. Students receive their scientific training and usually complete their research apprenticeships within a single discipline. New scientists earn their academic credentials and reputations as individuals with reference to a particular scientific community. Thus one scientist is usually associated with one discipline.

However, interdisciplinary efforts may be either collaborative or individual. Not all investigators restrict their scientific inquiry to their field of initial specialization. Some scientists, indeed it has been argued the better scientists (Ziman, 1980), enlarge their agenda of research questions, their research approaches, their visions of issues through interaction with scientists trained in other disciplines. Through study and experimentation, they expand their initial perspectives. They then incorporate these new scientific strategies into their research approaches.

Whether one scientist or more than one scientist is involved in a research effort has significant implications for the conduct and administration of that research. Cooperation, competition, motivation, leadership, power, teamwork, division of responsibility, and allocation of rewards are all important considerations which are influenced by the number of scientists involved in a research effort. Briefly then, these variables form the parameters of the people dimension of collaboration in an interdisciplinary activity.

Disciplinary Dimension

The second dimension of collaborative activity is the number of disciplines involved. Typically, scientists' research questions and avenues of inquiry follow the lines established by their training. Professional development and contacts with colleagues refine and expand these established approaches. The usefulness of those refinements is demonstrated in addressing questions or in identifying new questions for which the approach may be appropriate. The exceptional contribution identifies new tools (experimental and/or theoretical) with which questions can be either asked or answered (Price, 1982). All of these contributions are legitimized by disciplinary approval.

Another type of disciplinary influence occurs in the recommendation by scientific peers or disciplinary representatives that a project be funded. These disciplinary peers are considered the best qualified to advise on the scientific quality of a proposed activity (assuming that the activity falls within the disciplinary parameters represented by those peers). Identifying peers to review interdisciplinary activities depends either upon the availability of individuals who themselves have established expertise in several disciplines or upon the availability of a group of experts who are willing to fully acknowledge the legitimacy of each other's expertise (Russell, 1982) and together focus on the overlapping frontiers of the proposed activity.

Scientific peers also play important roles in encouraging or discouraging the identification of new research efforts, in acknowledging or ignoring contributions to new knowlege, and in accepting or denying new members. In these roles, the disciplinary group strongly influences what its members will study (Ziman, 1980). The criteria established for quality by the disciplines are "what make the disciplines disciplined" (Boulding, 1977). One can easily understand that such standards are maintained by adhering to them rather than by making exceptions.

Another disciplinary influence is the scientific literature. Publication is the means by which research is evaluated. It signifies that the scientific community of peers has judged the work acceptable by its standards. These disciplinary standards for review in publications are based on consensus regarding what questions, methods, and interpretations are deserving of the attention of that scientific community. Again, the academic respectability of a discipline is based on maintaining those standards rather than on making exceptions to them.

University research support is typically allocated to scientists on the basis of their abilities to make contributions to their respective disciplines. External support for research is awarded, on the other hand, according to scientists' ability to make substantial contributions to the concerns of the sponsor. When these concerns reflect problems which are complex the expertise from several disciplines is frequently required.

Styles of interdisciplinary collaboration have been defined using attributes of disciplinary paradigms, disciplinary literature bases, disciplinary methodologies and group identity, as well as goals, methods and outcomes (Birnbaum, 1981; Nagi and Corwin, 1972). These perspectives have developed definitional categories of disciplinary collaboration which are differentiated by the scientific methods which are used, the questions which are asked, the extent and characteristics of collaboration, and the extent of agreement among the collaborators on project goals. These perspectives view collaboration along a continuum of disciplinary autonomy and integration.

Peer review is the central disciplinary issue for research efforts involving more than one discipline. This issue resides in legitimizing the questions, the methods and the interpretations which make up the activity (Russell, 1981). Science is based on this type of critique. Recommendations for project funding and allocation of rewards, as well as decisions regarding professional recognition and quality control of scientific inquiry, are all influenced by the number of disciplinary perspectives involved in the activity.

Administrative Dimension

The third dimension of collaboration which is critical for both collaborative and interdisciplinary efforts is the number of administrative units involved in the project. These administrative units can be construed as the number of departments, institutions, states, or countries involved. Just as scientific communities are the professional and the ideological homes of scientists, administrative units are the financial and physical homes. These administrative units are accountable for the funds, both internal and external, allocated in support of scientists' research efforts.

Two primary considerations in this dimension are: "How many funding/administrative units are involved with the overall activity?" And "How many separate facilities/locations are a part of the total effort?"

This is the dimension of administrative accountability. Firstly, administrative units must account for research funds. These units are accountable for ownership of equipment purchased for the research and for the appropriate disbursement of funds for personnel and supplies. Secondly, they are also responsible for the products of research conducted with those funds, specifically degree recipients, new knowledge, intellectual property or potentially patentable innovations. Thirdly, administrative units are the sources of institutional recognition, incentives and rewards.

These responsibilities held by administrative units directly influence collaboration among scientists and collaboration among disciplines. Administrative policies provide the basic framework of the research environment within which the research activities are conducted. Unit goals and objectives provide the benchmarks against which the importance and relevance of scientists' efforts are evaluated. Policies regarding indirect costs, consultative arrangements, and reciprocity between units reflect the value of research activities and of collaboration to the administrative unit (Currie, 1978). Such policies influence the incentives or lack of incentives for collaboration (Ikenberry and Freedman, 1972). They influence the economic and noneconomic costs to departments of permitting and encouraging faculty members to participate in collaborative endeavors.

Collaboration between scientists from different administrative units is influenced not only by the policies of their individual units, but also by the policies of the central administration to which those individual units report. Policies which discourage collaboration between scientists from different administrative units, at any level of the organization, constrain the effectiveness of incentives at other levels (Russell and Sauer, 1983).

In summary, the classification of research according to personnel, discipline and administrative unit provides a three-dimensional framework for identifying and understanding the nature of collaboration and interdisciplinary activity (see Table 1), and the issues which characterize each dimension (see Table 2).

Table 1
Eight Types of Research Activities

People Dimension	Disciplinary Dimension	Administration Dimension
1. One scientist	One discipline	One administrative unit
2. One scientist	One discipline	More than one administrative unit
3. One scientist	More than one discipline	One administrative unit
4. One scientist	More than one discipline	More than one administrative unit
5. More than one scientist	One discipline	One administrative unit
6. More than one scientist	One discipline	More than one administrative unit
7. More than one scientist	More than one discipline	One administrative unit
8. More than one scientist	More than one discipline	More than one administrative unit

Table 2
Administrative Issues by Dimension of Collaboration and Interdisciplinary Activity

People Dimension	Disciplinary Dimension	Administration Dimension
Rewards and incentives Communication Division of responsibility Interpersonal conflict Problem focus Teamwork Motivation Leadership	Acknowledging new knowledge Admitting new scientists Professional recognition Maintaining standards Legitimizing the questions, methods interpretations Peer approval Graduate students	Accountability for funds Ownership of facilities and equipment Appointment of scientists Institutional recognition, rewards and incentives Unit goals-planning and prioritizing Unit prestige Research environment Accountability for intellectual property and patents

Several examples of research projects at the Microelectronic and Information Sciences Center (MEIS) illustrate these dimensions. MEIS is an interdepartmental center at the University of Minnesota. It is sponsored by state and federal funds and by several major compnies. MEIS sponsors several research programs, all of which have additional support from sources other than MEIS.

Examples. More than one scientist, more than one discipline.

On one MEIS project several faculty from one interdisciplinary department work together on a project. This project addresses the conceptual, theoretical and analytical innovations and the modeling and fabrication technologies which are important in producing a monolithic 3-D all semiconductor device. Seven faculty--three full professors, one associate, one assistant, and one emeritus--make up this team, which has a history of collaboration on education, as well as research, interests.

Their backgrounds include semiconductor physics, fabrication technology, circuit design, system architecture and testing. A core of journals represents their mutual professional identification and publication preferences. Since they share an administrative home, the primary considerations are the disciplinary and people dimensions:

How are graduate students apportioned?
How is authorship shared?
How is the project leader identified?
How are team members added or subtracted?
How will new courses developed out of the research be assigned to various faculty members?

The management of interpersonal issues is primary. Conflict, competition, cooperation and collusion may influence the collaboration. The maintenance of team work, leadership and appropriate recognition are major considerations for assisting this team in its efforts.

Summary

By identifying the various dimensions of collaboration which describe a research activity, the research management concerns of a particular project can be identified. The administrative issues of one scientist studying an issue from a single disciplinary perspective using funds from one administrative unit are surely different from those involved in financing a group of scientists from several administrative units who are combining several disciplinary perspectives. Each dimension of collaboration adds another set of management, scientific and/or administrative concerns.

Collaborative research is not necessarily interdisciplinary. Interdisciplinary research is not necessarily collaborative. The administrative assistance needed from a research administration office differs according to the character of the project. A single investigator who uses a synthesis of several disciplinary approaches and is funded by multiple sources requires different attention than a group of investigators from one multidisciplinary department who conduct their research in a laboratory sponsored by that department. When collaborative interdisciplinary research is interdepartment interfaces between administrative units are layered on top of interpersonal and disciplinary interactions.

Many of the highest priority research programs encompass problems that are broader than the expertise of any one scientist, discipline or department. Collaboration between scientists, between disciplines, and between administrative units has been and continues to be essential to address those problems.

Collaborative efforts among specialists have been recognized as the philosophical underpinning of the earliest community of scholars (Gusdorf, 1979). Historical documentation has shown, in fact, that scientific collaboration was the basis for the professional development of both scientists and disciplines (Beaver and Rosen, 1977). Dialogue and cooperation among scientists with different disciplinary training not only refine the foci of disciplinary or departmental efforts but have also formed the embryonic stages of subdisciplines and new disciplines (Cotterrell, 1979; Russell, 1981).

As the need for research expertise to address complex issues has grown, so also has the pool of knowledge on which such inquiries are based. The ability of any one scientist to fulfill professional responsibilities, continue investigations, and maintain currency with professional associates is limited by time constraints. In order to stay at the research front of an ever expanding pool of knowledge, a scientist's share of the total shoreline has decreased. In order to maintain the structure of scientific inquiry and its reliance on validation by peers, the concerns, the perspectives, the paradigms, the methods and the theoretical frameworks of each discipline have been specialized. Subdisciplinary communities characterize this emphasis on specialization.

Scientists have organized themselves into departments in order to instruct students and train successors. Departmental consensus on what should be taught has grown out of the disciplinary training of academicians and in turn shapes the development of their sciences (Brubacher and Willis, 1968). The increasing subdisciplinary specialization of scientific communities has strengthened the discipline of these communities. It has also led to the need for interdisciplinary research to provide new life blood into the disciplines and to utilize the strength of the disciplines in addressing real issues (Russell, 1982).

The people, disciplinary, and administrative aspects of collaboration provide a three-dimensional framework for identifying and understanding the nature of interdisciplinary research. The requirements of research administrators are specific to each dimension. By understanding these issues, the services of research administrators, can be effectively and efficiently targeted to their points of greatest impact. By understanding which dimension(s) are involved in the management of a research project, the assistance can be delivered by the administrative unit closest to and responsible for the issue.

REFERENCES

Beaver, D. deB., and Rosen, R. 1978. Studies in scientific collaboration. Part I. The professional origins of scientific co-authorship. Scientometrics 1(1):65-84.

Birnbaum, P.H. 1981a. Progress report on the organization and management of interdisciplinary research. Interstudy 2(6):1-31.

Birnbaum, P.H. 1981b. Contingencies for academic interdisciplinary research performance. Presentation at the Annual Meeting of the American Association for the Advancement of Science, Toronto, Canada, January 3-8.

Boulding, K. 1977. Peace research. International Social Science Journal 29(4):601-613.

Brubacker, J.S. and Willis, R. 1968. Higher education in transition. A history of American colleges and universities, 1936-1968, 3rd ed., rev. and enl. New York: Harper and Row.

Cotterrell, R. 1979. Interdisciplinarity: the expansion of knowledge and the design of research. Higher Education Review 8(3):47-56.

Currie, D.M. 1978. Interdisciplinary research and the university reward structure. Unpublished Ph.D. dissertation, University of Southern California.

Gusdorf, G. 1979. Past, present and future in interdisciplinary research. International Social Sciences Journal 24(4):580-599.

Ikenberry, S.O. and Friedman, R.C. 1972. Beyond academic departments. San Francisco: Jossey-Bass, Inc., Publishers.

Kuhn, T.S. 1970. The structure of scientific revolutions, 2nd ed. Chicago: University of Chicago Press.

Lewis, D.R. and Kellogg, T.E. 1979. Planning and evaluation criteria for allocating departmental and collegiate resources in a university setting. In D.R. Lewis and W.E. Becker, Jr., eds., pp. 131-143, Academic rewards in higher education. Cambridge, MA: Ballinger Publising co.

Nagi, S.Z. and Corwin, R.G. 1972. The social contexts of research. London: Wiley-Interscience.

Orlans, H., ed. 1968. Science policy and the university. Washington, D.C.: The Brookings Institutions.

Price, D.J. de Solla. 1982. Notes toward the philosophy of the science/technology interaction. Sociology of Science Yearbook.

Rossini, F.A. and Porter, A.L. 1981. Interdisciplinary research: performance and policy issues. Interstudy Bulletin 2:5, August.

Roy, Rustom. 1977. Interdisciplinary science on campus--the elusive dream. C&EN Aug. 29:29-40

Russell, M.G. 1981. Peer review in interdisciplinary research: flexibility and responsiveness. Presented at the Second International Conference on Interdisciplinary Research management, Manchester, England, July 1981.

Russell, M.G. 1982. Evaluating interdisciplinary research. In M. Russell, R. Sauer and J. Barnes, eds., Enabling interdisciplinary research. University of Minnesota: Agricultural Experiment Station Misc. Pub. 19.

Russell, M.G. and Sauer, R.J. 1983. Creating administrative environmental for interdisciplinary research. Journal of the Society for Research Administrators 14(4) Spring:21-31.

Saxberg, B.O., Newell, W.T. and Mar, B.W. 1977. Interdisciplinary research: a dilemma for university central administration. Project Report #9. Research Management Improvement, University of Washington, Seattle, May.

Sinaceur, M.A. 1977. What is interdisciplinarity? International Social Science Journal 19(4):571-579.

Ziman, J. 1980. What are the options: social determinants of personal research plans. Paper presented at the Conference on Scientific Establishments and Hierarchies, July 5, Oxford, England.

MANAGING HIGH TECHNOLOGY
An Interdisciplinary Perspective
B.W. Mar, W.T. Newell, and B.O. Saxberg (Editors)
Elsevier Science Publishers B.V. (North-Holland), 1985

IMPLEMENTATION OF A MODEL TO IMPROVE PRODUCTIVITY OF INTERDISCIPLINARY GROUPS

Harvey J. Gold[1] and Shirley E. Gold[2]

[1]Department of Statistics, Box 8203
North Carolina State University, Raleigh, NC 27695

[2]North Carolina Department of Natural Resources
and Community Development

At the 1981 Conference in Manchester, England (Gold and Gold, 1984; we will refer to this as Paper I), we described a model of the process that individuals from different disciplines engage in as they attempt to carry out interdisciplinary research or problem solving activities. The model was designed to serve as the basis for the development of tools which might help us cope with the interdisciplinary barriers that often lower the productivity of interdisciplinary research and problem solving groups.

REVIEW OF THE MODEL

In Paper I, we described an interdisciplinary research or problem solving group as a system made up of interdependent subunits which can be either individuals or groups, and which receive inputs from each other. For the purpose of this paper, the subunits will be taken to be individuals, though more generally, they may be composed of several individuals from the same or different disciplines.

Our focus is on the interpersonal transactions between these subunits, which we model as comprising four elements:

1. The specification of the input needed
2. Production of the specified input
3. Evaluation of the input provided (evaluation was not included as a separate element in Paper I)
4. Use of the input to produce an output

It should be noted that evaluation may be performed on an intermediate input and may cause the input specification to be altered.

The transactions can take place in a variety of modes depending on how responsibility for these four elements is divided or shared. Table 1 describes two extreme modes: contracting and partnership.

In the contracting mode, the potential user of the input has sole responsibility for specification, use, and evaluation of the input, while the supplier has sole responsibility for producing the input according to specification. In this mode, as long as the input conforms to the specification, the supplier has no responsibility for insuring that the specified input will satisfy the need, or indeed that it ever is used. This is the mode which characterizes the stereotypical New England farmer who answers yes when he is asked if he knows the time. The answer fully and accurately answers the question but does not address the need. If the user (the questioner in the example) determines through evaluation that the input supplied meets specifications but not the need, he may either redraw the specification (ask another question) or as frequently happens in interdisciplinary research, give it up altogether.

Table 1
Relative Division of Responsibility in the Pure Modes

Responsibilities	Contracting		Partnership	
	User	Supplier	User	Supplier
Specification	1	0	1/2	1/2
Supply	0	1	1/2	1/2
Evaluation	1	0	1/2	1/2
Use	1	0	1/2	1/2

In the other extreme, which we have termed partnership, and which is rarely to be expected in interdisciplinary work, all responsibilities are equally shared.

Between these extremes is a range of modes which we have termed consulting, in which each of the responsibilities is shared, but not equally. The user would normally have lead responsibility for using the input, the supplier would have lead responsibility for its production, while the specification and evaluation might be shared more equally. The sharing of responsibility for making the input useful is especially important in activities of low predictability because it may not be possible to specify in advance what the actual product will look like.

In an actual interdisciplinary group the interaction pattern is usually somewhat more complex, since the subunits often require inputs from each other, either at different times or simultaneously. The individuals may therefore become co-contractors or co-consultants.

In Paper I, we indicated that the distinction between contracting and consulting parallels, at an operational level, the distinction often made between multidisciplinary and interdisciplinary work. While multidisciplinary work can be carried out effectively in the contracting mode, effective interdisciplinary work usually requires the consulting mode. Unfortunately, the array of differences between disciplines that gives interdisciplinarity its creative, synergistic potential, also prevents many consulting (and therefore interdisciplinary) interactions from developing successfully. Even when the consulting mode is called for, the contracting mode will often be used. Consulting, with its shared responsibiltiies and high levels of interaction, may seem too difficult to be worth the effort.

ADDITIONAL UNDERLYING CONCEPTS

The processes of input specification, supply, evaluation and use take place in the framework of a consulting relation which evolves over time. In Paper I, we drew upon the consulting models of Bell and Nadler (1979) and of Lippitt and Lippitt (1978), which represent the consulting process as a time sequence of phases (see Table 2 for a summary). We used the Bell and Nadler model as an organizing framework, to show how the consulting relation moves through time. As it does so, a variety of factors operate as interdisciplinary barriers to the processes of specification, production, evaluation, and use of inputs that individuals from different disciplines require from each other.

Table 2
Phases of the Consulting Process

Bell and Nadler (1979)	Lippitt and Lippitt (1978)
ENTRY	Initial contact.
ENTRY	Formulating a contract and establishing a helping relationship.
DIAGNOSIS	Problem identification and diagnostic analysis.
DIAGNOSIS, RESPONSE	Setting goals and planning for action.
RESPONSE	Taking action and cycling feedback.
DISENGAGEMENT, CLOSURE	Contract completion: continuity, support, termination.

A discipline may be regarded as a body of knowledge, as a collection of methods, as a way of viewing the world, or as a scientific community. The factors that characterize a discipline also operate as barriers to interdisciplinary work. Building principally on the work of Hagstrom (1966), Kuhn (1970; 1977), Pelz and Andrews (1966), we grouped these factors into five main categories:

1. Cognitive framework consists of underlying theory and generalizations; of idealized models and analogies; and specific examplars. Taken together, these structure the

way in which an individual receives and interprets information.

2. Subject-matter factors consists of characteristics of the system under study which limit our ability to interact with, to observe, to manipulate, and to understand the system.

3. Normative factors are disciplinary value judgements about the types of questions which are valid subjects for investigation and about criteria for validity of solutions; the mind-set as to what is good science.

4. Community services include services offered by the disciplinary community, such as exchange of information, recognition, critical evaluation and feedback.

5. Personal identification is the association with the disciplinary community from which the individual draws his/her professional identity.

IMPLEMENTATION

We have used the concepts summarized in the preceding sections as a basis for:

1. planning how to carry out an interdisciplinary transaction

2. carrying out an interdisciplinary transaction

3. figuring out how to correct an interdisciplinary transaction that is not going well

These concepts can be used by individuals involved in the transaction, or by a third party, such as the director of a research project, to analyze the situation and provide guidance. Over the past several years, we have used these concepts in several ways:

1. Retrospective analysis of projects with which we had both been associated, and of case histories in the literature.

2. Advising graduate students involved in interdisciplinary collaborative research

3. Prospective analysis of the likely success of proposed interdisciplinary collaboration. In one case, the analysis lead to rejection of the collaboration, since one of the parties felt that a consultative mode was essential for success, while the other party was unwilling to expend more than contract-level effort.

4. In at least one case, drawing explicitly on the model to help guide an interdisciplinary research project

In the following, we discuss some of what we have learned, focusing on the entry and diagnosis phases of the consulting process. The most pervasive observation is that unsuccessful interdisciplinary collaborations tend to get stuck in a contracting mode. A researcher operating as a contractor is producing to someone else's specification. Such an activity is likely to be low in innovative content, and low in terms of reward and peer recognition.

In the remainder of this section we concentrate on the identification of those barriers that seem to be most important in the initial phases of entry and diagnosis, and suggest ways of dealing with these barriers.

ENTRY PHASE

Principle Barriers

In the projects we examined, the principle barriers that seemed to operate during the initial or ENTRY phase of the collaboration were:

> Differences in cognitive framework. Each of us receives and interprets facts and information in terms of our own disciplinary world view, or cognitive framework. Exchange of information is of minimal value if there is not a shared cognitive framework.
>
> Professional identification. Each scientist is convinced that his/her own disciplinary cognitive framework is the appropriate world view (and it is too bad the other scientist is having such a tough time seeing that).
>
> Normative factors. Disciplinary value judgments inflate regard for one's own scientific goals and lead to minimal toleration of the goals of other disciplines.

Each of these types of barriers were reviewed in some detail in Paper I.[1] An additional factor, however, emerged from our analysis, which had not been previously considered.

> Power sharing. Collaboration in a consultative mode involves the mutual granting of power, which is necessary if there is to be shared responsibiltiy. During the entry phase, each potential collaborator is faced with the prospect of being asked to surrender some degree of control over his/her own activities. We now feel that this is a major negative force, causing potential collaborators to hold back during entry phase, and later driving the collaboration to an inappropriate contracting mode.

Recommendations

The most pressing need during entry phase seems to be the development of a shared, or interfacing cognitive framework, which will underlie the collaborative work. Although our discussion has been in terms of pairwise interactions, it will normally be necessary for any group of closely collaborating individuals to develop a common interfacing cognitive framework. In the development of such a framework, it is often helpful to proceed backward; that is, starting from articulation of the group goal, and then while working backward to determine what is needed to satisfy that goal, identifying and developing common understandings. The tangible outcome of the process is a shared conceptual model of the system being studied and of the process by which the group goal is to be satisfied.

It is helpful if each member of the group articulates his or her understanding of the model in verbal as well as written form, for scrutiny by other group members and for mutual concept building. Attempts should be made to develop analogies, metaphors and exemplars which are meaningful to each of the participants. If, as sometimes happens, the primary task of articulation is assumed by a single individual, adequate care must be exercised to insure that the cognitive framework of that individual does not become dominant simply by default. The cognitive framework developed should not belong to any of the contributing disciplines, nor should it attempt to encompass them.

Relational input-output system diagrams are widely used as an effective medium for articulation of such conceptual models. Figure 1 is taken from a report (Gold, 1983) developed primarily as an internal working paper in a project dealing with agricultural pest management. The conceptual framework upon which a portion of this work is based is made explicit through several such diagrams, together with a good deal of explanatory text material. A knowledgeable scientist reading the report would have no difficulty recognizing that it could not have been based solely on the cognitive framework of any one of the contributing disciplines. Skills and techniques for the formulation of simple explicit relational models can be easily learned. Specialists in such modeling techniques may also play a third party facilitative role (see Walker et al., 1978; Holling, 1978).

The development of the interfacing cognitive framework is the process whereby the perspectives of the participants expand allowing their own world views to be altered and permitting new relevancies to be recognized. Given what has been written about the role of theory, models, and metaphors in scientific innovation (DeMey, 1980; Knorr, 1980), we believe it is no exaggeration to claim that this process is close to the source of the creative synergistic potential of interdisciplinary work.

During entry phase, it is desirable to explicitly recognize the need to address two types of goals: group goals and individual goals. The early discussions should include an attempt to understand each other's individual goal structures. Collaborators are more likely to invest in a consultative mode if they feel that their own goal structures will benefit thereby. Depending upon the benefits they see in the collaboration, some modification of individual goal structures may be negotiated.

Certain danger signs during this phase may signal that it is best not to proceed. First, inability to articulate a group goal which each participant accepts as valid and worth the effort is such a signal. Second, during the process of building the interfacing cognitive framework, some preliminary indication can be obtained of each participant's willingness to enter into needed mutual learning relationships and to share power in the sense previously discussed. Finally, this time may also be used by participants to satisfy themselves as to the competence of potential collaborators to provide inputs that will be needed and to use inputs provided to them. We could cite several collaborations which became frustrating

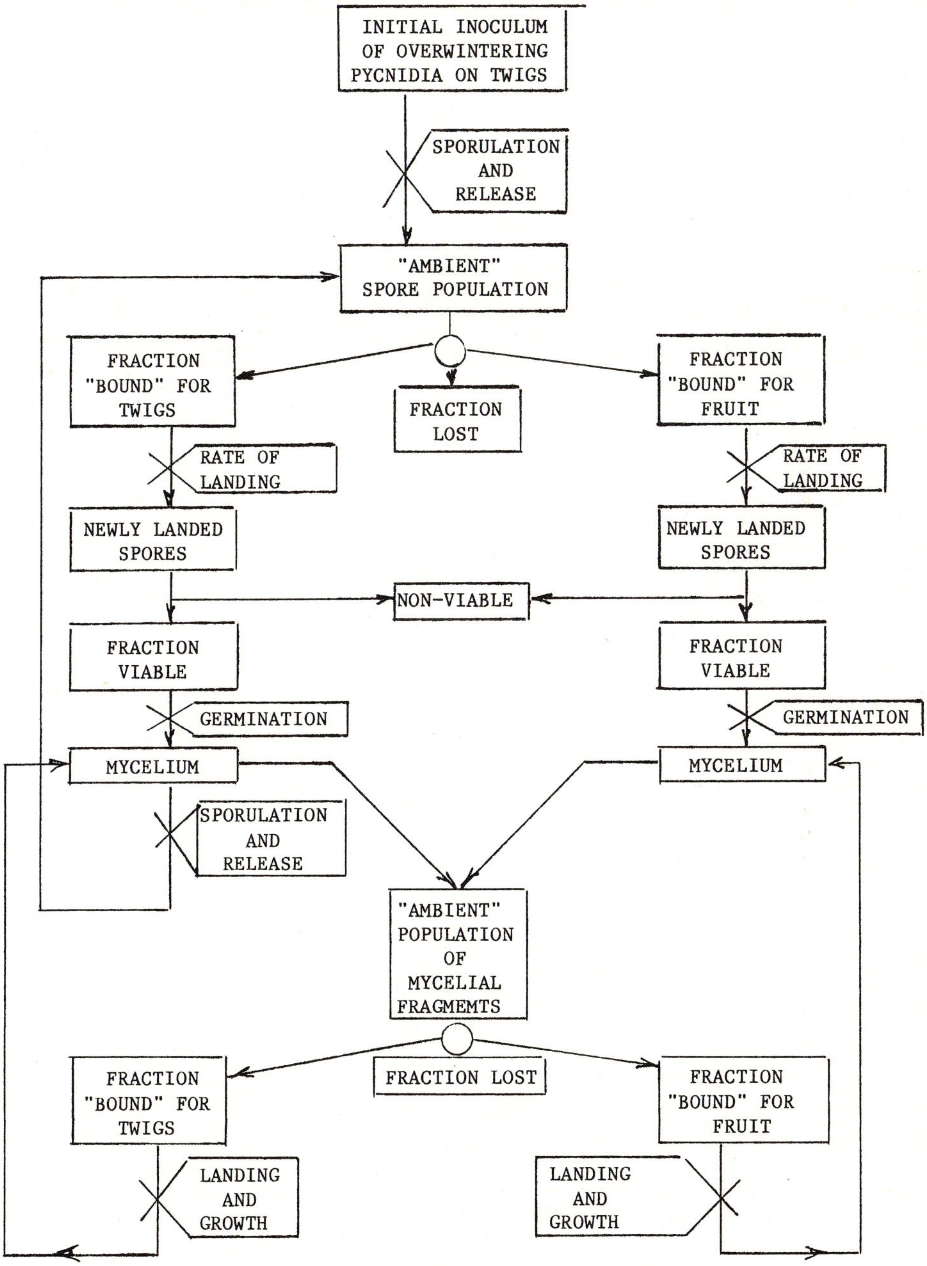

Figure 1
Example of a diagrammatic model which is part of the interfacing cognitive framework in a project on agricultural pest management (from Gold, 1983)

failures on this point alone. Since interdisciplinary research is a co-consulting relationship, the same care should be exercised as if a consultant were being hired.

DIAGNOSIS PHASE

Barriers

Once a basis for collaboration has been established, the collaborators are ready to begin specification of needed inputs. During this process, the mode of interaction is at least tentatively set. Understandings need to be reached concerning the sharing of responsibilites, the establishment of an ongoing mutual learning relationship, and for continued nonthreatening, but in-depth feedback and evaluation. To the extent that problems involving narrow professional indentification, value structures and differences in cognitive framework have not been previously dealt with, they must be addressed at this stage. We have studied several experiences in which the collaboration was allowed to proceed, trusting that these things could be handled along the way. The results were often a great deal of frustration, lost time, and ill will.

Two additional problems warrant discussion. The first is related to subject-matter factors. This was defined as those characteristics of the system under study which limit our ability to interact with, manipulate, and observe the system. The consultant in a given interaction must learn enough about the specifics of how his input will be used, so that he can see how best to fulfill the need. The client must learn enough of the consultant's discipline to enter into the evaluation of the input provided, and to develop an appreciation of the feasibiltiy of providing various types of inputs, and of the effort involved. An often heard complaint is, "He doesn't know what he's asking. He thinks it's an afternoon's work, but it will take at least a month." As the project goes through successive phases, a basis should be established for an on-going mutual learning relation, such as might be provided by formal as well as informal discussions, suggestions for reading, and visits to each other's laboratories or workplaces. Discussions should emphasize but not be limited to, the production, evaluation and use of inputs that participants are providing for each other. This mutual learning process is also important for the maintenance and updating of the interfacing cognitive framework.

Second, at this stage, the question of power sharing may become a major obstacle. Through the mutual learning relation, the supplier is granting the user the right (that is, the power) to evaluate the supplier's product. The user is also granting the supplier the right to enter into the determination of how that product will be used. That is, each one is granting the other the right to evaluate and to influence his/her own work. There are, in addition, the practical questions relating to who makes decisions concerning such things as resource use and technical support.

These problems may be exacerbated if, as the work goes through various stages, new objectives are defined and new inputs specified. In such cases, the roles of user and supplier may become reversed, and new modes of interaction may become necessary.

Recommendation

The problem of anticipating a time stream of activities is, of course, a common one in project management, and involves such commonly used tools as Gantt charts and PERT diagrams. In the interdisciplinary setting, we have found it useful to go one step further, and develop a time-flow chart for each of the project objectives, showing how responsibility is expected to be shared for input specification, production, evaluation, and use. This is illustrated by Figure 2, taken from the pest management project. The particular objective here is the development of a simulation for the pest system to be controlled. The work involves direct collaboration between scientists in biomathematics and either plant pathology or entomology. For tasks lying in the outside columns, lead responsibility (but not sole responsibility) lies with the indicated discipline. For tasks in the middle column, responsibility is shared on a more equal basis. Note that as we indicated in the description of the consultative mode, those tasks which have to do with producing and using inputs lie in the outside columns, while those dealing with specification, evaluation and concept development lie in the middle column. This "ping-ponging" of responsibilities continues through various phases of the work (Figure 3).

Diagrams of this type have helped us to deal in explicit and unembarrassed fashion with questions related to what each expects from the other, and therefore what each needs to learn from the other, as well as with questions of how we influence each other's work and with questions of how resource allocation decisions will be made. The diagrams also help in dealing with the very sensitive but crucially important question of critical evaluation, since the points at which evaluation will take place are made explicit.

SUMMARY AND CONCLUDING REMARKS

We have used the interpersonal transaction model described in Paper I for retrospective analysis of a number of interdisciplinary projects, for evaluation of the appropriateness of proposed collaborations, for guiding graduate student collaborations, and as an explicit guide in the initiation and on-going work of an interdisciplinary project. In the light of this model, several generalizations and recommendations emerge.

During entry phase of the collaboration, it seems to be especially important to begin development of a shared interfacing cognitive framework, which provides a context for further learning processes, and to take explicit account of individual as well as of group goals.

During the diagnosis phase, the stage should be set for the way in which responsibility will be shared for input specification, supply, evaluation, and use. We have developed a diagramatic aid which explicitly addresses the sharing of responsibility and depicts the way in which the mode of interaction may have to shift as the work proceeds.

It should be stressed that no formal structure or protocol will insure successful or creative interdisciplinary collaboration. Fundamentally, the success depends upon the competence and motiva-

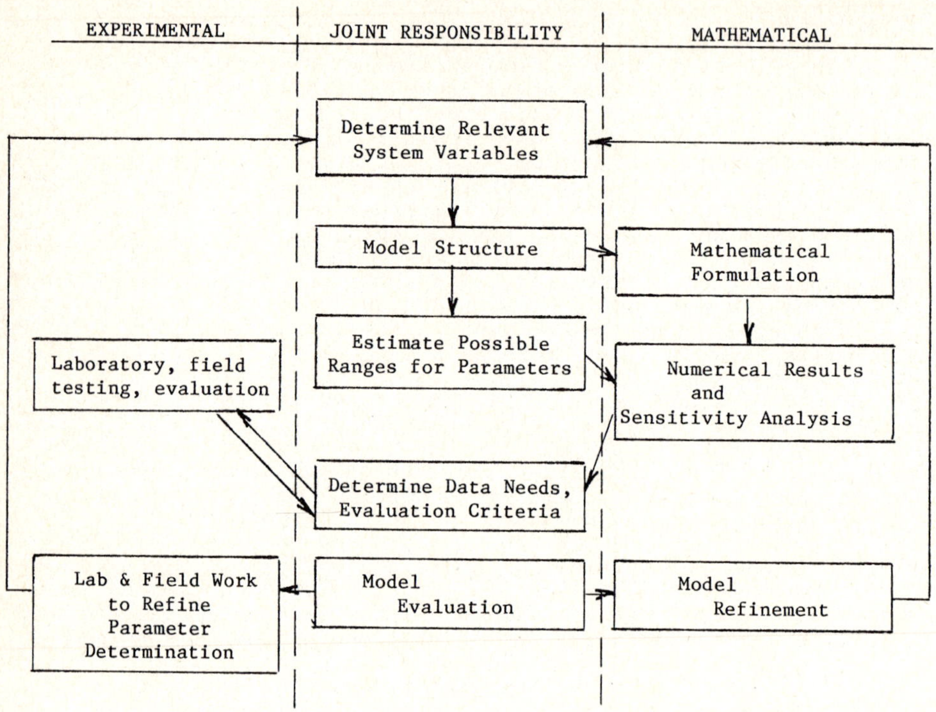

Figure 2
Work flow, indicating where lead responsibility lies for different phases of the work.
Objective: Use of simulation model for pest control

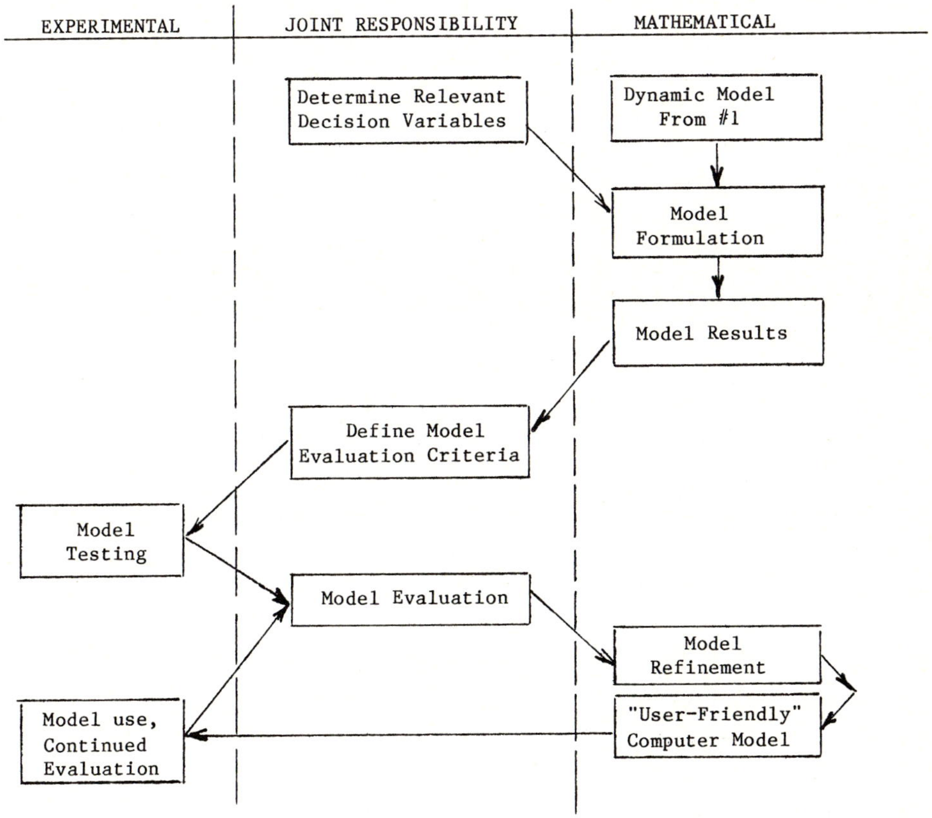

Figure 3
Work flow, indicating where lead responsibility lies for different phases of the work.
Objective: Use of simulation model for pest control

tion of the collaborators. Some who are successful at such Figure 3 collaboration may have the intuition to recognize and deal with the kinds of issues we have been describing, without the need or inclination or even ability to engage in formal analysis of the process. Studies such as ours, and conferences such as this one are designed to make the successes more frequent and more reliable. For most of us, the successes don't just happen--they require work.

FOOTNOTE

1. An interesting discussion (not available when Paper I was written) concerning the role of cognitive framework in the interpretation of observation is provided by DeMey (1980).

REFERENCES

Bell, C.R., and Nadler, L. eds. 1979. The Client-Consultant Handbook. Houston, Texas: Gulf Publishing Company.

DeMey, M. 1981. The Interaction Between Theory and Data in Science. In The Social Process of Scientific Investigation, ed. K.D. Knorr; R. Krohn; and R. Whitley, pp. 3-24. Boston, Mass.: Reidel Publishing Company.

Gold, H.J. 1983. Framework for Modeling the Epidemiology of Sooty Blotch and Flyspeck Diseases of Apple and Their Control. Biomathematics Series No. 16, Institute of Statistics Mimeo Series No. 1635, Raleigh, NC: NCSU.

Gold, H.J. and Gold S.E. 1984. Some Elements of a Model to Improve Productivity of Interdisciplinary Groups. In Managing interdisciplinary research, eds. S.R. Epton, R.L. Payne and A.W. Pearson, pp. 86-101. Chichester, U.K.: John Wiley & Sons.

Hagstrom, W.O. 1966. The Scientific Community. New York: Basic Books.

Holling, C.S., ed. 1978. Adaptive Environmental Assessment and Management. New York: John Wiley & Sons.

Knorr, K.D. 1981. The Scientist as an Analogical Reasoner: A Critique of the Metaphor Theory of Innovation. In The Social Process of Scientific Investigation, eds. K.D. Knorr; R. Krohn; and R. Whitley. pp. 25-52. Boston, Mass.: Reidel Publishing Company.

Kuhn, T.S. 1970. The Structure of Scientific Revolutions. Second Edition, Chicago: University of Chicago Press.

Kuhn, T.S. 1977. The Essential Tension. Chicago: University of Chicago Press, Chapter 9, p. 225, Chapter 12, p. 293.

Lippitt, G. and Lippitt, R. 1978. The Consulting Process in Action. Lajolla, California: University Associates.

Pelz, D.C. and Andrews, F.M. 1966. Scientists in Organizations. New York: John Wiley & Sons.

Walker, B.H.; Norton, G.A.; Conway, G.R.; Comins, H.N.; and Birley, M. 1978. A Procedure for Multidisciplinary Ecosystem Research: With Reference to the South African Savannah Ecosystem Project. J. of Applied Ecology 15:481-502.

MANAGING HIGH TECHNOLOGY
An Interdisciplinary Perspective
B.W. Mar, W.T. Newell, and B.O. Saxberg (Editors)

THE SHARED A PRIORI MODEL AS FRAMEWORK IN EXAMINING INTERDISCIPLINARY PROCESSES

Charles W. N. Thompson

Department of Industrial Engineering
and Management Sciences
Northwestern University
Evanston, IL 60201

To the extent that interdisciplinary processes are necessary or desirable in the identification and solution of problems, such as some or many of those in high technology areas, an understanding of how these processes are similar or different from other problem solving processes will be useful. As a framework for examining interdisciplinary processes, a shared a priori model is presented, and its use illustrated by examination of some of the underlying problems (and potential solutions) introduced by the use of interdisciplinary processes.

INTRODUCTION

From a somewhat simplified point of view, we may consider man as a problem solving animal or decision maker, and his organizations as problem solving organizations. From an evolutionary development (or epistemological) point of view we can consider the progressive discovery that some problems are similar, and some ways of looking at them and solving them work better, are more effective, and more efficient (Campbell, 1975, p. 16). In biological evolution, over long periods, we can see this in sensory processes, in various powerful reflexive processes, in emotions, and, more generally, in those characteristics psychology and physiology find common to most or all of man, and which are generally strong and stable. In non-biological evolution (or cultural evolution), and over a considerably shorter but still a range of time, we can see this in several ways (see Figure 1), but particularly in those processes we describe as disciplines or professions or vocations. Each of these, at least in part, arose in response to a set of problems and solutions which matched one another.

To the extent a problem is well (or correctly) defined and fits within or matches one of these disciplines, it may be argued that using a well developed and tested and successful problem solving process is to be preferred for most if not all purposes. However, over time, conditions change and new problems appear, and we find that at least some of these problems are not well matched to a single discipline, and may require matching to several disciplines in sequence, or simultaneously, or to processes not within any discipline, or it may not be clear which of these applies. Fortunately, these kinds of problems appear again and again, and the disciplines have some overlapping and common processes, and practitioners often overlap different disciplines, and methods for dealing with these nonsingle-discipline problems have been deve-

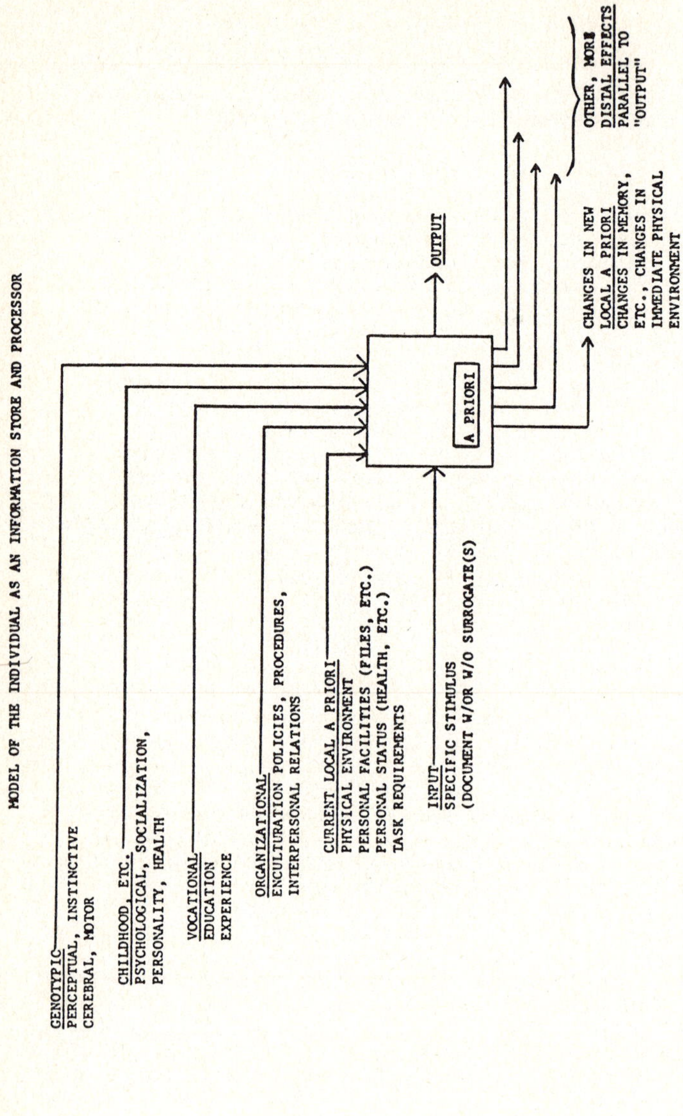

Figure 1

(From Thompson, 1969, p.48)

loped. For example, sequential disciplinary problems appear in the commonly observed research-development-production-marketing process; and simultaneous disciplinary problems are observed in teams, special forms such as matrix organizations, and in various coordinating and interfacing processes, such as planning in project management (Thompson, 1981). Without proposing a rigorous definition, it is suggested that these and other related processes may be considered interdisciplinary.

A note should be made of several processes which may be difficult to categorize as single discipline, interdisciplinary, or otherwise. First, sets of single discipline processes which are "added" together with little interaction; this is sometimes referred to as multidisciplinary (Birnbaum, 1981, p.1280). A second case is the interaction of two disciplines (or, more likely, very small parts of two disciplines) into a new, hybrid discipline (or a sub-discipline of one discipline), e.g., biochemistry. Third, the combining of several disciplines in the person of a single researcher, the "renaissance man" or Aristotle or a "specializing nonspecialist" (Sturgeon, 1963, p.250). Fourth, at least that small part of the systems world which considers itself transdisciplinary (Thompson and Rath, 1973, pp.155-6). None of these quite fit the interdisciplinary sense of two or more individuals from different disciplines working together on a common problem.

To the extent problems (and solutions) associated with high technology and the management of high technology are not found within or matched with a single discipline, it would appear that interdisciplinary processes are of importance (Robinson, 1983, p.1207; Thompson, 1978, p.329). Because problems and solutions, and the disciplines related to them, are not sets which are stable, finite, mutually inclusive and exclusive, and clearly and commonly observed (Thompson, 1975, pp.384-7), it would appear, further, that any specific problem may well include combinations of subproblems (and solutions and disciplines) (Thompson and Rath, 1974, p.251), some of which will match with a single discipline and other calling for interdisciplinary processes. This complexity suggests the potential usefulness of a framework which may be used in examining the matching or lack of matching between or among problems, solutions, disciplines, and the decision maker or makers.

THE SHARED A PRIORI MODEL

The matching of problems and solutions and disciplines and decision makers presents a number of combinations, any one (or all) of which might provide a focus for the discussion to follow. While it might be argued that the discussion applies directly or indirectly to all, for convenience in presentation, the focus will be on the match between two decision makers, i.e., communication, with limited reference to the closely related areas of decision making and/or problem identification and solution.

It is proposed to draw upon a model proposed some years ago by Warren Weaver (1949) which described communication in terms of problems at three levels, the technical (referred to here as syntactic), semantic, and effectiveness (referred to here as pragmatic). Without offering an analysis of subsequent discussions, or attempting to describe possible similarities to other triads, e.g., DMUC,

DMUR, DMUU; deterministic, probabilistic, heuristic; etc., it is proposed that most, if not all, extensions suggest that these different levels are differences in kind, for example, Hyvarinen (1968, pp. 2-6).

The framework model to be described here asserts that the three levels are continuous, and that the apparent differences in kind, or qualitative differences, are solely a convenience in describing parts of the whole which differ only in the number of symbols or sets of symbols involved in the process. This model will be further described in terms of Figure 2 and outlined with examples from the three levels, and some comments on how they relate to one another in the communication process.

For communication to take place between two decision makers, A and B, at the syntactic level the following must occur:

> A and B must have a shared (or common) set of symbols which can be communicated, i.e., sent by A in some physical sense and received by B, for example, (.-,-..., etc.)
>
> A must select less than all of the set, and send it to B, e.g., (--.,---)
>
> B must receive the sent set
>
> B must match it to those in his set of symbols to determine shich part of the set A communicated.

At least in the sense intended by Weaver, the sets of symbols here are finite, denumerable, unambiguous, quantitative, etc.

For communication to take place at the semantic level, the following must occur:

> ALL OF THE ABOVE, PLUS
>
> A and B must have a shared (or common) set of symbols which relate the above symbols to other symbols, or sets of symbols, including, at least, symbols for concepts such as words or sentences, and other symbols for referents such as objects, events, and relationships (and confidence in those relationships) in the real world, e.g., (--.,---) equals "GO" equals "(you) move from a particular place"
>
> A must select the syntactic symbols which match the semantic message he wishes to send
>
> B must match the syntactic symbols received to the semantic symbols to determine which part of the set A communicated.

In practice, the set of symbols here includes at least the set of all referents, some redundancies and ambiguities, and combinations, suggesting at least an order of increase in size over the syntactic level.

For communication to take place at the pragmatic level, the following must occur:

SHARED A PRIORI MODEL

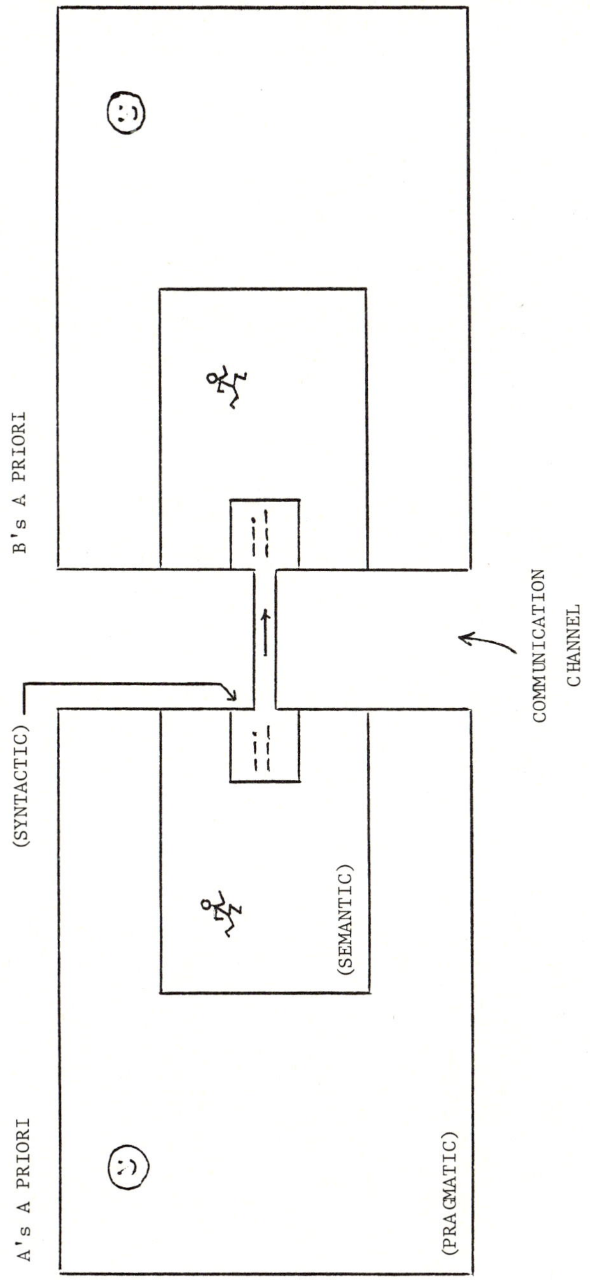

FIGURE 2

ALL OF THE ABOVE, PLUS

A and B must have a shared (or common) set of symbols which relate the above symbols to other symbols, or sets of symbols, for referents which relate to purpose, goals, preferences, value structures, etc., e.g., "GO" equals approaching a reward or avoiding a punishment (a "smile on one's face")

A must select the semantic symbols which match the pragmatic message he wishes to send

B must match the semantic symbols (which were matched to the syntactic symbols received) to the pragmatic symbols to determine the part of the set A communicated.

In practice, at least, the set of symbols here includes at least the set of all symbols relating all of the above to the set of all preferred states, with all of the possible combinations, direct and indirect, certain to uncertain, independent, inconsistent, etc., suggesting another increase in order.

The above descriptions and examples propose that the three levels are continuous, the parts of which are differentiated in terms of the number of symbols involved and which may be (but not necessarily are) distinguished by the kinds of symbols. In addition, it seems appropriate to mention briefly the relations betewen the levels and with some closely related uses.

First, all of the sets of symbols mentioned above in connection with A (or B) are part of A's a priori, in the sense of information available to A at or prior to the time of making some decision, e.g., to send a message to B. It should be clear that each decision maker (A, B, ... N) has an a priori, the sources of which are outlined in Figure 1. It should, further, be pointed out that the a priori may include information (symbols) not only in the brain (memory) of the decision maker, but also in his related knowledge processes, e.g., sensory systems, and in (some) external information stores, e.g., personal files, available to him. A more complete definition is in Thompson (1974, pp.335-6).

Second, the sense of shared or common a priori is something more or less than identity. For example, for communication as intended by A, A's a priori must contain symbols which describe, directly or indirectly, B's symbols, although A may not agree with B's symbols, particularly with respect to confidence (belief structure) and utility (value structure). Conversely, B may receive a communication even though unintended by A if B's a priori contains symbols describing A's a priori.

Third, purposeful activities require all three levels whether cognitive or behavioral (including the behavior called communication).

Fourth, the sense of communication from A to B applies also to the case of B to A, and, to a considerable degree, to information transfer to and from sources or sinks other than other human beings, including for example, the process of identifying a problem and/or solutions.

Fifth, although there appear to be systematic differences between the levels, all parts of the a priori are subject to variations in the degree of certainty and/or availability, analogous to particles rising and sinking in a turbulent stream.

Sixth, various ongoing processes, including newly inputted information, make changes in the a priori at all levels.

Seventh, the several matches between levels in both sending and receiving a prioris and between the received message and the receiving syntactic a priori, are or may be complex decision processes involving, among others, search and retrieval processes, various comparison processes, may operate at levels different from the match level, and may "create" new symbols and sets of symbols (e.g., patterns, chunks, etc.) as intermediate or terminal processes.

PROBLEMS

To the extent one can distinguish interdisciplinary problems, it seems clear that interdisciplinary processes (and/or sets of decision makers or problem solvers) provide a potentially better match between the problem and the solution. It seems also clear, particularly where communication and cooperation between decision makers is required, that much of the advantage in the single discipline process for the single discipline problem has been lost. Instead of a high degree of shared a priori, at least at the syntactic and semantic level, interdisciplinary processes introduce differences which may include both nonintersections and conflicts. Some brief examples will be provided.

At the syntactic level, we may find differences between the cultures i.e., verbal versus mathematical (versus conceptual or symbolic), and within verbal, e.g., English versus Swedish, and within mathematical, etc.

At the semantic level, we may find differences which may be extensions of the above, e.g., vocabularies, but also differences on several bases, including phenomenological, i.e., how different disciplines view the same object; terminological or conceptual, i.e., how different disciplines describe or theorize about the same referent; methodological, i.e., how different disciplines solve the same problem. A compounding problem is that relations between or across phenomena, etc., may be outside the a priori of both disciplines.

Finally, at the pragmatic level we may find differences across the disciplines in several areas, including, in relevant peer sets, e.g., "add to man's knowledge" versus "add to man's wealth;" goal hierarchies, e.g., time horizons, preferences; uncertainties, i.e, cause and effect relationships accepted; authority relations, e.g., organizational, family, psychological, cultural; risk preferences, e.g., low risk/low payoff versus high risk/high payoff.

The above proposes that most if not all of the disadvantages (and advantages) associated with the use of interdisciplinary processes can be usefully described in terms of the match or lack of match among the a prioris of the individuals from the several disciplines (and the match or lack of match between the problem and solution and the a prioris of the set of decision makers).

SOLUTIONS

It seems obvious that interdisciplinary processes should not be used for single discipline problems; and, as a corollary, it would appear that for those multidisciplinary problems which can be decomposed into single discipline problems and then combined through simple interface processes, a similar caution applies (Birnbaum, 1981, p. 1281). For interdisciplinary problems, by definition, the advantages require incurring the disadvantages, and it is proposed that solutions can be found in describing and understanding the disadvantages, and, in some cases, minimizing or compensating for their effect.

Describing and understanding interdisciplinary processes is a large and ongoing research activity. For the purposes here, it is proposed that it would be useful to use the model as a reference framework.

At the syntactic level, much work to improve communications at this simple level has been undertaken. Obvious areas for improvement include understanding differences in the symbolic languages of the several disciplines, and the development of improved common languages, ease in translation, and perhaps improvements in indexing and retrieval.

The semantic level presents greater problems and greater opportunities. With respect to phenomena, it may be possible to develop methods to ensure that individuals from different disciplines or points of view can see the same object or event and, in the process, see the other's point of view (Lewin, 1984; p. 8). With respect to concepts and terminology, in addition to translation and transforms, it may be possible to develop common models or framework models which provide a stable and neutral middle ground for several disciplines (Thompson, 1974). With respect to methodologies, reference to underlying epistemological processes and an understanding of the different requirements and uses of different methodologies may, at least, increase the tolerance for diversity in methodology if not also lead to an awareness of the necessity for matching the method to the problem (Thompson and Rath, 1974, p. 243).

At the pragmatic level, the problems may be either very difficult of very easy. Some parts of this a priori are shared very widely, e.g., almost all of us react the same way to a poke from a sharp object; others are not only not shared, they may be undecipherable. At least for those value structures which can be identified and which may be widely shared, e.g., wanting to get paid or wanting to be identified with or associated with a successful project, it may be possible to introduce a powerful reward-and-punishment structure which will encourage and/or force the several different disciplines to work hard to overcome the differences in a priori (National Institute of Neurological and Communicative Disorders and Stroke, and the National Institute of Mental Health, 1984; Ulsamer, 1984). This is the flagpole or lifeboat model of group cooperation, providing a common fate for the group.

SUMMARY

It has been proposed that a useful approach for examining the advantages and disadvantages of interdisciplinary processes is a shared a priori model, that it provides a common framework for examining different kinds of problems in terms of the same concept of matching different sets of symbols. Examples and references are provided to illustrate the concept and its use.

REFERENCES

Birnbaum, P.H. 1981. Contingencies for interdisciplinary research: matching research questions with research organizations. Management Science 27(11):1279-1293.

Campbell, D.T. 1975. On the conflicts between biological and social evolution and between psychology and moral tradition. American Psychologist 30(12):1103-1126.

Hyvarinen, L.P. 1968. Information theory for systems engineers. Berlin: Springer-Verlag.

Lewin, R. 1984. Ancestors worshipped: paleoanthropologists have been discussing their agreements and disagreements in the presence of most of the world's hominid fossils. Science 224(4648):477-479.

National Institute of Neurological and Communicative Disorders and Stroke, and the National Institute of Mental Health. 1984. Multidisciplinary research center(s) for the study of the neurological basis of disorders of language, behavior and learning during infancy and early childhood, (announcement). NIH Guide for Grants and Contracts, 13(6):13-14.

Robinson, A.L. 1983. Theory center awaits NSF word on renewal. Science 219(4589):1207-1208.

Sturgeon, T. 1963. Never underestimate. In 17 X Infinity, ed. Groff Conklin, pp. 244-262. New York: Dell Publishing Co., Inc.

Thompson, C.W.N. 1974a. Administrative experiments: the experience of fifty-eight engineers and engineering managers. IEEE Transactions on Engineering Management EM-21(2):42-50.

Thompson, C.W.N. 1974b. Model of human performance reliability in health care systems. Annals of Assurance Sciences 7(2):335-339.

Thompson, C.W.N. 1969. The initial screening of technical documents by the user. A dissertation, Northwestern University.

Thompson, C.W.N. 1975. Technology Utilization. An Annual review of information science and technology, ed. Carlos Cuadra, pp. 383-417. Washington, D.C.: American Society for Information Science.

Thompson, C.W.N. 1978. Procuring systems: experiences from civilian and defense agencies. In Proceedings of the seventh annual DOD Procurement/Acquisition Research Symposium, Hershey, Pennsylvania, May 31-June 2, 1978, pp. 196-200.

Thompson, C.W.N. 1981. A dimensional approach to uncertainty in the systems acquisition process. In Proceedings of Management of Risk and Uncertainty in the Acquisition of Major Programs, eds. M.D. Martin, A.J. Rowe, and H.A. Sherman, pp. 152-161. Colorado Springs, Colorado: United States Air Force Academy.

Thompson, C.W.N., and Rath, G.J. 1973. Planning solutions aided by management and systems technology. The Annals of the American Academy of Political and Social Science 405:151-162.

Thompson, C.W.N., and Rath, G.J. 1974. The administrative experiment: a special case of field testing or evaluation. Human Factors 16(3):238-252.

Trivers, R.L., and Hare, H. 1976. Haplodiploidy and the evolution of the social insects. Science 191(4224):249-263.

Ulsamer, E. 1984. In Focus...SDI settles into stride. Air Force Magazine 67(5):27-31.

Weaver, W. 1949. Recent contributions to the mathematical theory of communication. In The mathematical theory of communication, C.E. Shannon and W. Weaver, pp. 93-117. Urbana: The University of Illinois Press.

MANAGING HIGH TECHNOLOGY
An Interdisciplinary Perspective
B.W. Mar, W.T. Newell, and B.O. Saxberg (Editors)

INDIVIDUAL PSYCHOLOGY AND INTERDISCIPLINARY RESEARCH

Ivan T. Robertson

University of Manchester,
Institute of Science and Technology,
P.O. Box 88, Manchester M60 1QD, ENGLAND

One of the main targets of any interdisciplinary research program is to combine and develop the sometimes diverse knowledge and expertise of research team members. Thus an area of central interest for interdisciplinary research management involves the psychological problems and opportunities presented when individual researchers attempt to collaborate. For example Mar (1984, p. 6) has noted that "A critical issue in the management of high technology or other innovative efforts is the need to understand and control the (psychological) conflicts between the individual's relationship to the home unit, their current assignment, and their personal goals." Other opportunities for either conflict or cooperation arise in terms of the individuals' relationships with each other. The issues involved here form the focus for this paper.

The paper is divided into two parts. First, there is a brief discussion of the psychology of group processes followed by consideration of the factors that might improve creativity within interdisciplinary research teams. The second part of the paper concentrates on individual differences in thinking style, the relevance of such work to research and problem solving activities, and finally a consideration of the implications that this work has for interdisciplinary research.

Interdisciplinary research conducted within organizational settings can be examined at various levels of analysis. In terms of increasing specificity, four significant levels are: the organization as a whole, the subgroup/department, the working team or group and finally the individual. This categorization corresponds more or less with that proposed by Payne and Pugh (1971) and various other organization theorists (e.g. Herbert, 1981). Researchers in the interdisciplinary research field have shown considerable interest in the two most general levels of analysis, i.e. the organization and the work group (e.g., Mar, Newell and Saxberg, 1983; Mar, 1984).

This paper in keeping with previous work (Robertson, 1981) focuses on lower levels of analysis (ie. groups and individuals).

GROUPS

The processes of group dynamics and development and their relevance within the wider organizational context have been studied for some time (see for example Schein 1980; Duncan, 1981; Smith et al., 1982; Robertson and Cooper, 1983). This work has produced a number of guidelines that may help to develop an effective team. A fundamental condition for group efficiency is to produce and maintain a cohesive (though not necessarily conflict-free) group. Achievement of this aim is related to a number of factors. For example, if a group is difficult (or at least perceived as difficult) to join, members are likely to value their membership more and this has implications for their level of motivation and effort. Groups threatened from the outside or competition with other groups tend to be more closely knit and cohesive. As far as interdisciplinary research teams are concerned a factor related to group cohesion that seems particularly apt is the extent to which members need to cooperate with each other. In groups where the need for cooperation is high, cohesion is better. By explicit recognition of the need for cooperation and the complementary nature of different individuals' expertise, interdisciplinary research teams and their managers may be in a position to influence group cohesion and thus, in the longer term, effectiveness.

Another concept worth mentioning is the distinction between a person's functional role in a group and his team role. Within an interdisciplinary group an individual's functional role is likely to be related to his home discipline, e.g., they are included in the group because they are a physicist. An appropriate mix of functional roles is an important ingredient for success; but team roles are important too. A person's team role relates to the role played in relation to the interpersonal dynamics within the group. Several researchers have demonstrated the importance of such roles for team success. Belbin (1981) for example has proposed a set of specific team roles (Figure 1) and shown that a team within a balanced set of roles is more successful (on certain criteria) than a team with a less balanced set.

ROLE	OBSERVED CONTRIBUTIONS
SHAPER	1. Shaping roles, boundaries, responsibilities, tasks and objectives/ 2. Finding or seeking to find pattern in group discussion. 3. Pushing the group towards agreement on policy and action and towards making decisions.
MONITOR/ EVALUATER	1. Analyzing problems and situations. 2. Interpreting complex written material and clarifying obscurities. 3. Assessing the judgments and contributions of others.
OTHER ROLES CHAIRMAN	
PLANT	
COMPANY WORKER	
TEAM WORKER	
RESOURCE INVESTIGATOR	
COMPLETER	

Figure 1
Bebin's (1981) team roles

Creativity and problem-solving

Successful research of any kind involves producing original, unusual or creative solutions to problems. Without wishing to get too entangled in a discussion of what creativity is, I should like to draw attention to a definition proposed by Ackoff and Vergara (1981), which emphasises the effect that self-imposed constraints have on limiting creativity.

Various techniques and procedures such as brainstorming, morphological analysis, goal orientation etc. (see Rikards, 1974; Prince, 1970) can be used to help modify people's self-imposed constraints. As an illustration of the power of self-imposed constraints, consider the problems given in Figure 2.

Puzzle A

Join the dots with four straight lines, without removing your pen from the paper.

Puzzle B

AB=?

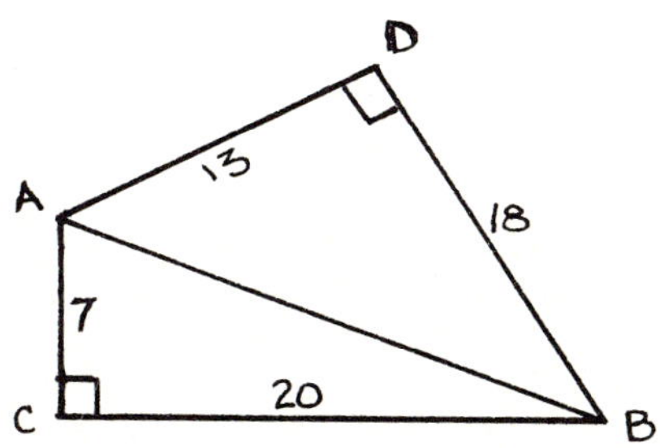

Figure 2
Some Puzzles to Solve

One extremely successful way of minimizing the influence of self-imposed constraints is to involve nonexperts in the problem solving process at appropriate stages. Nonexperts are not limited by the kinds of mental sets that limit the thinking of experts within a discipline. (Often routine ways of perceiving problems are so deeply internalized that we do not recognize that there is any alternative.) Although there some material in the literature to demonstrate the value of nonexperts, my personal conviction of this comes from experience of running a series of creative problem solving courses over the last three years for mixed discipline groups of engineers and scientists.

Within an interdisciplinary team everyone has access to nonexperts. In a sense, using nonexperts represents something of a reversal of our traditional ways of looking at research teams, i.e., members are valuable because of what they do not know rather than what they do know.

INDIVIDUAL ASPECTS

Individuals joining an interdisciplinary research team bring along with them their own skills and personal characteristics and like their finger prints, no two people's psychological make-up will be identical. The individual differences likely to distinguish one person from another can be summarised in three categories:

1. ability/aptitude/technical expertise

2. personality and attitude

3. thinking and information processing styles

There are clearly differences between people from different disciplines in terms of their abilities/expertise and in their personality (see Cattell, Eber and Tabuoka, 1970 for some examples). The remainder of this paper, however, focuses on the area of thinking (cognitive) style.

Cognitive style has been defined as, "Differences in information processing which are stable across a variety of situations and are reflected in qualitative differences in the processing, storing and recalling of information" (Robertson, in press).

Cognitive Style

The literature on cognitive style is fairly large and a variety of cognitive styles have been proposed and investigated, indeed Messick (1976) provides a glossary of nineteen different terms.

Perhaps the best known and most widely used cognitive style dimensions has resulted from the work of Herman Witkin and his colleagues (Witkin, 1976-1977).

Witkin's research began with an interest in perceptual factors but developed into an attempt to conceptualise and measure an aspect of cognitive style known as field dependence-independence.

The difference between the field dependent (undifferentiated) and independent (differentiated) style lies in the extent to which the individual structures and analyzes information. The field independent style involves a more global, less analytical processing of stimulus material. The styles can be identified by use of Embedded Figure Test, where the respondent is required to identify a specific figure embedded in a more complex figure. Field independent people have less difficulty in finding the embedded figure whereas field dependent people have great difficulty even with simple items.

Other cognitive style dimensions identified include (a) Harvey, Hunt and Schroder's (1961) conceptual systems approach and the later related work of Driver and Struefert (1969) and Driver and Mock (1975) which argues that "decision making differs among people in two key dimensions: ... complexity or amount of information used and ... focus or number of alternatives generated from the information" (Driver and Rowe, 1979, p. 151). (b) Pask's work on holist and serialist strategy (Pask 1976; Pask and Scott 1972). Briefly a serialist strategy is exemplified by a person who progresses in linear fashion through learning or problem solving material, taking a step by step approach and increasing understanding in small increments. Holist strategy involves a more global approach. Pask and his colleagues have observed holist and serialist strategies in a wide range of people and situations, although in more recent work they have developed the concepts of operation and comprehension theory. The basic distinction between a detailed analytic approach and a more global holistic approach to information processing is maintained. Although Pask's work has had significant impact within educational psychology, little use seems to have been made of his ideas to date in other applied fields. (c) Convergent and divergent thinking. As Guildford (1967) has pointed out, many tests of human information processing ability (e.g. IQ tests) call for a convergent mode of thinking; i.e., the components of each question point to one single, correct solution. Many problems of course, particularly within real life settings, do not have one single, unambiguously correct solution and often a more divergent mode of thinking is needed where the problem solver is prepared to free-wheel and adopt a broader more intuitive perspective for the problem under consideration.

In general the available evidence does seem to support the validity of divergent thinking as a distinct mode of thinking. There is however considerable uncertainty concerning the extent to which divergent thinking, intellectual ability (intelligence) and other human factors can be separated.

It is also worth noting that as the study of cognitive style expands, the need for examining patterns of relationships separately for each sex may increase. The results of Hyde et al. (1975) and Thompson, Mann, and Harris (1981) point to a unitary cognitive style for men but not for women.

A-Synthesis

Despite the difficulties involved in the conceptualization of cognitive style, it does seem, that there are striking similarities between many of the existing STYLES.

Entwistle (1981) has reviewed much of the work on cognitive style and pointed to the similarities between various dimensions.

> Holists tend to be divergent thinkers, which might imply that they use wider conceptual categories and are more ready to accept thematic as opposed to analystic links between concepts. It might also imply that they are likely to be impulsive, emotionally inhibited but not anxious, and to opt for courses in languages or humanities. Serialists are more convergent in their thought processes. They are thus likely to be cautious, conser-

vative, analytical, and tend to reject possible connections by using narrow categorizations. Emotionally they might be expected to be inhibited and their area of study likely to be mathematical or scientific". (Entwistle, 1981, p. 217).

Entwistle then goes on to produce, essentially, a two category grouping which places descriptively similar strategies under two headings, style A and style B (see Table 1). Empirical evidence supporting this is limited, despite its intuitive appeal.

Table 1
Styles A and B
(Adapted from Entwistle, 1981)*

Style A	Author*	Style B
(Right hemisphere)		(Left hemisphere)
Holistic	Pask (1976)	Serial
Field dependent	Witkin (1977)	Field independent
Divergent	Hudson (1966)	Convergent
Impulsive	Kagan (1966)	Reflective
Cognitive complex	Bieri (1966)	Cognitively simple

Left brain - right brain

In addition to pointing to various similarities between different cognitive strategies Entwistle goes on to suggest a possible link between styles A and B and the neurological activity of the human brain.

Although the brain is usually thought of as a single structure, it is actually divided into two halves, joined by a bundle of connecting fibers - the corpus callosum. The left side of the body is controlled mainly by the right side of the brain, and the right side of the body by the left side of the brain (for most people). Despite the fact that both hemispheres of the brain share the potential for many functions, in the normal person the two hemispheres do tend to specialize to some extent. Ornstein (1977) expresses it thus:

> Although each hemisphere shares the potential for many functions and both sides participate in most activities, in the normal person the two hemispheres tend to specialize. The left hemisphere (in those people--the majority--in which it is dominant) ... is predominantly involved with analytic, logical thinking, especially in verbal and mathematical functions. Its mode of operation is primarily linear. The hemisphere seems to process

> information sequentially. Since logic depends on sequence and order, this mode of operation must of necessity underlie logical thought ...
>
> If the left hemisphere is specialized for analysis the right hemisphere ... seems specialized for synthesis. Its language ability is quite limited. This hemisphere is primarily responsible for orientation in space, artistic endeavour, crafts, body image, recognition of faces. It processes information more diffusely than does the left hemisphere, and its responsibilities demand a ready integration of many inputs at once (quoted in Entwistle, 1981, p. 218-9).

In turn Entwistle suggests that cognitive style A may relate to neurological activity in the right hemisphere and style B to activity in the left hemisphere. This parallel between certain modes of thinking and brain activity has been suggested by other authors (e.g. Mintzberg 1976, Robey and Taggart 1981) although as Entwistle acknowledges, there is little direct evidence that cerebral dominance can be used to interpret cognitive strategy and style differences.

Despite the apparent similarities between various measures of cognitive style/strategy and the possible connection between cognitive styles and neurological functioning, it is worth emphasizing Entwistle's point that the evidence to support such a tidy view of the issues involved is not extensive. It is clear for example that the two hemispheres of the brain do tend to specialize: left hemisphere language, right hemisphere visual/spatial (e.g. Sperry, 1968; Gazzaniga, 1970). It is far from clear however that cognitive strategy/style factors can be related to brain functioning in any straightforward way (e.g. Bogen et al., 1972; Ornstein and Galin, 1974) although some authors argue strongly that there are links (e.g. Robey and Taggart, 1981). There is also little evidence available to show how the various cognitive style/strategy factors relate to each other (see Berlin and Languis, 1981; Springer and Deutsch 1981). Research in our own laboratory (Gardiner, 1983) investigated links between cerebral dominance, cognitive style, discipline and occupational choices. No evidence to support any clear links was produced.

Thus we are left with a popular and influential, but as yet unsubstantiated theory suggesting that there might be significant differences in terms of thinking style between people from different disciplines based on differences in neurological functions--with artists at one extreme, physical scientists at the other and the remainder ranged out in between.

Implications

Before considering implications in any detail it makes sense to examine the basic elements of the processes that are likely to be involved when members of an interdisciplinary team work together on a research problem. Drawing on the work of a variety of authors (e.g. Gordon, 1961; Prince, 1970) it can be argued that the basic building block for creative problem solving involves alternate phases of divergent and convergent thinking (or style B and style A thinking). The divergent phases involve things such as removing constraints, redefining the problem; generating a wide set of ideas, ideally all taking place under conditions where judgment of feasibility/practicality/relevance, etc., ... is suspended). The convergent phases involve (style B) thinking where evaluation, logic, reality, feasibility etc. are to the fore. For successful problem solving, convergent and divergent thinking activities need to be combined.

Within the context of interdisciplinary research the convergent phases of the process draw on the subject-matter expertise, knowledge and experience from the participants' home discipline and specialist knowledge is very useful.

By contrast, during the divergent phases being a nonspecialist may be particularly helpful in overcoming (perhaps simply by being unaware of them) self-imposed constraints.

A further possible implication is that during any interdisciplinary research endeavor it is important for the research team members to have some awareness of the current phase of problem solving (i.e., divergent or convergent) and thus the type of contribution that is appropriate.

The ideas discussed earlier in this paper point to two important potential features of interdisciplinary research, both of which imply that interdisciplinary research is significantly different from single discipline research and furthermore suggest that it may produce more creative thinking on the part of participants. The first feature involves recognition of the fact that a person from another discipline does not only possess a different set of knowledge and expertise but may also be in a better position to have creative ideas about one's own discipline by removing self-imposed constraints. Secondly someone from another discipline may place more emphasis on different and perhaps valuable modes of thinking and information processing. Careful planning and management of these special features may produce considerable benefits within interdisciplinary research settings.

ACKNOWLEDGEMENTS

I am grateful to Reg Talbot for many fruitful discussions of the topics raised in this paper and for introducing me to the puzzles (Figure 2).

REFERENCES

Ackoff, R.L. and Vergara, E. 1981. Creativity in problem solving and planning: a review. European Journal of Operational Research 7:1-13.

Belbin R.M. 1981. Management teams why they succeed or fail. London: Heinemann.

Berlin D.F. and Languis. 1981. Hemispheric correlates of the rod and frame test. Perceptual and Monitor Skills 52:35-41.

Bogen J.E., Dezure, M., Tenhouten, W.D., and Marsh, J.F. 1972. The other side of the brain. Bulletin of the Los Angeles Neurological Society 3F:49-61.

Cattell R.B., Eber, H.W., and Tatsuoka. 1970. Handbook for the sixteen Personality Factor Questionnaire. Windsor: N.F.E.R. Publishing Company.

Driver, M. and Mock T. 1975. Informaton processing decison style theory and accounting information systems. Accounting Review 50:490-508.

Driver, M. and Struefert, S. 1969. Integrative complexity. Administrative Science Quarterly 14:272-285.

Duncan, W.J. 1978. Organizational behaviour. New York: Houghton-Mifflin Company.

Entwistle, N.J. 1981. Styles of learning and teaching. Chichester: Wiley.

Gardiner I.A. 1983. Left-brain-right-brain, thinking and behaviour: an empirical test of a popular theory. Unpublished dissertation, U.M.I.S.T., Department of Management Sciences.

Gazzaniga, MS.. 1970. The bisected brain. New York: Appleton-Century-Crofts.

Gold, H.J. and Gold, S.E. 1984. Implementation of a model to improve productivity of interdisiplinary groups. In this proceedings.

Gordon W.J.J. 1961. Synectics: the development of creative capacity. New York: Harper and Row.

Guilford, J.P. 1967. The nature of human intelligence. New York: McGraw-Hill.

Harvey, O.J., Hunt, D., Schroeder, H. 1961. Conceptual systems and personality organization. New York: Wiley.

Herbert, T.H. 1981. Dimensions of organizational behavior, 2nd ed. New York: Collier-macMillan.

Hess, S. 1984. Rational ranking of development candidates. In this proceedings.

Hyde, J.S., Geiringer, E.R., and Yen, W.M. 1975. On the empirical relation between spatial ability and sex differences in other aspects of cognitive performance. Multivariate Behavioural Research 10:289-310.

Lewis, B.N. 1976. Avoidance of aptitude - treatment trivialties, In S. Messick, ed., Individuality in learning. San Francisco: Jossey-Bass.

Mar, B.W. 1984. Summary of related literature. Interstudy Bulletin 5:5-9.

Mar, B.W., Newell, W.T., and Saxberg, B.O. 1983. Matrix organizations: the key to effective interdisciplinary/interdepartmental management? Interstudy Bulletin 4.

Messick, S. and Associates. 1976. Individuality in learning. San Francisco: Jossey-Bass.

Mintzberg, H. 1970. Planning on the left-side and managing on the right. Harvard Business Review July-Aug. 54:49-58.

Ornstein, R.E. 1977. The Psychology of Consciousness, 2nd ed. New York: Harcourt, Brace, Jovanovich.

Ornstein, R.E. and Galin, D. 1976. Psychological studies of consciousness. P. Lee, R.E. Ornstein, D. Galin, A. Deikman and C. Tart, eds. Symposium on Consciousness. New York: The Viking Press.

Pask, G. 1970. Styles and strategies of learning. British Journal of Educational Psychology 46:128-148.

Pask, G. and Scott, B.C.E. 1972. Learning strategies and individual competence. International Journal of Man-Machine Studies 4:217-253.

Payne, R.L. and Pugh, D. 1971. Organizations as psychological environments. In P.B. Warr, ed., Psychology at Work. Harmondsworth: Penguin, 1971.

Prince, G. 1970. The Practice of creativity. New York: Harper and Row.

Reeves, E.I. 1984. Spacecraft design and development at TRW: matrix management applied. In this proceedings.

Rickards, T. 1974. Problem solving through creative analysis. Ann Arbor, MI: Gower Press.

Robey, D. and Taggart, W. 1981. Measuring managers' minds: the assessment of style in human information processing. Academy of Management Review 6:375-383.

Robertson, I.T. In press. Individual differences in information processing strategy and style. Behaviour and Information Technology.

Robertson I.T., and Cooper, C.L. 1983. Human behaviour in organizations. Plymouth: MacDonald and Evans.

Robertson, I.T. 1981. Some factors associated with successful interdisciplinary research. Journal of the Society of Research Administrators 13:44-50.

Schein, E.H. 1980. Organizational psychology, 3rd ed. Englewood Cliffs, New Jersey: Prentice-Hall.

Smith, M. Beck, J.E., Cooper, C.L. Cox, C.J., Ottoway, R.N., Talbot, R.J. 1982. Introducing organizational behaviour. London: MacMillan.

Sperry, R.W. 1968. Hemispheric deconnection and unity in conscious awareness. American Psychologist 23:723-733.

Springer, S.P. and Deutsch G. 1981. Left brain, right brain. San Francisco: W.H. Freeman.

Thompson, E.G., Mann, I.T., and Harris, L.J. 1981. Cognitive complexity, sex and spatial task performance. British Journal of Psychology 72:249-256.

Witkin, H.A. 1976. Cognitive style in academic performance and in teacher-student relations. In S. Messick, ed., Individuality in learning. San Francisco: Jossey Bass.

Witkin, H.A., Moore, C.A., Goodenough, D.R., and Cox, P.W. 1977. Field-dependent and field-independent cognitive styles and their educational implications. Review of Educational Research 1-64

Section III :

ALTERNATIVE MANAGEMENT ACTIONS

MANAGING HIGH TECHNOLOGY
An Interdisciplinary Perspective
B.W. Mar, W.T. Newell, and B.O. Saxberg (Editors)

MANAGEMENT ALTERNATIVES

Editorial Perspective

Classic solutions to management problems have been to attempt to redirect the behavior of the parties, to alter the perceptions and values of the parties by education and improved communication, to reallocate resources to alter the stature or power of the parties, or to realign the structure of the organization to form units with more common behavior, perceptions and ability to use resources. These solutions have been suggested as needed improvements to improve interdisciplinary processes in the academic setting. The need to improve communications between individuals from different disciplines, the need for organizations that protect the interests of interdisciplinary units, the need to provide administrative support for interdisciplinary units, and the need for common office space are typical recommendations. High technology efforts appear to face similar types of problems and the potential solutions are similar.

MANAGING HIGH TECHNOLOGY
An Interdisciplinary Perspective
B.W. Mar, W.T. Newell, and B.O. Saxberg (Editors)

THE IMPORTANCE OF LONG RANGE PLANNING IN RESEARCH AND DEVELOPMENT

B.G. Tweedy, Ph.D. and Steve Banegas

CIBA-GEIGY Corporation
Box 18300, Greensboro, North Carolina 27419

Long-range planning has been with us for many, many years; one of the first examples is the construction of the Great Wall in China. The Great Wall took over 1,000 years to complete. It was apparently a successful plan when the possible uses, the length in terms of miles and numbers of years to complete the structure are considered.

Long-range planning has different meanings to people. To some, planning and forecasting are synonymous terms, and to others planning and strategy are synonymous, but planning is really different from either of these meanings. Simply expressed, planning is taking a hard look at the organization, the environment, the opportunities and the potential risks, and then developing a set of actions to confront these future opportunities. Good planners recognize that no one can accurately forecast the future and that luck favors the prepared mind. Good planning establishes an overall direction that is clearly understood and accepted by the appropriate responsible personnel. It is not an elaborate document to be filed away, but rather a commitment made by the personnel within an organization or group to work toward accomplishing a common goal or mission. This commitment is arrived at through give and take interactions by individuals and groups addressing their future. Long-range planning is important to academic, governmental, and industrial organizations. The primary objective of planning is to determine for an organization a specific course or courses of action uniquely appropriate to: (1) the opportunities available in the current and future environment, (2) the risks associated with each of these opportunities, (3) the resources available to the organization, and (4) the personal values of the organization's leadership as well as their preference for risk taking.

Long-range planning requires careful thought, experience, creativity and a great deal of communication. Planners need to be somewhat optimistic in order to live with the inevitability of change particularly when these changes cannot be foreseen. As stated earlier, change is inevitable and when a change occurs, organizations with good planning are in a much better position to simply modify their plans and meet these changes successfully. When we develop long-range plans, we frequently give much consideration to the past, but this is not always to our benefit. Edward Burke once wrote that "you can never plan the future by the past." These words are certainly true, but successful planners tend to draw heavily upon the past as well as project the future.

Basically, planning approaches can be classified into three groups: the proactivists, the reactivists, and the inactivists. The proactivists contemplate future challenges and develop a plan to meet these challenges. They build their own bandwagons and others jump onto them. The reactivists wait for the challenges to represent themselves and develop a plan to meet it. They watch the bandwagon being built by someone else and then jump onto it. The inactivists continue with business as usual and just deal with things on a day to day basis. Someone builds a bandwagon and it passes by them, but they simply watch it go by.

Strategy is one very important ingredient of any good plan. George Allen, of the Washington Redskins, is one individual who most people consider had a very well developed plan for building a winning football team. When he went to the Washington Redskins, he looked over his talent and decided that he would need some new talent if he was to have a winning football team. He considered many alternatives and decided that his strategy would be to trade the draft selections for veteran players. By doing this, he was looking to each Sunday or each season; he wanted quick results with little regard for the future. Tom Landry, of the Dallas Cowboys, on the other hand also had a very well developed plan, and he supposedly went through the planning stage in a very similar manner; but his strategy was quite different than that of Allen's. He decided to scout new talent at the college level, draft new players, and train them by bringing them up through the Dallas system. Both strategies have been very successful. Both coaches have had winning teams and are well known throughout the football arenas. George Allen planned for the short-term, and eventually time caught up with him. And Tom Landry on the other hand, planned for the longer-term. There are many similar examples that can be used today for universities and for industries for building various programs, many of which are successful and many of which are not successful.

Long-range planning helps to keep the management of an organization the master of its fate, rather than groups going in so many different directions and being driven by special interest groups, competition, and other factors. Long-range planning helps an organization deal with its future decisions that are made or are chosen not to be made. The most difficult questions that face the executives today are concerned with developing strategies for 1990 and beyond. Some of these are: (1) what should be done to get ready for the unknowns of the 1990's, (2) to what extent, if any, are todays decisions impacting on our freedom to act in 1990, and (3) are the decisions that we do not choose to make today circumventing our freedom to act in 1990. These are very pertinent questions to all executives or large organizations whether academic, governmental, or private.

Long-range planning helps us to feel more comfortable in taking risks. Risk is rewarded by an appropriate return. Long-range planning is undertaken to make sure that the risks that an organization is taking are the right kind of risks, and that lead to end results which are worthy of taking these risks. In long-range planning, we are seeking to learn enough about the future consequences of today's risk-taking to be reasonably satisifed that we are doing the right thing today. Peter Drucker stated that "long-range planning is the process of making current risk-taking decisions with the best possible knowledge of their future consequences."

Planning is important in any organization in order to successfully complete the objectives in a timely manner. When I changed from academic world to the industrial world, I learned that one big difference is that academics are more apt to be late rather than wrong, and industry wants a timely action regardless of whether the action taken was the right or wrong one. However, some universities have well developed plans that have enabled them to be highly successful. A vice president of agriculture of a large university drew up a plan to develop expertise for his university for a bio-technology program. This plan called for a certain time frame at a defined level of cost. The state legislature appropriated funds consistent with his plan, and now they have done as George Allen has done and have acquired the necessary talent for an excellent research program.

In order to develop a good plan, one must have a clear set of objectives. In CIBA-GEIGY, we develop long-range objectives which are periodically reviewed. We have developed action plans for the short, intermediate, and long-range to meet these objectives. Short-range is considered to be one to two years, intermediate is considered to be three to five years, and long-range is considered to be ten to twenty years. For example, we have developed part of a plan for the year 2000. We have received inputs from all over the world to project what is going to be the agricultural situation for the larger crops in the year 1990 and the year 2000. We've looked at world trends in terms of economics in the developed and developing countries.

We have evaluated the productivity of many countries in terms of their food and fiber production. The better the data base, the better we can define our opportunities and take advantage of them. For example, ten years ago it was projected that the developing countries would be growing at a more rapid rate economically speaking than some of the more developed countries. They would require increasing use of our products compared to the past several decades, and now we are finding that this is actually true. To meet these challenges, our parent company in Switzerland developed a strategy to take the greatest advantage of this opportunity. We have many different departments and many different disciplines. In order to accomplish our goals and objectives, it is imperative that each one of us complete a portion of the major plan and finish it in a timely manner. It is important for us to keep informed about developments in other disciplines which impact our work, and to keep our colleagues informed on developments in our area of expertise. Thus, communications across departments, countries and languages are the glue that holds all our plans and activities together.

Research is an essential ingredient of enabling CIBA-GEIGY to accomplish the goals and objectives defined in our long-range plan. There are many disciplines involved in research and most disciplines are in different departments. Thus, planning and communication of this plan between departments becomes imperative. When a new product is identified and developed for the market, we must first define the biology, i.e., how it can be used to control crop pests. A second aspect is to investigate the biochemistry for the purpose of defining the fate of the chemical in crops, feed, and the environment. Another area of investigation is to determine the safety of the product in terms of man and wildlife. While these aspects are being investigated, the marketing groups must project markets and evaluate alternatives for placing the product in the marketplace.

Another very important group is production. This group must investigate what formulation can be produced, what is the best route for synthesis, if we can use existing facilities or build a new one, and where we can obtain the raw materials. This process takes six to ten years. The interdisciplinary activities and coordination efforts are imperative. All of the research, marketing, and production areas must be completed in a timely manner; otherwise, one group could finish one or two years ahead of another. Because of the complexity of the product development and the length of time required, changes in one area, such as the discovery of performance or customer preference for different formulations, can change production plans or biochemistry projects at any given time and our responsibiltiies are delegated on country bases. Thus, you can readily see that good planning on a short and long-range basis is essential to the success of our business.

In summary, it is foolish to overlook the impact of chance, of sudden initiative, of accident or of stupidity. But by the same token, it is also foolish to bet your organization's future on the premise that the entire environment of public and private enterprise is completely capricious and random. Successful long-range planners recognize that unplanned events will always occur despite good plans and that they will do so at the most embarassing and inopportune times. There is little or no formal administrative defense to prevent all of these disruptions.

We can say that successful interdisciplinay efforts require special attention to the principals of planning. Briefly stated, these principles center around:

> Making a plan which has goals which are understood and supported by the key parties
>
> Working the plan to add your share of the efforts toward the goals
>
> Communicating, especially between disciplines, to take into consideration the predictably unpredictable changes which impact the plans

But having taken this position, the successful organizations size up their environment as being more orderly than random, more structured than dissarrayed, more benign than malevolent, more knowable than inscrutable, and thus they are more likely to respond in a positive and profitable manner to intelligent, consistent well planned strategies than simply shots in the dark.

MANAGING HIGH TECHNOLOGY
An Interdisciplinary Perspective
B.W. Mar, W.T. Newell, and B.O. Saxberg (Editors)

RATIONAL RANKING OF DEVELOPMENT CANDIDATES

Sidney W. Hess

Vice President, Atlas Aerospace Division
ICI Americas, Inc.
Valley Forge, Corp. Park
Valley Forge, PA 19481

DIVISION BACKGROUND

In 1980 the Aerospace Division was the smallest of eight divisions of ICI Americas. ICIA with sales of about $1 billion is the wholly owned subsidiary of Imperial Chemical Industries of the United Kingdom. The latter at the time had world sales of over $10 billion. Most of the Aerospace Division sales were miniature explosive actuating devices for military uses (for example, missile components). In addition, the Division had one commercial security product, a dye and tear gas money pack, which the teller would slip to would-be bank robbers.

The division was treated as a cash cow and was devoid of any significant new product development activity. We sensed the need to expand and diversify, particularly if we were ever to become a significant factor within ICI Americas.

BRAINSTORMING

To solicit new product and business ideas, an internal brainstorming session was organized. Approximately twelve representatives of the Division's engineering, development, production, and marketing departments plus three representatives from Corporate Research and Corporate Marketing Research participated. Figure 1 summarizes the panel's definition of the Division technology and what makes us distinctive. Fifty-three individual new products and businesses were identified. At the conclusion of the session each was rated according to general level of attractiveness and our market position. For each of these criteria, the ideas were ranked high, medium, and low. One idea was high in two categories, 21 ideas were high in one or the other of the categories.

At a subsequent date, the brainstorming exercise was repeated by interdisciplinary teams from Arthur D. Little. The little panels were thought to have less of a parochial view. This turned out to be the case although approximately half of the 40 ideas they suggested were common with our own 53. the balance of this paper, however, is concerned only with the ideas generated within the Division.

Science/Engineering of controlled use of explosives in small devices

Electronic/Explosive matching (electromechanical)

Miniature/Reliable

Glass-to-Metal Seals

DISTINCTIVE
(Why do customers come to us?)

Customer needs/application/service

Fast prototype development/new product design testing capability

Marketing
make/buy decision and response

Reliability/Precision parts/high QA

Production/miniature production

Routine handling of explosives/small

Machine design capability

Hi/lo volume capability

Components - (vs. systems)

Figure 1
Aerospace Division Technology

EVALUATION CRITERIA

At a subsequent meeting, the ICI Americas panel developed the following eight criteria for screening the 53 ideas:

technical risk
high growth
sales in excess of 1 million per year (after five years)
profitability of 20 percent on sales
development time of less than one year
development costs less than 10 percent of sales
timing
capital costs

The group rating of each idea against the criteria was rapid and subjective; no attempt was made to be specific on any of the eight criteria but just highlight whatever anyone happened to say regarding the idea opposite any of the eight criteria. By consensus, the 53 ideas consisted of the best eight ideas that should receive immediate attention. Group 2 were fair and might be reexamined for a potentially useful idea at a later date. Group 3 were ideas that should be examined only if the ones and twos failed to produce any significant opportunities.

Some of us on the panel were dissatisfied with the casual ranking since the eight criteria seemed to be treated, but are not, of equal importance. How should they be weighted?

ECONOMIC MODEL

We propose a simple expression for the expected net present value which would place each of the criteria in a true relationship to the others and its contribution to the project value. The technique is similar to ones used before in screening Corporate Research projects.

The expression for expected net present value is similar to a Chemical Research example (Figure 2) but deletes the sanction question between development and commercialization. The complete expression we used is:

$$E(NPV) = -R + P_t(-D + P_c kS \int_T^{T+10} e^{-it} dt) \qquad (10)$$

$E(NVP)$ = Expected net present value
R = Research costs; i.e., the cost to prove technical feasibility
P_t = Probability of technical success; i.e., demonstrating technical feasibility
D = Development costs; i.e., the cost of moving from technical feasibility to commercialization
P_c = Probability of commercial success; i.e., achieving the forecast profit level
k = Trading profit as a fraction of sales
S = Average annual sales over first 10 years
T = Years to commercial introduction
i = Annual discount rate

$\int_T^{T+10} e^{-it} dt$ = Cumulative continuous discount factor

Essentially all of our eight criteria are incorporated in equation (1) except capital; I suggested it be treated as a lease change influencing the numerical value of k. Ideally parameter estimates for each product/business would be used calculating E(PVN). Independent projects would be ranked on the ratio of E(PVN)/R; in other words, selecting projects which would give the highest return per dollar of research.

SHOULD RESEARCH ON 2ND GENERATION PROCESS CONTINUE?

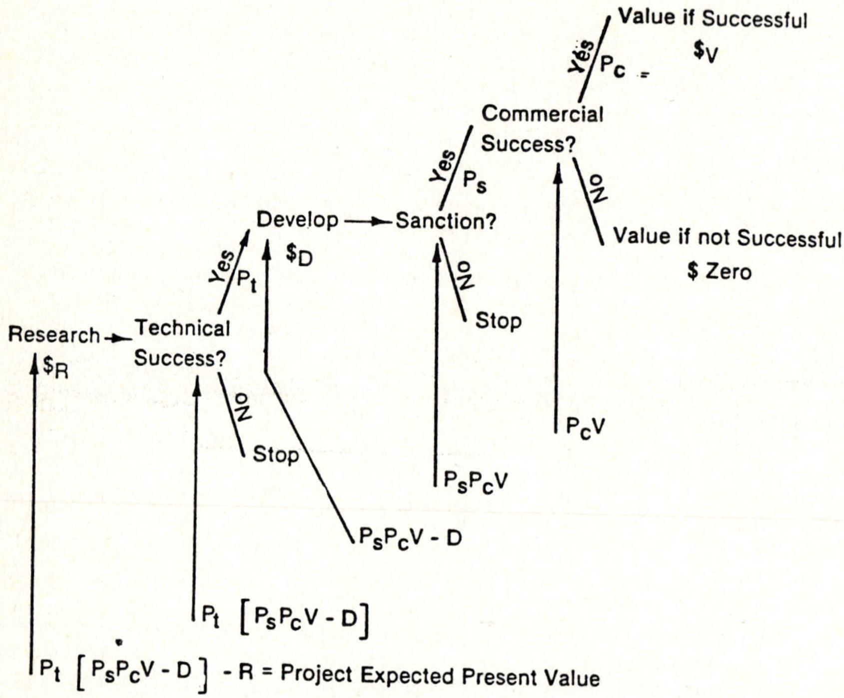

Research should continue if Project Expected Present Value > 0

i.e. if

$$P_s > \left[\frac{R}{P_t} + D\right] \frac{1}{P_c V}$$

Figure 2

PARAMETER VALUES

Estimation is a thankless and nearly impossible job. Our need, however, was not precision but the ability for gross screening. We needed to separate the good projects from the also ran. We utilized a three-point scale for each parameter, assigning a high, average, and low value of the parameter for three points. While it is difficult to assign specific numerical values to each parameter for each project, we could make reasonable guesses as to whether the particular value was about average or significantly higher.

Except for average sales volume and development expense, the parameters are independent. The higher sales volume products will sustain higher development expenses and we paired those parameters accordingly. A table of values is in Figure 3. One approach was to use the nominal value unless we felt strongly otherwise.

Parameter	Nominal Value	Significantly Better	Significantly Worse
S;D very correlated	\$10 mil.; 1 mil.	\$100 mil.; 3 mil.	\$1 mil.; 0.3 mil.
k	.25	.35 (e.g., patent protection)	.15 (e.g., government sale)
$\int_{T}^{T+10} e^{-it}dt$ i=.15	2.5 (T~5 yrs.)	3.5 (T~3 yrs.)	1.5 (T~8 yrs.)
R	\$.1 mil.	\$.03 mil.	\$.3 mil.
P_t	.5	.8 (sure thing)	.2 (difficult)
P_c	.6	.8 (confident)	.4 (risky)

Figure 3

RESULTS

Each of the panel members was given the opporunity to independently estimate all the parameters for all the 53 projects. many of the panelists declined to estimate all the parameters. Those that were estimated generally followed the numerical values of Figure 3. The averages for each parameter were calculated, and the ratio of expected net present value to research dollars was used as a ranking parameter. The top 20 percent (11 products) are tabulated in Figure 4.

RATING AND RANKING OF "BRAINSTORM" PRODUCTS

Economic Rank Order	Identification Name	Subjective Rating: Attractiveness Market Position	Subjective Rating: 1, 2, or 3	Approximate Economic Ranking (See Fig. 6)
1	Theft deterrents	Medium-low	3	8
2	Lead, booster, det. cord mfg.	Low-medium	3	4
3	Sensing and actuating systems	Medium-low	2	11
4	Pipeline protector	Medium-low	3	6
5	Fire protection devices	High-low	1	2
6	Explosive plugging nuclear steam generator	Low-medium	Not listed	30
7	Process vessel protection	Medium-low	Not rated	1
8	Public building safety devices	Medium-low	3	3
9	Oil well service explosives	High-medium	1	19
10	Small business protection systems	High-low	Not listed	9
11	Explosion suppression	High-low	3	14

Figure 4

Also on Figure 4 are the subjective rankings, both attractiveness--market position and 1, 2, or 3. Recall that 21 of the products were high in either attractiveness or market position; that is about 40 percent of the total. Eight or about 15 percent were rated 1, of best products.

Only four of the 50 percent, which include a high, matched any of the top 20 percent by rank order. Further only two of the top 15 percent by the 1, 2, or 3 criteria matched the 20 percent by economic rank order no better than random.

We felt more comfortable with the economic ranking and therefore rejected the subjective ratings.

ESTIMATION

As the input values are no better than one significant figure, precise calculations can be approximated. Note that the range of the parameters--integral, d_c, P_t, and K-- are approximately + or - 40 percent. Since these four parameters are multiplicative, the effect of one parameter's being at the high level is the same as if another were at the high level. Further, a high parameter nearly balances a low parameter. I assumed the following kind of approximation.

$$(1 + 0.4) \times (1 - 0.4) = .84 \cong 1 \tag{2}$$

For the four parameters, integral, P_c, P_t, and K, I count the net number of highs minus lows, use that as an abscissa and read E(NPV)/R on the ordinate of Figure 5.

Figure 5 shows the dominant influence that average sales volume has on project value. Within the range of parameter estimates any $100 million project will always be worth working on; a $1 million project will be worthwhile only if nearly all other parameters are at their high level.

In Figure 6 pluses and minuses depicting significant favorable or unfavorable values from average are displayed together with E(NPV) and ranking for each of the eleven products/businesses with an E(NPV) of $1 million or better. Eight of the eleven products in Figure 6 match the top eleven in Figure 4. The approximate ranking has also been added in Figure 4 in the right hand column. The agreement between the economic rank order and the approximate ranking are significantly better than chance.

CONCLUSIONS

Three years have passed since the brainstorming exercise. Of the top 20 percent by economic rank order, three of the eleven products are now commercial. These are the oil well servicing explosives, theft deterrents, and explosive plugging of nuclear stream generators. Only the oil well service explosives rated a high or a #1 by the subjective rating.

Two other products in the top eleven are still in development. One of these two is also rated high and #1 by the subjective ratings.

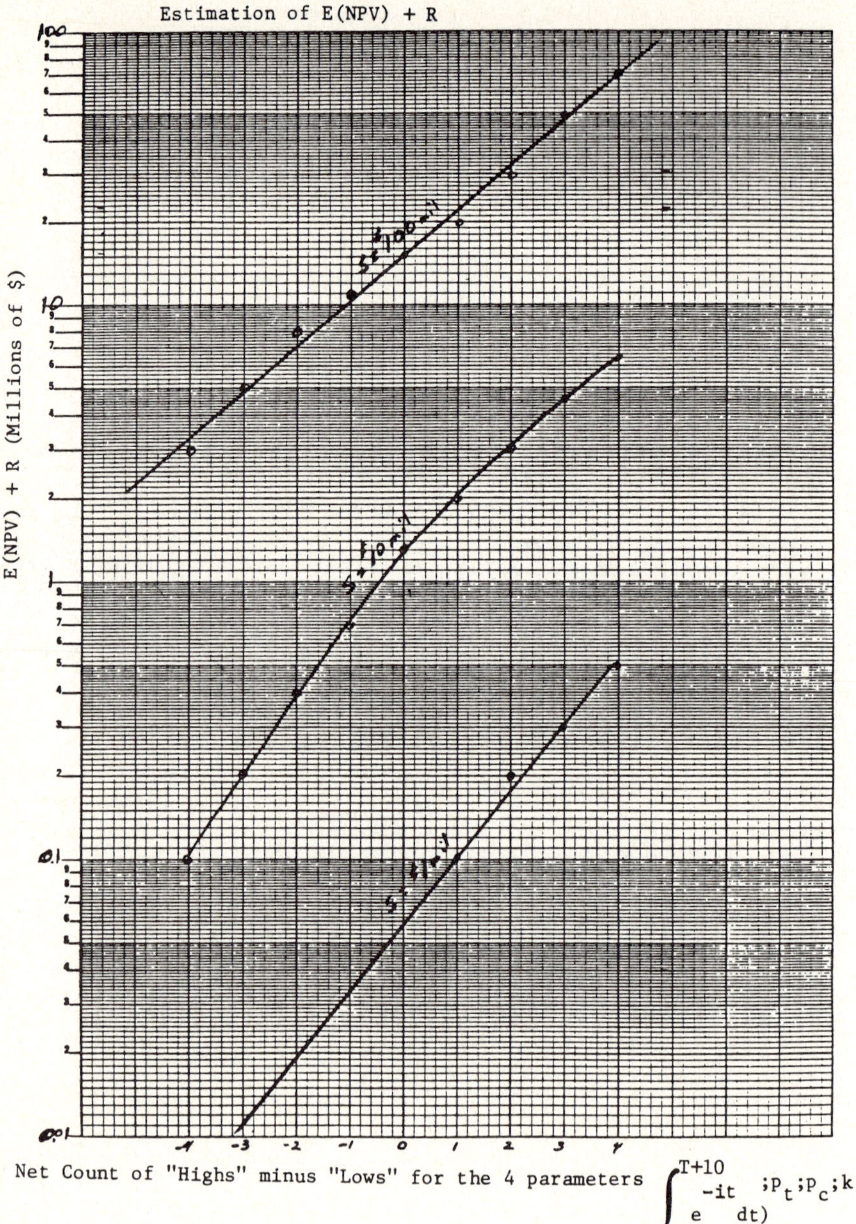

$\left(\int_T^{T+10} e^{-it}\,dt\right); p_t; p_c; k$

Figure 5

TOP 20% USING APPROXIMATE PARAMETER ESTIMATION

Rank**	Description	E(NPV) \$ Mil.	R \$ Mil.	S^*	T	k	P_c	P_t
1	Process vessel protection	4.5	.03			+	+	+
2	Fire protection devices	3.0	.03			+		+
3	Public bldg. safety devices	7.9	.1	+	-	+	-	-
4	Expand expl. line/det. cord, lead and boosters	2.0	.03					+
5	Industrial energy conservation	.7	.03					+
6	Pipeline protector	1.9	.1			+		
7	Home security	1.9	.1			+		
8	Theft deterrents	1.9	.1			+		
9	Small business protection systems	1.2	.1			+		-
10	Runaway braking	1.2	.1		-	+		
11	Sensing and actuating system	1.2	.1			+		-

*In last five columns, plus means significantly better than average; minus means significantly worse.

**Ranked on E(NPV)/R

Figure 6

Finally, one other development product not in the top eleven is still in the works.

The economic rank ordering was easy to implement and very helpful in guiding our future growth. More important, the entire brainstorming/ranking exercise signaled the Division and the ICI we were serious about revitalizing the Division with new products, technology and businesses. Sales in 1984 will be double the level of 1980, a 20 percent growth rate.

MANAGING HIGH TECHNOLOGY
An Interdisciplinary Perspective
B.W. Mar, W.T. Newell, and B.O. Saxberg (Editors)

TECHNOLOGY FORECASTING

Aant Elzinga

Science Council of Canada
100 Metcalfe Street
Ottawa, Ontario
Canada K1P 5M1

INTRODUCTION: TECHNOLOGY FORECASTING AND INTERDISCIPLINARITY

Technology forecasting today is methodologically somewhat different from what was practiced under this heading fifteen or twenty years ago. The 1960's was a decade when systems theory and rationalistic planning models were in vogue. Forecasting was frequently associated with linear trend extrapolation and simplified consensus surveys of expert opinion. Today technology forecasting involves not only a broad base of interdisciplinary information; it also requires the polling of extrascientific opinion.

Jean-Jacques Salomon in his review of the history of science policy has noted how the Brooks report of 1971 was one of the documents that articulated the distinction between tactical and strategic choices. The former deal with scientific problems in a given field, and as such a tactical choice involves internal scientific criteria. Strategic choices on the other hand determine priorities between broad areas of endeavor on the basis of external criteria relating to social, economic, and political relevance expressed in national or regional goals. Interdisciplinarity has to extend to social scientific expertise, and nonscientific practitioners also come into the picture.

> As soon as these strategic choices are determined by objectives of a social nature, the formulation and elaboration of research programs can no longer fall within the competence of the representatives of a single field or even of connected fields: it is necessary, at the very least, that the specialists of natural sciences and those of social sciences learn to work together. But this in itself would not be enough to encourage public opinion and, through it, the public powers, to foster basic research; at that level of decision making the participation of the future users' representatives (trade unions, consumers, local authorities) is no less useful than that of the representatives of administration and industry (Salomon, 1977).

The involvement of nonscientific, or end user expertise in forecasting has led to greater self-awareness on the part of the forecaster, an interest in historical perspective, and also modifications in terminology. All these reflect a shift away from the technocratic naivity of some of the earlier forecasting literature.

INFLUENCE OF THE INSTITUTIONAL CONTEXT

There exists a great variety of forecasting techniques. Long-term trends can be identified by trend extrapolation, projection of historical cycles, or pipeline flow analysis (based on the presumption of normal progress with respect to efforts currently in the works).[1] More elaborate methods involve Delphi forecasting, scenario construction and model building or simulation. In the literature there is no significant consensus as to the most adequate techniques for science and technology forecasting. The energy shock in the early 1970s was one major event that prompted a considerable amount of rethinking and self-criticism. As the inaccuracy and variability of past energy forecasts became obvious, some writers suggested that the problems presented by forecasting are not necessarily solved by increasing the sophistication and quality of the data and its analysis.

One problem is the complexity of interdependent relationships. Although each relationship may be explicit in a model, their interactions may produce emergent properties that have not been foreseen. This is even the case with complex models used to simulate reality. "Methodological sophistication buys little, if anything, in terms of forecast accuracy," is the verdict of Brunner (1982). Rescher after a review of different methodologies, says: "Model building is virtually useless for making forecasts about science and technology--at this stage of the game, at any rate." He points to the importance of retrospective studies and providing for them in the present. There is a tendency today also to recognize the legitimate role of social scientists, historians, and humanists in futures studies. This is because of an increasing awareness of the influence values and institutional arrangements have on outlook.

Institutional forms influence the way in which problems are defined and what solutions are reached. Hooker and van Hulst (1977; 1979) have pointed out how several studies made by institutions in the energy and health fields in Western societies embody certain assumptions that ultimately derive from a market paradigm based on neoclassical economic theory. In a similar manner bureaucratic hierarchy and functional division of responsibility tend to reinforce sectorialization and fragmentation of perspectives (Elzinga, 1979). Robinson (1982) argues further that when an institution uses analytic techniques of forecasting which assume the value postulates implicit in the organization, one gets a semblance of value free inquiry and neutrality, and a tendency to favour the kind of policy orientation already adopted. "Forecasting is not a neutral method of analysis, but a means of imposing in the future a perspective determined by the framework within which we presently operate." If this is true, even to some extent, it would reinforce the suggestion of bringing in some social scientists and particularly historians, e.g. economic history and history of science and technology in society, and others with a general training (Chandler, 1977; Noble, 1979).

FROM FORECASTING TO FORESIGHT

A shift from the reductionist perspective of the 1960s to an emergent historical perspective is evident in concepts like outlook, foresight and issues management, which have in part come to replace or soften up terms like prediction and forecasting. Foresight gives sense of perspective, ranging from a look at problems immediately ahead, through trends (5-10 years hence), to longer range emerging issues. The term foresight or outlook seems to be more often associated with forecasting in governmental settings, while the private sector has a preference for the terms issue management and strategic thinking. Also, it is sometimes noted how foresight activities of a broad scope function better in the private sector than they do in the public sphere. In the private sector business environment, strategic planning has two main objective. One is to provide management with guidelines, based on a scan of emerging issues and factors in rapidly changing political, socio-economic and cultural (life-style) environments. The other is more specifically to identify future product opportunities and the backward linkages from emerging (high) technology to R&D (Thomas, 1984).

In the case of the private sector the distant early warning of intelligencing information is usually directly transmitted to the Chief Executive Officer, whereas in the governmental setting the information from environmental scanning and issue identifying activities have to pass through several filters before they reach levels of authority that can act on it. Thus the distance between awareness and action is much smaller in the private sector than in the public. This is one of the reasons why in the private sector context foresight is often referred to as issue management. As such it involves scanning of economic, social, political, and technological environments to devise a number of scenarios and alternative lines of action by which the corporation can act on/adapt to a rapidly changing environment. In the governmental setting, by contrast, this type of global scanning and issue identification often tends to become marginalized or else lives on in relatively ineffective groups of frustrated persons concerned with strategic thinking/planning. The alternative in this setting is to refocus the exercise with a short-term planning horizon which attaches to, say, a two-year budgetary cycle.

A NOTE ON DIFFERENCES OF NATIONAL STYLE[2]

In each country the institutional and political contexts of forecasting activities are important determinants in the shaping of organizational arrangements. Where there is a considerable degree of decentralization and multiplicity of decision makers, this is reflected in a looseness of organizational structure. The European program for Forecasting and Assessment in the Field of Science and Technology (FAST) started up in early 1979 within the Council of Ministers of the European Communities. Here we have an enterprise that serves a loose knit voluntary association of countries with different resource bases, economies and cultures. Foresight activities are given a strong research component, and an effort has been made to anchor it throughout the member states of the Community.

Two contrasting models exist for foresighting activities in official agencies in the U.S. One is in the work of the Congressional Research Service (CRS) at the Library of Congress. It has put out a handbook entitled Survey of Science and Technology Issues Past and Present, which reviews the national problematic in various functional areas like environment, health, communications, science education/human resources, science policy, etc., and it looks briefly at the U.S. in a global context. The other model is the

OSTP-NSF-NAS Science and Technology: A Five Year Outlook, which has taken a selective look at specific aspects of science and technology, and prospects in various field, science in the U.S., the institutions of science and technology, the changing global context, and so on. The CRS studies are in-house efforts, whereas the NSF Outlooks have involved farming out topics to a variety of experts who have written background papers. The National Academy of Science for its part has gone on to scan areas of science and produce sets of Research Briefings, providing more of an internalist perspective on trends and prospects.

Compared to FAST, the exercise in the U.S. appears to have a somewhat tighter structure, with a division of labor which reflects the diverse interests involved, mainly scientific communities and federal agencies. In countries like Sweden where there is a strong tradition of concerted government effort, this division of labor is much more systematic and forecasting activities have much closer ties with the centers of political and administrative powers in the state apparatus.

In the Netherlands there is a dual model of both government based and university based projects which complement each other. The strategy of implanting foresight activities regarding S&T into a university milieu seems to reflect a long-standing tradition of consultative interchange between government and academe. The orientation and development of the Science Dynamics program at the University of Amsterdam is sometimes taken as an example of this type of implantation of an advisory function. The present orientation of the program, as you might know, is towards the study of the steerability of the growth of scientific knowledge production.

In France the creation of a Center d'Etudes des Systemes et des Technologies Avancees (CESTA) under the Ministry of Research and Technology provides yet another model. In keeping with the task of raising public awareness, CESTA has come to emphasize its role as a forum for interregional encounters and public colloquia on issues of science and technology. This constitutes a concerted governmental effort to mobilize broad participation for S&T assessments, economic ones as well as assessments from sociopolitical and cultural perspectives.

Sweden, with a different tradition of concerted government effort (dirigism), reveals a pattern where one has a systematic division of labor, with forecasting activities located in several agencies simultaneously. Sometimes the perspectives they present are complementary, and sometimes they are mutually divergent, thus prompting national debates.

THREE BASIC MODES OF FORESIGHT ACTIVITIES

Underneath differences of national style there is a level where one might compare modes of foresight on a more general basis. It seems possible to identify at least three such modes: the responsive mode, the reflective mode, and the mode of participatory awareness. The first is a mode of operation usually associated with the traditional role of commissions, studies and task forces in response to crises and pressures for immediate action. The reflective mode in its most typical form is associated with think tanks and critical reflection in the academic community. Foresight activities in this context have a stronger research component. The participatory awareness mode, finally, is the one I associate with an action orientation towards public awareness in a broad sense. Examples may be found in groups involved in action oriented research connected with social movements and public interest organizations (e.g., pollution probes).

Each of these three modes has its own distinct characteristics and difficulties, which I summarize in Table 1.

A MATRIX APPROACH

I think it is important to somehow combine all three modes of foresight which are important in North America because of the many conflicts that have arisen around central areas of science and technology during the 1970s. Also there is the unequal distribution of benefits and opportunities that often flow from science and technology, because of differences between industrial sectors, geographic regions, and layers in the population. These aspects are particularly significant in Canada with its mosaic of cultural diversity, contradictory interests, and regional loyalties. For this reason the Science Council of Canada is employing a matrix approach to identifying emerging science and technology (EST). The basic idea is to set up consultative panels that will bring together experts in specific S&T fields with various end user categories. The panels will be convened around a select number of what might be called intersection-points between potential S&T applications and end-user areas (Table 2).

The foregoing is only the beginning of a matrix that can be constructed, for example with the help of input from a broad survey of expert opinion by questionnaire. The questionnaire asks respondents in industrial laboratories, universities, governments, and high technology associations to give their perception of what are, say, the five major emerging technologies over the next five years. This, together with literature studies will provide input to the matrix map. Workshops organized around emerging areas of scientific and technological interest, such as for example, the subject of Advanced Industrial Materials can also be used to further identify strategically important fields. At the same time, if carried out properly, both the questionnaire survey and the workshops will themselves serve as vehicles in the consultative process.

The next step is to use ranking criteria on the multitude of intersecting points in the matrix to arrive at a select number of emerging fields for further investigation. An important factor here will be the criteria chosen: e.g., domestic needs, international market opportunities, existing technical possibilities or research problems still to be overcome to bring the emerging technology to the market place, differential national advantages, considerations of national integrity, self-reliance, as well as social relations and quality of life. It is evident that such deliberations require interdisciplinary as well as extrascientific consultations. The composition of panels in such an exercise therefore would illustrate what Jean-Jacques Salomon states as quoted in the introduction to this paper (1977). In the final analysis, the mode in which this is done will depend upon the context--in our case we would like to obtain a

Table 1
Three Modes of Foresight

Summary

Mode Operation	Responsive	Reflective	Participatory Awareness
primary orientation	an arm of decision-making	think tank in-depth research	public interest organizations, corporations, trade unions, etc.
primary format of reporting	reports to gov'ts. executives and other high level decision-making bodies, politi-cians	inter- and intra organiz-ational, research papers and quality research reports	popularization, thematic issue reports and vention in mass media
timetable	tied to that of legislature machinery	own time-table, dependent on intraorganiza-tional needs	related to flux of publc issues and debates in society
dangers	drift to purely reactive mode, uncritical, short-term focus	drift to insularity and academization	drift to pupulism and reactiveness to the flux of popular debate

Table 2
Matrix of S&T and End-user Sectors

	End-user areas / S&T	Mfg*	Forestry	Mining	Health	Transport	Agr.
BIOTECH	fermentation tech ...						X
	tissue culture		X				X
	bioreactors ...				X		
	bioleaching			X			
MICROEL	remote sensing		X	X		X	X
	CAD/ Robotics	X	X	X			
	optical fibres	X		X	X	X	
ADVINMAT	composites ...	X			X	X	
	fine ceramics ...	X			X	X	
	glassy metals	X		X		X	
ENERGYPROC	coal liquefaction ...				X		
	solar and other renewables ...					X	
CHEMPROC	catalytic process			X			

* (e.g., Primary Metals Industries, Fabricated Metal Products Industries, Non-Metallic Mineral Products Industries, etc. and Other Manufacturing Industry categories)

balance, which should be possible for an organization operating at arms length from government.

CONCLUDING REMARKS

Despite differences in national style and basic modes of foresight or technology forecasting, it is striking how efforts in various countries have led to a convergence of opinion and analysis around a few key areas, such as microelectronics/telematique, biotechnology, and advanced industrial materials. A longer list would also include energy production technologies, food processing technologies, waste management technologies, etc. One might ask whether this convergence reflects similarities pertaining to international market pull, endogenic science and technology push, or both. For my part, I think it is a question of both. Furthermore, this raises the question of the rationale behind a resurgence of interest in technology forecasting.

On the one hand, generally speaking, we have to consider the generic nature of some of the new and emerging technologies, in view of their vast transformative potential vis-a-vis many sectors and groups in society. Also the new technologies have a close symbiotic relationship to the fundamental research into physical, chemical, and biological processes. The great variety of the emerging technologies, and the financial as well as social risks associated with some of them thus require more systematic approaches to scanning, monitoring, and identifying opportunities, risks and constraints.

On the other hand, as the finalization theorists in Germany have argued, modern science and technology are embedded in societies characterized by strong relevance and accountability pressures on the communities of scientific investigators. They are linked to national political agendas by hybrid communities involving scientists, politicians, civil servants and experts representing industry as well as other nongovernmental interests. In this context of a political direction of science, there are conscious efforts to transform political goals into science policy programs, and further, into concrete research strategies. In this context technology intelligence and forecasting has an important function.

A final reflection I want to make pertains to an imperative to which technology forecasting itself seems to be a response or perhaps is it a manifestation?

Today, industrial nations, whether their leaders like it or not, are being pushed even more strongly in the direction indicated by Salomon in his comments on the Brooks report. The uncertain environment of intense global economic and technolgoical competition and the strategic role of emerging science and technologies in this context seems to prompt three things: (a) increasing government involvement in the economic life--despite frequent rhetoric to the contrary; (b) increasing pressures on both governments and private sector actors to think and act strategically on issues of economic and technological development; (c) and, as a consequence of this, the need to more systematically facilitate strategic planning, to gain wide ranging bases of information, and to implement technology and economic policies, there is a need to develop broad consultative building mechanisms around issues of technological change.

This is particularly important in view of the regional, cultural, and economic differences and inequalities which in many industrial countries are a source of serious tensions.

FOOTNOTES

1. For a critical review see Nicholas Rescher, "Methodological Issues in Science and Technology Forecasting: Uses and Limitations in Public Policy Deliberations," Technological Forecasting and Social Change, 20 (1981), 101-112. An interesting review of methods used by a forecasting institute may be found in Andrew Lipinsky and Denis Loveridge, "How we forecast", Futures 14(1982), 205-239. It is the experience of the Institute of Future's Studies of the U.K., its identification of sources eliciting information, quantifying, scenario construction, "modelling".

2. Jamison uses the term "national" style to refer to differences in scientific growth in different countries. See A. Jamison National Components of Scientific Knowledge. A Contribution to Social Theory of Science (Research Policy Institute, Lund 1982).

REFERENCES

Brunner, R.D. 1982. The policy sciences on science. Policy Sciences 151:115-135.

Chandler, A.D., Jr. 1977. Visible hand: the managerial revolution in American business. Cambridge, MA: Harvard University Press.

Elzinga, A. 1979. Swedish science policy: sectorialization and adjustment to crisis. Research Policy 9.

Hooker, C.A., and van Hulst, R. 1977. Institutions, counter-institutions and the conceptual framework of energy making policy in Ontario. Report for the Royal Commission on Electric Power Planning. Toronto: University Press.

Hooker, C., and van Hulst, R. 1979. The planning of environmental problems in public policy institutions. In Ecology and politics in Canada, ed. W. Leiss. Toronto: University Press.

Noble, D.F. 1979. America by design: science, technology and the rise of corporate capitalism. Boston: Galaxy Books and Oxford University Press.

Robinson, J.B. 1982. Backing into the future: on the methodological and institutional biases embeded in energy and supply forecasting. Technological Forecasting and Social Change 21:229-240, p. 234.

Salomon, J.J. 1977. Science policy studies and the development of science policy. In Science, technology and society: a cross-disciplinary perspective, eds. D. Spiegel-Rosing and D. de Solla Price, p. 65. London: SAGE.

Thomas, L.J. 1984. Technology and business strategy - the R&D link. Research Management 27(3):15-19.

MANAGING HIGH TECHNOLOGY
An Interdisciplinary Perspective
B.W. Mar, W.T. Newell, and B.O. Saxberg (Editors)

TECHNOLOGY EXCHANGE IN LARGE DIVERSIFIED CORPORATION: A CASE STUDY

Dr. Julianne H. Prager and Dr. Bruce A. Langager

Corporate Technical Planning & Coordination
3M Center, 220-4E
St. Paul, MN 55144

Optimum use of the combined technical abilities of any technical staff, be it large or small, is realized only if staff members exchange ideas and keep abreast of what their colleagues are doing. In larger industrial corporations it can become quite easy to miss this exchange as a result of organizational structure and immediate job responsibilities which can promote the isolation of individuals within their own laboratory.

Failure to maintain communications outside one's own laboratory can result in two very significant shortcomings for any vital technical organization. First, the individual misses out on the stimulation that accompanies exposure to indisciplinary groups. Secondly, the company misses out on the synergism that occurs when receptive technical minds come together. This synergism frequently results in clues to solve one's own problems as well as sparking new ideas which can lead to major new technology and product developments.

3M is not unique in that it faces many of the same problems that other large industrial corporations must overcome in promoting technology and idea exchange. Perhaps what does make 3M unique in this area is the fact that it recognized the need for the exchange, as well as the importance of creating a climate which would foster the desired interaction, several decades ago. Since that time 3M has maintained a steadfast commitment to the task, exploring new ways of promoting the exchange and modifying older mechanisms to more adequately address today's needs.

THE COMPANY

Before discussing the mechanisms which 3M employs in promoting technology exchange, it is informative to take a brief look at the company. From its founding in 1902, 3M has grown into a large, multidisciplinary corporation having manufacturing operations in 31 foreign countries. It is in the Fortune magazine "Top 50" list with 1983 worldwide sales in excess of $7 billion. It has more than 45 major product lines with nearly 50,000 products which are supported by over 85,000 employees. Throughout its growth, it has employed an evolutionary spin-off philosophy, developing unrelated business in established groups or divisions and when these businesses have become large enough to be self-supporting, splitting them off from the parent division and creating a new division. This philosophy has fostered a strong sense of sharing, while at the same time it encouraged innovation and diversity.

The evolutionary spin-off philosophy has led to today's organization which consists of four business-oriented sectors that are organized into more than 40 divisions, which are the basic operating units of the company. This organizational structure might seem much like a conglomerate with 40+ small companies operating under the 3M umbrella, but the fact of the matter is that the organization is the antithesis of a conglomerate. Extensive technology sharing and interdisciplinary cooperation provides the "glue" which in effect unifies the company and essentially creates a single large technical organization.

Research, development, and engineering activities within 3M are pursued by 5000 employees in over 50 laboratories. The laboratories include staff sponsored laboratories, sector laboratories, and divisional laboratories with R, D & E activities encompassing virtually everything from long-range fundamental research to shorter range applied research.

Longer range fundamental research activites are concentrated in staff sponsored Central Research Laboratories. In addition to R&D activities, these laboratories are also charged with maintaining contacts with academic institutions to insure that, as a company, 3M keeps abreast of technical developments within academic research circles. Staff sponsored laboratories also provide specialized technical services (i.e., analytical, process engineering, etc.) to complement services offered in individual laboratories.

Intermediate range R&D activities are the primary responsibility of the Sector Laboratories. These laboratories are charged with providing the fundamental undergirding of each sector's key technologies and to broaden the sector's technology base to provide the building blocks needed for the development of new products and new businesses.

The major share of 3M's technical effort is concentrated in the divisional laboratories where over 75 percent of the compnay's R&D staff pursue activities directed at individual market segments. This work ranges from quite basic to very applied and typically is of a shorter term nature.

THE PROCESS

Unlike many other large industrial, technically diverse corporations that have widely dispersed R&D activities, the majority of 3M's R&D activities are situated in the 3M Center complex. This concentration of technical personnel with diverse areas of expertise creates a natural setting for the company's technical exchange programs.

As with most good programs, they don't just happen, but rather they are the result of considerable thought, planning and effort. 3M's technology exchange program is no exception to this rule. The organizational structure and concentration of technical employees are insignificant elements of its technology exchange program but they, in and of themselves, are not the reason for the success of the program. More realistically, one can point to the numerous formal and informal activities, functions, organizations and services within the company, all of which are directed toward stimulating and enhancing technology exchange and making this exchange an effective tool in the company's overall R&D effort, as the principal reasons for the success of the program.

An underlying theme of the technology exchange program at 3M is that the maximum effectiveness of the program will be realized only when all levels of the technical community are active participants in the exchange. To this end, various elements of the program are targeted toward specific audiences, some to management, some primarily to the staff, and some to both management and staff.

TECHNICAL COUNCIL

Interdivisional technology exchange at management levels is a principal objective of the Technical Council. This organization, which consists of the Technical Directors of the various divisional, sector, and staff laboratories, holds regular monthly operational meetings as well as two special off-site theme meetings each year. To many, this may seem like a very high and possibly unnecessary meeting frequency. To the contrary, however, 3M has found that this high frequency builds a trust and rapport among Tech Council members which stimulates a free and sustained exchange of information.

Technology exchange is promoted at all of the meetings through formal technical presentations as well as informal discussions before, during and after meetings. The off-site meetings particularly emphasize the informal exchange as they provide a more relaxed social atmosphere in which to discuss technical subjects. A hidden advantage of these informal interactions is that significant technology exchange occurs without lengthy meetings and "telephone chases" typically associated with more formal exchange processes.

TECHNICAL FORUM

The most widespread and effective mechanism for technical exchange within 3M is the Technical Forum, an organization run by and for the technical community. It was organized in the early 1950's for the purpose of providing middle management and the senior technical staff a vehicle to assist in idea and technology exchanges that could help both groups improve their respective performance. Forum membership grew from a few hundred initially to several thousand today, with all members sharing the commonality that they are engaged in technical work at 3M.

Today, the Technical Forum serves 3M's technical community in two very important ways. First, it promotes the interchange of technology and ideas across divisional lines. Secondly, it stimulates the professional development of the technical employee. The Forum has become a significant factor in helping to combat technical obsolescence as well as promoting technical excellence in all phases of research, development, and engineering as a result of its many programs.

Annual Technology Fair

The premier event for promoting idea and technology exchange which the Technical Forum sponsors is the Annual Technology Fair. This event features divisional and staff displays of emerging technologies or new products. An informal atmosphere permeates these events which encourages detailed technical discussions and ample opportunities for idea exchanges. These events are very popular with the technical community as is evidenced by the fact that they are attended by a majoity of the Technical Forum members. A secondary benefit that is realized with these events is the fact thay they are also a social function where friends and acquaintances of the 3M technical community can visit and catch up on recent events over a cup of coffee and cookies.

There are a number of other Technical Forum sponsored activities which complement the Annual Technology Fair in promoting idea exchanges, including in-house technical symposia, special program series, co-sponsored conferences with academia, and specialized educational offerings.

In-House Symposia

The in-house technical symposia series provides and opportunity for 3M technical personnel to present their work to the 3M technical community as a poster, talk or a formal presentation even when business considerations might preclude a presentation outside 3M. Typically, at least one of these events is held off-site each year in conjunction with a dinner and social hour.

Lecture Series

A series of six to eight general interest programs are also coordinated through the Technical Forum each year. The principal objective of these programs is to expand the knowledge and perspective of the 3M technical community.

Academic Relations and Educational Programs

The Technical Forum also promotes interactions with academia by sponsoring conferences involving prominent academic researchers and company experts in informal settings where a mutually beneficial sharing of ideas and technology can occur.

Another effective vehicle for idea and technical exchange is the specialized educational offerings which the Technical Forum coordinates. As instructors for these courses, company experts share their experiences with other members of the 3M technical community who are interested in learning about a technology and applying it to their specific areas.

Technical Forum Chapters

A notable special feature of the Technical Forum is its organization into discipline focused chapters. These chapters, which range in size from less than 50 to more than 500 members, provide an opportunity for idea exchange in a smaller group having a common interest. Typical chapter programs include lectures by outstanding scientists from outside 3M, lectures by 3M experts, panel discussions, or vendor presentations. The diversity of interests of the 3M technical community is reflected in the fact that there are currently 20 active chapters including: Adhesives/Adhesion, Analytical, Computer, Electronics, Engineering, Imaging, Inorganic Materials, Interface Science, International, Life Sciences, Non-Wovens,

Organic Chemistry, Patent Liaison, Photochemistry, Physics, Polymer, Product Design, Quality Control, Technical Service and Thin Films.

International Technical Forum Organizations

Formation of Technical Forum organizations in technical operations remote to 3M Center has been encouraged as a result of the success that the Forum has had in promoting idea and technology exchange on the 3M main campus. In response to this encouragement, over ten international operations have organized or are exploring the possibility of organizing their own Technical Forum. Because most of the international Forum organizations are relatively new, the bulk of their efforts to date have been focused on their own local activities. A longer term objective, which some of the older international Forum organizations are now addressing, is to develop a stronger dialogue and sharing between the parent and international forums.

INFORMATION LIAISON FUNCTION

Information acquisition or exchange in today's large corporations represent an increasingly more complex problem whether dealing with internal or external information. 3M has addressed this problem by establishing an Information Liaison network which consists of an Information Liaison representative in each laboratory. The Information Liaison function is charged with the objective of contributing to the company's R&D productivity by promoting increased usage of internal and external information and increased communications between laboratories. Specific areas where the Information Liaison function helps promote the flow of information include being aware of and referring people to appropriate information resources, providing leadership in information gathering, sponsoring seminars and other educational functions for their respective laboratories, and maintaining a familiarity with their home laboatories as well as other 3M research functions.

INFORMATION SERVICES

Duplication of effort or "reinventing the wheel" is a very costly mistake which every research organization must diligently strive to avoid. One of the best ways to avoid this mistake is to have a good working knowledge of the literature applicable to any given research program. This knowledge of published literature should include literature generated outside the company, as well as internally generated information.

3M, like virtually all other large industrial corporations engaged in technical research and development activities, maintains an extensive library and information collection and retrieval system. Information generated outside the company is accessible through a number of libraries and information service centers, including:

1. Technical Library System, which consists of a central library that serves as the corporation's basic source of scientific and technical information as well as a back-up for six subject collections located throughout the company

2. Business Information Service, which provides a collection of books, journals, and reference materials oriented toward 3M's business needs

3. Engineering Information Service, which maintains a basic collection of journals, texts, standards and handbooks in all engineering fields applicable to 3M

4. Patent Information Center, which provides retrospective and alerting patent searches for both U.S. and foreign patents

5. Current Awareness Service, which coordinates requests for information on a continuing basis that will keep the 3M researcher up to date on developments in a specific field

6. Marketing Information Services, which coordinates access to externally generated marketing studies in support of the corporate marketing effort.

Another service, the Technical Communications Center (TCC), serves as a central clearinghouse for all internally generated research and development information. Working from its files which contain over 100,000 technical reports, the TCC can provide the 3M researcher with an historical as well as a present status report on a technology or product. The TCC also provides biweekly current awareness updates on personalized profiles.

In addition to its technical report clearinghouse activities, the TCC also maintains a 3M Chemical Registry. This Registry lists novel compounds which have been prepared at 3M. Information on these compounds is accessible through structure and substructure, chemical name and property attributes.

The TCC also encourages 3M researchers to submit special interest items to them for inclusion in their files. These items, which can include technical literature items, scientific meeting attendance, 3M publications and patents, and marketing studies, are communicated to the 3M technical community on a monthly update basis in the Technical Awareness Gazette.

TECHNICAL AUDIT FUNCTION

A discussion of technology exchange at 3M would not be complete without mentioning the Technical Audit function and the role it plays in the company's overall technology exchange effort.

The Technical Audit function was organized several years ago to provide internal appraisals of major research and development programs within the company's various laboratories. The audits consist of a critical examination and analysis of the technical and business aspects of major programs within the laboratory by a group of 10-16 auditors consisting of a core group from the audit function, the division's operating committee, and a rotating group of "outside" technical directors and senior level scientists. The audit team critiques the programs relative to their strengths and weaknesses, attempts to judge the probability of success on both technical and business considerations, as well as a number of other criteria which experience has identified as being critical to successful programs.

The audit team also makes recommendations to the division operating committee on changes for specific programs as well as total laboratory programs. While these recommendations are not binding, they are carefully considered by division management in developing their strategic plans. A direct result of the activites of the Technical Audit function is that as a company, 3M has improved its ability to identify earlier, those programs which have a higher probability of succeeding and making a more significant contribution to the company's profits.

In addition to realizing its primary objective of providing management with additional information for decision making, the Technical Audit program has been very successful at creating a higher awareness of 3M technologies and resources by promoting the sharing of technologies. Technology sharing is "built into" the Technical Audit process by the composition of the audit committee. Initially, the audit committee is exposed to the technologies and products of the laboratory being audited. Secondly, the audit committee with the majority of its members typically coming from ouside the division, brings a broad knowledge of other 3M technologies and resources to the audit. The auditors call attention to technologies that would have potential applicability to programs which are being audited, as well as suggesting specific contacts for additional information.

SUMMARY

Optimum use of the combined abilities of a technical staff can be realized only when there is a free and sustained exchange of ideas and technologies between staff members. The approach of one large, diversified corporation, namely 3M, to the problem of promoting this free exchange of information is described.

A basic premise of 3M's approach in promoting information exchange is that the process will be successful only if all members of the technical community, management and technical staff alike, are active participants in the process. The 3M program has activities which are targeted toward different audiences to insure their involvement in the information exchange process (i.e. the Technical Council targets its activities principally toward management, the Technical Forum and Information Liaison activities aim their efforts toward the technical staff and Information Services and Technical Audit fu
staff).

MANAGING HIGH TECHNOLOGY
An Interdisciplinary Perspective
B.W. Mar, W.T. Newell, and B.O. Saxberg (Editors)

FACILITATING INTERDISCIPLINARY RESEARCH WITH INTEGRATED SOFTWARE SYSTEMS

Peter V. Norden

IBM Corporation and Columbia University

INTRODUCTION

The explosive developments in computer technology recently chronicled in the literature have centered largely about these three themes:

1. Miniaturization and increase in computing speed of hardware

2. Local area (and general) networking

3. Artificial intelligence

The first brings with it massive increases of raw computing power available to the individual researcher. The second allows widely dispersed users to be welded into what has become known as on-line communities. The third is attempting to extend the discrete and deterministic logic of computers into the complexities and subtleties of human reasoning.

My talk today will attempt to describe a small subset of all of the above: The attractiveness of an integrated software system to facilitate communication among individual researchers and to enhance their problem-solving abilities. I will use, as an example of how this notion has been reduced to practice, IBM's recently announced Engineering/Scientific Support System (E/S^3) and at the end make a few short comments concerning the evolution of scientific software.

HISTORY

In an announcement of a short course entitled "Developing Computer Solutions to Scientific and Engineering Problems," offered by the School of Engineering and Applied Science of the George Washington University, the instructors, W.E. Schiesser and S.H. Johnson, make the following comments:

> The modeling of scientific and engineering systems is gaining increased acceptance as an important adjunct to experimental studies. In science, theory can be extended to complex problem systems through computer analysis; in engineering, modeling is used to avoid serious and costly design errors and proceeds efficiently to an optimized design with a minimum of time-consuming experimental testing.

> As models grow in complexity to encompass more detail, however, they present formidable requirements for their solution. The availability of a high-performance scientific computer does not in itself guarantee success in solving the model equations. Careful consideration must also be given to algorithm selection and implementation.
>
> Most scientists and engineers do not have the time or inclination to learn the numerical analysis and computer programming skills necessary for approaching large-scale problems. In fact, the scientist or engineer would typically rather concentrate on the problem at hand and not become enmeshed in these details of computation.
>
> The most efficient approach to the solution of large-scale problems is to use mathematical software written by experts. Libraries of quality mathematical software are now available at modest cost, representing major commitments by numerical analysts and computer scientists that could not be duplicated with any reasonable effort and cost by nonexperts.

Past efforts to develop computer aided systems usually resulted in many discrete and separate systems in a given organization. Each functional organization will seek to develop its own computer system which is suboptimized to that special function. An engineer or scientist faced with the need to use many functions will find a complex set of computer systems which are in various phases of development. Each system will be different, and information exchange between systems will be difficult. Some systems will be outdated, while others will require long lead times to become operational. The current strategy appears to be the development of workstations with enough hardware and software capability to communicate with these existing computer systems, but will be easy to install by the user, will be easy to learn and use, will be modular and adaptable to each user's need.

The IBM E/S^3 system is a reduction-to-practice of these aims. It is a general purpose interactive facility designed to put powerful, responsive computing tools in the hands of engineers and scientists. E/S^3 contains tools and examples which can improve productivity in many areas of engineering and scientific work.

E/S^3 is intended to reduce the time and effort for implementing selected programs by providing pregenerated products with preapplied service in a VM/SP environment. It is based upon and significantly enhances the VM/SP System IPO/E by eliminating many of the implementation tasks which would normally be required.

The user may elect to delete programs from E/S^3, or may choose to add additional programs that serve the specific needs of the user's environment. The user should consult VM/SP documentation and the E/S^3 System Administrator's Guide for assistance in this area. When such changes are made, the integrity of E/S^3 is the user's responsibility.

E/S^3 has extensive library facilities for storing and executing production programs or other packages. Users can add the production programs or packages to E/S^3-supplied menus or create their own menus and prompts. In fact users are encouraged to tailor the menus to their own requirements.

E/S^3 offers a VM/SP based LOAD-and-GO system with facilities for performing some activities that an engineer or scientist may expect to encounter as part of the daily routine. These activities might include:

Design and presentation graphics
Analysis/simulation
Interactive and personal computing
Support of non-IBM devices
Engineering and scientific administration

Design and Presentation Graphics

Very powerful interactive graphics tools provide a way of communicating information, and large quantities of information can be reduced to diagrams and charts, so that meaning and inferences can be drawn quickly. These tools can help a designer visualize a design on the computer before building a model, could reduce the time to complete a design, and could also reduce the problems inherent in the design. Presentation graphics can be used to distribute the most recent data across organizational boundaries quickly, to improve the discussion about the design, and to automate the extraction of data for documentation and analysis.

E/S^3 contains the Computer-Graphics Augmented Design and Manufacturing (CADAM) system which provides the ability to zoom, translate and copy drawings, the ability for dimensional and textual annotation, the ability to extract data for BOM and analysis, the analysis of three dimensional piping, the ability to exchange graphics data, numerical control and APT and COMPACT II interfaces for numerical control. The Circuit Board Design System 2 (CBDS2) provides the ability for entry of schematics, automatic gate assignment, component placement, and printed wire routing, the ability for logic simulation and test pattern generation, and the generation of component data bases.

Analysis and Simulation

Analysis and simulation activities can evaluate a concept or design before expensive prototypes are constructed, and can accurately simulate conditions which are costly to produce in a laboratory environment. These activities can also permit the evaluation of many more concepts than can be done when only manual techniques are available. These evaluations will reduce the problems and errors in the development phase, and reduce the number of potential problems that could arise in the prototype and production phases.

E/S^3 provides for a range of simulation tools including software for finite element analysis, structural analysis, continuous and discrete simulation, library subroutines, simulation languages, and a wide variety of application packages from non-IBM sources. Software to analyze piping, structures, industrial/social/economic situations, digital circuits, integrated circuits, and power systems are also available.

Interactive and Personal Computing

A computer aided system for personal computing provides an individual with the ability to be more productive and be less frustrated in the conduct of daily and special tasks. Such a personal system must be able to access larger data bases to obtain input data and to deliver outputs, and must be capable of providing the special calculations and data processing unique to each individual. In addition, such a system must facilitate interactive problem solving, the application of development tools and the presentation of data in graphic formats.

The Conversational Monitor System of E/S^3 permits interaction with other work stations, and permits each work station to use the full computer resource. Each work station is provided with a full screen editor, the ability to operate in concurrent, online and batch mode, and the ability to communicate not only with other work stations, but also with other computers. E/S^3 uses VS/FORTRAN as the major programming language, and supports FORTRAN 77 ANSII standards. An interactive debug capability is provided and a high-accuracy arithmetic subroutine library is available, as well as an elementary math library.

Engineering and Scientific Administration

The management and administration of engineers and scientists can also be facilitated with a computer aided system. Software is available that can assist the planning, monitoring and control of engineering and scientific activities. Technical organizations can be examined as separate units and the scheduling and project management of these units can be documented, planned and controlled with such software.

FUTURE

As the use of the E/S^3 system increases and more experience is available, the need for other software packages can be clarified. One major concern for the future of E/S^3 systems is the general direction of packaged software development. Software packages seem to fall into two categories. Either the software has been designed for ease of use and has sacrificed features, function and performance to become user friendly, or the software is sophisticated and offers many features, functions and high performance. The system environments (operating systems) also follow this pattern and trade ease of use for fewer features. One of the new developments that may reverse this is the development of artificial intelligence systems that may reduce the need for the user to have knowledge of how to use the software. The artificial intelligence system will guide the user through the use of the software and will react to the inputs made by the user.

If the trend to develop sophisticated software for sophisticated users is not reversed, then the ability to enhance interactive computing and interdisciplinary problem solving on computers will decrease. The disciplinarity of computer aided problem solving will be perpetuated and fewer participants would enjoy the advantages of such systems.

MANAGING HIGH TECHNOLOGY
An Interdisciplinary Perspective
B.W. Mar, W.T. Newell, and B.O. Saxberg (Editors)

BRIDGING RESEARCH AND DEVELOPMENT

Israel Dror

Rafael
Defense Research and Development Authority
Haifa, P.O.B. 2250

ABSTRACT

The paper refers to policies of managing R&D in high technology organizations. Different R&D categories and phases are surveyed, with special concern to interdisciplinary research and development. An intermediate predevelopment stage between research and development is introduced, and results of implementing such an idea in one large organization are given. It was found that usually four to five years had passed from the start of a predevelopment project until its success was apparent. In the failure cases, the failure could have been seen after only two years.

Understanding the different dynamics of the R&D phases, and funding and controlling them accordingly, renders an overall more effective R&D, an R&D that matches better overall strategic goals.

INTRODUCTION

Some internal aspects of R&D activities will be defined and analyzed in the following chapters, trying to understand the internal dynamics that lead complex ideas from a research stage up to a fully developed product or system.

A perspective is first set by showing relative R&D expenditures in some industries and in U.S. Department of Defense. Following this, different R&D categories and phases are discussed briefly.

Main attention is given to the interface between a research and its successive development task, and analyzing a case in which an intermediate stage between research and development was implemented. Finally, some aspects of coordination R&D with company goals are discussed.

R&D SHARE THE INDUSTRY

Industrial R&D expenses over sales is commonly used as an indicator for R&D intensity.

This indicator is based on the assumption that some cause/effect relation exists between R&D expenses and its benefit to the industry concerned. The assumption has not (yet) been proven, however it is still a plausible one. Some correlation may even be found between R&D expenditure and future sales--but not with future profits (Parasuraman, 1983). This fact coincides with Rubenstein's observation (1976) that the R&D end-product is actually only an intermediate objective. The ultimate goals (e.g., profit) is attained after accomplishing other activities beyond R&D.

An example of typical R&D expenditures for different industry groups is given in Table 1 (I.B.W., 1983).

Table 1
Industrial R&D Expenditure
(expenditure as a percentage of sales)

Industry Group	Average Expenditure	Range
Information processing	6.4	0.7 - 26.7
Aerospace	5.1	0.7 - 12.7
Electronics	3.8	0.7 - 14.5
Automotive	3.2	0.6 - 4.8
Chemicals	2.9	0.4 - 5.4

Note that high technology companies spend, on the average, more on R&D than low technology companies.

However, the spread is very large. For instance, some of the aerospace companies are spending 12.7 percent of sales on R&D while others are spending as little as 0.7 percent of sales--and still doing good business.

Obviously each company uses its R&D resources differently. No attempt will be made here to analyze the differences in the R&D share across industries. A further discussion on this subject may be found elsewhere (e.g., Blake, 1978).

Our main concern will be the a priori break-down of R&D allocation to different R&D phases and the related intermediate objectives set within the R&D process.

Before going into more details, a reference is made to the stability of the R&D share in the U.S. defense budget outlays. During the period 1967-1982, the yearly budget outlays (in constant dollars) varied in the range of ±23 percent (G.A.O., 1983). But sure enough, each year the R&D share in the budget was quite stable: 10 percent ± 1 percent. At the same time, the procurement share varied between 19 percent and 31 percent of the yearly total budget. Obviously, allocations of total defense R&D are aimed at global strategic goals; they do not result from analyzing separate single future needs.

Industrial R&D is not totally different from defense R&D. Allocations for industrial R&D are essentially based on similar broad strategic objectives--and faith.

PHASES OF THE R&D WORK

Categories in Industrial R&D

Industrial R&D is usually seen as pursuing a new product idea from a laboratory up to the marketplace. This approach implies that the new idea is well defined and composed just at the beginning of the R&D process. According to this line of thought, a new product development cycle comprises the following six stages: exploration, screening, business analysis, development, testing, and commercialization.

This model overlooks some types of new ideas, those which are being shaped within and by the R&D process itself. Note that this is the case for most of the research ideas that intend to comprise several disciplines. Another deficiency of the model is the fact that it does not allow for technology R&D, work that is intended to trigger a line of products, rather than developing one particular product. A broader view on R&D was suggested in a recent symposium (Parker, 1983). Three categories of R&D were described. The first two conventional categories refer to R&D for existing business and for new business. The third category is defined as: "Exploratory and basic research to generate knowledge leading to new business areas, and keep abreast of developments in the basic technology of the business." The important place given to that concept is somewhat new, as it is not commonly used in the business context.

Anyhow, it signifies the role of R&D in keeping a company viable and abreast in the market place. A survey of 105 different companies had shown that an average of 7 percent of the R&D was devoted to this exploratory category. See details in Table 2.

Table 2
Percentage of R&D Spent on Four Categories
(source: Parker, 1983)

Category	Percentage R&D Expenditure	
	average	range
Exploratory and basic research	7	3 - 13
R&D for new business	12	7 - 18
R&D for existing business:		
- Major improvements	47	40 - 61
- Minor improvements	34	21 - 44

Comparison With Defense R&D

Once again, we compare our findings with defense R&D. Table 3 is an example of the U.S. defense R&D in 1980, comparing allocations to different categories. (The information was drawn from Hearings, 1980).

Table 3 is typical to the defense R&D budgets for many years. It illustrates the fact that about 20 percent of a defense R&D budget is devoted to advancing a strong technology base. The basic idea is first to invest in building a broad technology base, and then use it to support the development of advanced specific products.

Table 3
Allotment in U.S. Defense R&D, 1980
(relative share of R&D categories)

	Percentage
Research	4
Exploratory development	13
Advanced technology development	5
Total science & technology programs	22
Development programs	78
Total R,D,T, & E programs	100

A similar logic is inferred for industrial R&D. Referring back to Table 2, the exploratory basic research category may be seen as a tool by which a company can continually advance its technology base. the other R&D categories are then seen as higher echelons in the R&D process. They comprise well defined projects for developing new products, based on that advanced technology.

AN INTERMEDIATE PREDEVELOPMENT PHASE

The Need for Predevelopment

Ideas for new products are formed by the interaction of technology push and market pull. After selection, a viable idea is carried on through R&D work to produce, hopefully, a new product, process or system.

In some cases the development of the idea is straightforward. This chapter refers especially to the other cases, namely ideas for complex products, which usually comprises several disciplines. In such cases, a general idea is not enough for describing what development work is needed, it is not sufficient for defining and launching a sound development project. Some work has to be done before the actual development starts, to enable a more directed and efficient development project.

It is suggested to define such preliminary work as a separate predevelopment phase within the R&D process. This phase includes feasibility study, as well as actual R&D work, to close possible gaps between disciplines and to define plausible configuration for a suggested product/system.

In the sequence of R&D, a predevelopment phase follows the development of a technology. It differs from technology oriented research by being product/system oriented research, i.e., a research with some specific application (product) in mind.

Following a successful predevelopment comes the full development phase. The resultant development project is thus well defined, and a likely sponsor will be more easily persuaded to finance the project.

A Case Study

The idea of defining predevelopment as a separate mission-oriented R&D stage was implemented by one large R&D organization, surveyed by the author. Some of the findings are presented below. Table 4 is a summary of results of 38 predevelopment projects carried out during nine years. An outcome was defined as a success wherever a predevelopment project had actually initiated a sure development project. (The arbitrary) threshold for surety was the point where actual investment in a development project becomes ten times larger than investment in its related predevelopment project.

Table 4
Results of Predevelopment Projects

Project Outcome	No. of Projects	Average Investment in a Project	Length of Project (years)	
			average	range
Success	7	1.3	2.6	1 - 7
Failure	11	0.8	3.4	1 - 5
Not concluded	20	1.0	3.1	7 <
Total	38	1*	3.0	---

* Total investments, divided by no. of projects, defines one "unit" of investment--and used as a reference point.

It was found that usually four to five years had passed from the start of a predevelopment until a resultant development project had reached the threshold decision criteria mentioned above. The survey found twenty (out of thirty-eight) cases which, at the time of the survey, had not yet reached their threshold point.

Out of the eighteen predevelopment projects that had been concluded, seven were successful. Eleven cases were tagged failure. They include also cases where a development project had actually been initiated, but aborted before the threshold point had been reached. Investment in a successful predevelopment project was, on the average, about 50 percent higher than investment in a project that failed. The difference is due to the fact that each project was reviewed annually, and a project that showed good advances was funded more generously. On the other hand, a retarded predevelopment project got a chance to continue, but with a tighter budget.

Regarding the lengths of projects, one can see just the opposite phenomena. A successful predevelopment project was terminated, on the average, after 2.6 years (only to give way to a full development project initiated by it), whereas a failure case was terminated after 3.4 years. In most of the predevelopment cases, the failure could have been seen after only two years.

However, the work went on, though at a slower pace, trying to find ways to overcome the difficulty. In some cases the main problem of the management was how to persuade the researchers that the project should indeed be terminated. It should be stressed here that even so, predevelopment involved less budgetary and personal commitments than full development. Thus, the termination act, though difficult, was easier than in a usual development termination case.

Cause/Effect Relations

The allotment for different R&D categories in the organization surveyed is given in Table 5 for a typical year. The research program account for 15 percent of the total R&D. In it, four percent were spent on predevelopment projects. The share of the development program is 85 percent and 15 percent in it were caused directly by previous predevelopment projects.

Table 5
Allotment by R&D Phases
(an example)

	Percentage
Phenomenon oriented research	7
Technology oriented research	4
Product oriented research (predevelopment projects	4
Total research program	15
Projects initiated by predevelopment	15
Development projects (other)	70
Total development program	85
Total R&D programs	100

The research program is divided into three distinct phases: phenomenon oriented research, technology oriented research, and product oriented research. All of it is applied research, but the phases differ by their distance from the application. The nearest is the product oriented research, which includes all the predevelopment projects. Results of this research are meant to initiate some product development, and thus cause/effect relations between research and development can be easily identified. Roles of the other research phases were also set according to the relations between phases. This enabled management to establish an elaborate project selection process, in which different criteria were used to select research ideas for each one of the three research phases.

The R&D process was thus better managed. Expectations from each R&D phase were more clear. And even more the volume of work in each R&D category was more easily regulated, to get an overall more effective R&D.

DISCUSSION

Coordinating Research Activities

The U.S. department of defense has a very elaborate system to manage R&D. Their R&D phase called "Exploratory Development" is the nearest to our predevelopment phase. Its definition is (DOD, 1974): "Exploratory development includes all effort directed towards the solution of broadly defined problems, with a view to developing and evaluating technical feasibility." The approach suggested here goes one step further, by defining for each effort some mission orientation. The main advantage of such predevelopment phase is establishing a common focus to diverse R&D activities. Such a focus has become extremely important in interdisciplinary research projects. It coordinates feasibility studies and combines technology development with engineering R&D. Without consistent orientation, different research trusts tend to diverge. Results may prove to be interesting, but have only slight impact on an intended product development.

Introducing Strategy to Research

In a recent study, Booz, Allen and Hamilton surveyed trends in new products management. One of their major findings was that industrial companies had begun to develop specific new-product strategies (Booz, Allen and Hamilton, 1982). Such a new-product strategy is introduced at the beginning of a new-product development process and is used to guide it through. Booz, Allen and Hamilton have found that this approach, together with other best practices actually raised the product success rate from 1 out of 58 ideas, to 1 out of 7.

The approach described here complies with the Booz, Allen and Hamilton findings, though with a somewhat different emphasis. The main similarity is in introducing strategic consideration at the very beginning of the idea screening process. but in addition, the predevelopment phase is used to actively shape a new idea to fit some specific strategic goal. At the end of a predevelopment phase and according to its outcome, management is in a better position to set precise and final goals to a full development project.

Concluding Note

Two of the important facets of the R&D process have been discussed here. One facet is the pursuit of an idea from an abstract concept stage up to a finished product. Another facet is the coordination of activities within each R&D phase, whether technology advancement, research or development. Different R&D activities, having both short and long term objectives, are regulated by management. The common guidelines to all these diverse activities are the company strategic goals. To optimize total R&D results towards strategic goals, a company has to have an overall concept combining the different facets of R&D. Implementing such an overall concept (especially) in the R&D funds allocation process, may render an overall more effective R&D, a better aid to achieving the company strategic goals.

REFERENCES

Blake, S.P. 1978. Managing for responsive R&D. San Francisco: W.H. Freeman and Co.

Booz, Allen and Hamilton. 1982. New products management for the 1980s. New York: Booz Allen & Hamilton, Inc.

DOD. 1974. Research and development in the Department of Defense. U.S. Government Printing Office 625-101/581; May, 1974.

G.A.O. 1983. Defense spending. GAO/PLRD-83-80; June 9, 1983.

Hearings. 1980. Hearing on military posture. H.R. 6495. FY 1981, Washington.

International Business Week. 1983. R&D scoreboard, 1982. June 20.

Parasuraman, A. and Zeren, L.M. 1983. R&D's relationship with profits and sales. Research Management 26(1).

Parker, R.C. 1982. The survival of industry R&D. Conferent report by R.C. Parker, R&D Management 13(4), 1983.

Rubenstein, A.H. 1976. Some observations on the effectiveness of federal civilian-oriented R&D programs. Priorities and efficiency in federal R&D, a compendium of papers submitted to the Subcommittee on Priorities and Economy in Government of the Joint Economic Committee, Congress of the United States, Washington; U.S. Government Printing Office, pp. 46-64, October 1976.

MANAGING HIGH TECHNOLOGY
An Interdisciplinary Perspective
B.W. Mar, W.T. Newell, and B.O. Saxberg (Editors)

THE KNOWLEDGE HOUSEHOLD: A NEW CONCEPT FOR COMMUNICATION AND PROGRAMMING WITHIN THE PHYSICAL PLANNING PROCESS

Hans Jacobs

Department of Architecture, Building and Planning
University of Technology
Den Dolech 2, P.O. Box 513
5600 MB Eindhoven, The Netherlands

INTRODUCTION

This paper discusses the relationship between research, planning and public policy making in the field of urban and regional development and concerns such issues as land use, settlement, and urban change. It is especially focused on the problems of communication and programming of policy research. Planning here is seen as an activity to increase the interrelations of future decisions by connecting knowledge to organized social action.

Although research, planning and public policy making within an organizational context are often seen as a continuum, we are interested in the distinction between research on the one side and planning and policy making on the other side. This distinction is often discussed as a gap which can be bridged by a better organization of the relationships between both sides, or by increasing the utility of knowledge through new research methods, or by improving programming activities of research as well as policy making. Though these adjustments sometimes are effective, they are criticized because of their traditional, instrumental view on the relationship between science and politics which neglects the influence of social relations.

In recent literature on policy research we have identified three important problems which are of crucial significance in understanding knowledge utilization in decision processes in public policy making.

1. Decisions are not made in one single moment but rather as a process of decision accretion. Decision making is often preceded by planning and followed by implementation, management, and control activities. In each of these activities successive decisions are taken.

2. Within decision making many actors are involved directly and indirectly both inside and outside governmental institutions, each with different interests, standards, and values, and different perceptions of the problems at stake. So decision making must be considered as a primarily political process, ruled by a political ratio and therefore often in sharp contrast with the scientific ratio.

3. Decision making is a practical process characterized by consultations, negotiations, confrontation, and bureau-

cratic organization of human actors. These processes require different types of knowledge (both fundamental and applied research findings; both scientific and casual everyday knowledge).

These characteristics and the problems related to these are some of the central issues within the debates on technological development as well as on urban and regional (physical) planning. The positions taken within these debates will be related to three fundamental models of the relationship between science and policy making within society: the decisionistic, the technocratic, and the pragmatic model. Within the pragmatic model all three characteristics mentioned above are integrated. This model is seen as a fruitful base to understand and improve the relationship between research on the one side and planning and policy making on the other. Within this model fits a new concept of policy research: the management of the "knowledge household." We will discuss this concept as a framework for communication between all political actors involved as well as a framework for programming policy research.

The ideas presented are the results of the reflection within our research group on current planning practice in the Netherlands rather than based on empirical research.

This paper is structured as follows. We will first draw some lines through the debates on technological development and physical planning. Next we will describe the three fundamental models on the relation between science and politics, resulting in some conclusions. We will then discuss some characteristics of policy research. The knowledge household will be introduced and finally we will discuss the need for pluriformity in knowledge production for public decision making.

THE DEBATES ON TECHNOLOGICAL DEVELOPMENT AND URBAN AND REGIONAL PLANNING

Technology and technological development have become one of the main foci of economic policy of governments in most industrial nations. It is seen as an important means to cope with the economic recession. This policy is generally based on the assumption of a direct relation between science and technology development on one hand and economic and social progress on the other. The role of government is seen as restricted to stimulating finances and to the creation of an infrastructure that allows technology to develop as smoothly as possible, e.g., by the creation of technology transfer centers, technology oriented education programs, etc.

Opposite to this is the view others are looking at the social problems which are related to technological development, such as degradation of natural resources, differing benefits for specific social groups, environmental damage, the increase of health risks, military applications, unemployment, etc. Within this view the social control on technological development and its effects and direction are emphasized as crucial. The question no longer just concerns whether technological development can be related to social needs and desires, but also whether technological possibilities are socially desirable as well (Tuininga, 1981; Rip, 1981).

These two approaches, creation (financial and technical) versus control, are in conflict very often. Policy makers and entrepreneurs must meet both of them nevertheless. Moreover, other actors have become involved in technological problems as, e.g., citizen groups organized on issues such as housing, environmental problems and consumer interests (Nelkin, 1978; Nelkin, 1979).

One of the ways in which governments are trying to bring those two approaches together is through planning and participation. However, communication fundamental for these activities is very often hindered by severe obstacles. Besides political and power barriers an important obstacle is constituted by the difference in access to data, information, and knowledge relevant to the issues at stake. In this paper we will discuss the knowledge obstacle. Technology Assessment has not been able to offer a simple medicine, as Smits and Geurts (1982) have put it: "the expectation of the first hour that Technology Assessment would contribute considerably to the social steering of technological developments seems not to have come true."

Pluralism and politics apart, policy research studies have made us aware of the following: "obviously, utilization (of scientifically produced knowledge) was a complex process involving bureaucratic, ethical, attitudinal and social considerations that took precedence over the information conveyed" (Rich, 1979, p. 20). Utilization as such in a clear instrumental way could not be taken for granted, the process itself had to be understood before knowledge could be translated into action. In understanding the process it was clear that utilization should be considered in the context of overall social problem solving (Rich, 1979).

So technological development essentially is a political issue with impact on all sectors of society; a field of public policy making like many others and a field in which a broad spectrum of knowledge and information is playing an important role. A policy field in which issues such as the utilization of knowledge and information, as well as the possibilities for effective control and democracy are at stake. In this perspective, technological development does not essentially differ from other fields of public policy making. Even when we consider the fact that in decision making on technological development the government is only one of the actors together with private and semiprivate firms operating on the market. Because as Rip (1981) points out: the central problem here is the possibility and capacity a society has, to control the social process in which technological developments take place. So some of the problems at stake as summarized above can be related to other areas like urban and regional (or physical) planning which will be discussed now.

During the last few years within Dutch planning theory and practice an intensive discussion has been taking place around the role of planning in society. This discussion is centered on the so-called paradox of planning related to the effectiveness of planning to initiate and guide social change. The enormous expansion of governmental planning activities in the 1960s and 1970s has not come up to the expectations based on those efforts. It displays a picture featured by more planning and less control (Jacobs, 1974; Castells, 1978; Hall, 1981).

The social sciences, particularly planning theory, have developed and tried out numerous models in order to overcome these problems, each from different conceptions about science, about the rationality of the policy process and about planning as a scientific way of policy making. Generally these models are apolitical or even technocratic. They pay much attention to the rational combination of means and given ends to the internal system characteristics, but much less attention is paid to the problems in the implementation of a plan as a social and political process. This has resulted in overrating of the potential planning, and as a consequence it is sometimes discarded due to lack of appreciation.

The pendulum swings back: deregulation, privatizing, less control. In doing away with the filthy bathing water the baby tends to be done away with at the same time. But a modern style of planning does harmonize with deregulation and privatizing. It requires a different style of planning: communicativeness, focus on both framework planning and flexible implementation planning (den Hoed, 1983; Edwards, 1984; van Houten, 1984), awareness of the pluriformity (many interests including participation), rejection of the central-rule-approach (van Gunsteren, 1976; Ringeling, 1981; Ringeling, 1983). In sum, the new planning style aims at bringing planning back to the stage of the actual actors.

Dutch planning practice has shown the parallel between public policy making related to technology development on the one hand and urban and regional planning on the other. To overcome the actual problems different authors (e.g. Van Gunsteren, 1976; den Hoed et al., 1983; Ringeling, 1983) are stressing the need for planning with a more open and incremental approach, oriented towards societal action and actors, based on the acknowledgment of conflicting goals. This planning gives a great weight to communication, participation, and collective learning. Planning is seen primarily as a political process, in which many actors are involved. Besides, governments, (as a complex whole of actors) interest groups, individual citizens, different types of institutions and organizations are playing a role within this process. Together with this emphasis on the political dimension of planning we see a reorientation towards a substantive planning (van Houten, 1984). When planning is inventing and willing the future (van Gunsteren, 1976) then the actual situation, its problems and the possible future developments and uncertainties have to be identified. Planning must offer a substantive direction and orientation towards the future and refer to factual societal practices of many actors. Procedural and substantive planning have to be integrated within this perspective.

Governmental planning is seen as integrated into actual social relations and therefore defined in its scope and effects within a pluriform society. Contextuality is one of the main features. Planning is an enterprise, including the awareness among planners of essential uncertainties about the environment, about related decision areas, and about changing values and standards (Hall, 1981; den Hoed et al., 1983). This statement brings new insights to the role of policy research. For no knowledge can give that certainty in advance. So this type of planning requires other types of knowledge and knowledge handling not based on a one and only scientific quality standard, nor with general claims.

THREE MODELS FOR THE RELATION OF SCIENCE AND POLITICS

The debates on technological development as well as on planning can be related to the three, generally distinctive, fundamental models on the relationship between science and politics (Habermas, 1968). The decisionistic model based on the ideas of Max Weber is characterized by the complete separation of science and politics. Standards and values are seen as irrational as they result from arbitrary choices or decisions. Facts on the other hand are perceived in terms of rational considerations. Decisions are made within the political domain. Experts and scientists are the servants of politics and deliver factual knowledge. The contribution of science to politics is not recognized.

Within the technocratic model science and technology have completely penetrated the political domain, in fact they replace politics. Political decision making is rationalized on a scientific basis. Social development is governed by its own separate laws. Standards and values on which they are based can be determined scientifically. Politics is ruled by experts. Habermas criticizes both models due to their exclusive accent on scientific rationality, strictly separated from social standards and values, or the neglect of these. Political practice, on the contrary, shows an uncontrolled mixture of social standards and scientific facts. The technocratic model sees scientific and technological development mistakenly as an autonomous process. According to Habermas this apparent autonomy is caused by the invisibility of the determining social forces and networks. He does not agree with the assumption (on which this model is based) that the technically best solution is also the socially best solution from a normative point of view. Social practice shows that a problem can be solved in many different technical ways according to the social standards and values of those involved. Both models are characterized by a narrow view of the relationship between science and politics in which only experts and politicians are involved. The role of society as a whole (stakeholders, the public) is neglected.

Habermas criticizes both models due to their exclusive accent on scientific rationality, strictly separated from social standards and values, or the neglect of these. Political practice, on the contrary, shows an uncontrolled mixture of social standards and scientific facts. The technocratic model sees scientific and technological development mistakenly as an autonomous process. According to Habermas this apparent autonomy is caused by the inevidence of the determining social forces and networks. He does not agree with the assumption (on which this model is based) that the technically best solution is also the socially best solution, from a normative point of view. Social practice shows that a problem can be solved in many different technical ways according to the social standards and value of those involved. Both models are characterized by a narrow view of the relationship between science and politics in which only experts and politicians are involved. The role of society as a whole (stakeholders, the public) is neglected.

Habermas then introduces the "pragmatic" model in order to solve these problems. The strictly functional separation of scientist and politicians is replaced here by a critical interaction. In a permanent dialogue social needs and political needs are related to technological possibilities and, the other way around, technological possibilities to social and political needs. Essential to this dialogue is that it is broadened to include society. The pluriform public opinion is of crucial importance, for the outcomes affect society as a whole. New institutions in the field of public information have to enable those who are interested (in both senses) to participate in decision making. Rational decisions within this model are based on a communicative rationality, of which technical or means-end rationality is only part. Communicative rationality is based on argumentation within a dialogue between all political actors. To enable this process of formulation of political will the powers at stake have to be made explicit in the dialogue. It means that science as well as political decision-making and implementation have to be related to everyday social life and the specific problems at stake.

Besides many authors confirm the need for a critical interaction or dialogue between researchers and policy makers both to contribute to a better mutual understanding as well as to increase the rationality of decision making. Their ideas are based on the experience and results of evaluation studies of the utilization of policy research, and the understanding of the impossibility to strictly separate the formulation of social ends and the rational selection of means for achievement. They reject the idea that objective science is exclusively able to trace and define social problems, needs and desires. Within a dialogue between science and politics, social ends and means have to be mutually adjusted. Rationality of decision making in this vision is to be found within this dialogue as a significant communication process. Within this process normative judgments can play an equal role and the public opinion gets an important place. In his recent study on the theory of communicative acting Habermas (1981) gives a further explanation and instrumentalization of this pragmatic approach based on a thorough analysis of the actual social problems and the state of sociological theory in Western society. We ascertain that the pragmatic model offers an interesting conceptual approach to the understanding of the problems in the relation between science and politics as shown in the former sections, and also, from a normative point of view, a desirable direction for finding solutions.

We see three important characteristics in the pragmatic model which can be directly related to the conclusions of the actual debate on knowledge utilization in public policy making. First, the pragmatic model departs from the very character of the decision making process. A political process is not linear in nature but involves a whole system of formal and informal practices, continuous adjustments, negotiations resulting in decisions which are usually compromises between conflicting interests, goals, values, and perspectives (Weiss, 1980; Ringeling, 1983). It is a process which is further characterized by partial views of public interest, short-term commitments and incremental and sequential ways of dealing with high interdependent and complex systems of human interaction (van Gunsteren, 1976). Second, the pragmatic model emphasizes the "three communities approach" (Smits and Geurts, 1982). Recent discussions on knowledge utilization show that the so-called linkage or gap-filling theories with their suggestions of a direct linear relationship between research findings and policy decisions do not fit in with the complex nature of the decision making process. Weiss (1982) has shown that the relationship is actually much more subtle or even problematic. Policy making and the social sciences can be said to occupy two angles of a triangle with society as a whole at the apex. Both research and policy making are component parts of a greater entity, i.e., society at large, and as such subject to its inherent forces. It is a process of society as a whole in which scientists as well as policymakers, bureaucrats, interest groups, and individual citizens are participating and have got to participate; "Democracy is too important to be left to technocrats" (Jansen, 1981). Third, in the pragmatic model the relation between research and politics is perceived as an interaction or communication process. The traditional linkage or diffusion model with its connotations of an active transmitter and a passive receiver has given way to a model in which different transmitters and receivers are playing an active interdependent role. Here utilization is not uniquely related to the various steps in the decision making process, rather it is seen as one of the fundamental sources for the communciation process among the actors in decision making. It is based on the thesis that the effect of social science research is conceptual rather than instrumental (Weiss 1982).

Different authors confirm the major conceptual use of knowledge, particularly in relation to decisions on major issues which affect society as a whole (Caplan, Weiss, Lindblom and Cohen, Rich). Rather than making a rational impact on clear cut decisions, research provides a background of data, empirical generalizations and ideas that affect the way that policymakers think about problems (Weiss, 1982).

Both the pragmatic model and the three characteristics discussed above require a broadening of the concept of policy research to be discussed in the next section.

POLICY RESEARCH

In the planning process of urban and regional matters each phase and each activity requires knowledge, either existing or to be acquired. The acquisition may be done either by production of new knowledge or by processing existing knowledge. Planning as the complex combination of substantive and procedural issues requires a great amount and a wide range of policy research, and the knowledge and information which result from it. Thus, policy research has to be inter- or even transdisciplinary in character as policy problems are not defined in terms of (academic) disciplines. Solving these problems therefore requires the combined involvement of several disciplines, resulting in integrated studies. These studies are characterized by linking and/or integrating theories, methods, cognitive structures, and processes, information and expertise gained in different disciplines.

Policy research is justified by the intended direct or indirect use of its results in policy processes. It therefore is utilization focused. The goal of a policy research process as well as its methodology should be adapted to this goal (van Lohuizen, 1981). Because the direct orientation in policy research is towards practice, the situational (time, space) character of the problem has a significant influence on the research outcomes. So the policy context must not be neglected[1].

Research is further expanded from mere production to the handling and management of knowledge (van Lohuizen, 1980; van Lohuizen, 1983a,b,c; van Lohuizen, 1984; Jacobs, et al. 1983). In this expansion lies an important characteristic of policy research which becomes more and more central.

First we see an enormous increase of information in society. Information overload and abuse are well known (Cuilenburg, 1982). In many cases it will not be necessary to produce new knowledge and information, but rather to find and process existing knowledge and information. Processing is done in terms of structuring data and information into knowledge, translating, synthesizing, applying and integrating knowledge in relation to the problems at stake, as fundamental activities for understanding, judging, and decision making.

Second, the debate on knowledge utilization shows that utilization in relation to public policy making must not be seen in the context of the so-called problem solving model in which social problems are solved by direct application of research findings. Research directly attached to the policy process is only one of the sources from which participants in the policy process derive their ideas and judgments on the problems at stake. In political practice a diffuse and conceptual use seems to be dominating, based on a permanent interaction of political actors and a stream of information and knowledge flowing through this process: "knowledge creep and decision accretion" (Weiss, 1980). So-called policy research findings are only part of this stream.

Taking into consideration this broader concept of knowlege utilization, the stress on communication and interaction in the planning process, the role of different types of research and knowledge, we conclude that a new concept of policy research is needed: the management of the knowledge household.

THE KNOWLEDGE HOUSEHOLD

In our view knowledge production, transfer and translation on the one side, planning and policy making on the other, and society as the third (Smits and Geurts, 1982; Weiss, 1982) must be seen as three different communities or subcultures (Caplan, 1975; Zaltman and Desphane, 1983). These have a different language and different action and perception frameworks, and therefore different specified responsibilities, although they have a common task in finding solutions for policy problems. To make the two or three communities meet, an intermediary framework is necessary, which sometimes may materialize in organizational forms. In this intermediary framework the problem of translation is a crucial one; it must consist of a mutual communication framework to create the possibilities for dialogue and interaction. It means that knowledge production, transfer, and translation as well as planning and policy making have to be mutually broadened without neglecting their specific responsibilities. The intermediary framework is to be found in the overlap which results from this broadening. Through an accurate definition of the characteristics and demands of the different communities, a common language can be developed, not only for mutual understanding and use of information but also for selection purposes, e.g., selection of information supply and demand (van Lohuizen, 1982). On the one hand this means explication of the characteristics and demands of the process of planning and policy making, the political character and the use of different types of knowledge. On the other hand, it means explication of the characteristics and demands of the research process. So this mutual broadening is a conceptual broadening of the different communities within the intermediary framework. This mutual explication and definition process constitutes both the communication framework and a common language.

This intermediary framework is called the knowledge household[2] of a planning process, which is defined as all the available and unavailable knowledge elements relevant to a planning process and particularly to decision making in that process. In the knowledge household the elements are handled in terms of management: efficiency and effectiveness, by both responding to the political character of the planning process, and representing the scientific requirements implied in handling knowledge. Activities in the knowledge household, among others, are: gathering, structuring, qualifying, marshalling, controling, analyzing, synthesizing, interpreting, translating, transferring, disseminating, utilizing. In this sense the knowledge household can be seen as a widening of the concept of policy research by including the handling and management of various forms of knowledge (van Lohuizen, 1983a,b,c; van Lohuizen, 1984).

The knowledge household must be understood as both a structure and a dynamic information process directly attached to the planning process: it serves as a framework for programming relevant research. It must be seen as a process in which knowledge and information elements are selected and translated into usable knowledge attuned to specified planning and policy problems at stake in order to understand and judge them, as well as their possible solutions. At the same time it must be seen as an information demand process: it helps translating planning and policy problems into suitable scientific problem definitions for policy research. As decision making and planning are seen as continuous processes, permanently adjusted to changing practical situations, the production, management and translation of knowledge within the knowledge household must be organized in a continuous way too. As concluded by Yin and Gwaltney (1981): "social science knowledge is most readily utilized in service settings when a human communications network has been built and maintained. The network must include both knowledge producers and knowledge users, and the network must operate continuously even as knowledge production is occurring" and as planning and policy making is progressing. They propose the use of linkers or field-agents within this network. They would have the same characteristics as the disseminators of Knott and Wildavsky (1980): translators, interpretors and gatekeepers for government decision makers. But it is questionable whether new roles should be introduced in the intricate network operating already.

A further characteristic of the concept of the knowledge household is that it need not and surely should not be limited to scientific knowledge. Policy making and planning depends to a large degree on everyday casual knowledge. Accepting both types is not only desirable but also necessary to create a suitable communication framework between policy research and planning and policy making.

Within this framework it becomes possible to trace and explicate also the limitations of knowledge and information in relation to a specific planning or policy problem. By structuring the information and knowledge supply the knowledge household may contribute to overcome two currently occurring problems in knowledge production for planning: information overload and information lack (van Cuilenburg, 1982; Knott and Wildavsky, 1980).

In this way the knowledge household gets an active role, not only in connecting supply and demand of knowledge and information, but also in tracing ignorance in planning and policy making. Both knowledge and the lack of knowledge are conditioning factors in planning and policy choices to be made. It is therefore required to explicate the knowledge which is lacking besides the knowledge which is available (Kruithof, 1975). The advantages of this activity are clear: signalizing these knowledge deficits can sometimes result in immediate new research projects, or in programming activities, or in transforming existing knowledge at other places. It also enlarges the transparency of the planning and decision making process and the possibility of making adjustments in the next phase. It results in flexible and effective programming. So the knowledge household is a center for the social dialogue on public policy making. Therefore, it has to be accessible for all political actors involved (the three communities); it needs different inputs and feedbacks. It therefore must be a public knowledge household. This implies that the knowledge household, although closely related to, and in interaction with, the planning process has to maintain an independent position.

THE PUBLIC KNOWLEDGE HOUSEHOLD

In programming, like in contracting and conducting policy research for planning and public policy making, government plays an important initiating role. Very often there is a governmental monopoly. For several reasons this is not a very desirable situation. First, there is a narrow relationship, as we said before, between the context in which policy research is initiated and done and consequently its result. Policy research is essentially influenced by the interests involved. One of the fields to which this applies is technology research. Policy making and planning are primarily political processes in which a great number of actors and often opposing interests are involved. Although government as one of the actors generally pretends not to serve a particular interest but, on the contrary, to balance interests, it very often occurs that governments, intentionally or not, favor particular interests. Moreover, governments may have their own interests at stake; e.g., employment, idle investments, landownership, competition among government agencies. So government itself, or more precisely the difference agencies involved, are participating in different roles and with different interests in the process of planning and policy making: all next to each other or even in conflict with, the other actors involved.

Thus it is advocated here that this type of research needs pluriformity both in problem definition and in conducting, not only within government agencies, but more broadly among the different actors and interests outside government.

This will bring the essential elements of planning and policy making again in the political arena where they belong. And at the same time it enlarges the possibilities for political control and criticism of the planning and policy making process by interested citizens, counter movements, etc., a crucial condition for democracy.

Knowledge production and utilization have become important instruments in politics. This must not be monopolized by government or any other interest group because participation can only be effective if specific goal- and interest-oriented policy research is conducted and made accessible for those wanting to participate (Jacobs, 1983). It is necessary on the one hand to accept the relative significance of research findings (Lindblom and Cohen, 1979; Weiss, 1977, 1980, Weiss, 1982, Weiss, 1983; Caplan, 1975, Caplan, 1981) and on the other hand to emphasize the great importance of research findings as a conditioning factor in a political debate. Industry and government aside, it is of particular interest to pay attention to citizen groups and organizations which, because of their social position often have limited access to available knowledge and generally have no means to contract research projects. Especially these groups often have major difficulties in getting their problems and interests put on the political agenda, particularly in periods of economic crisis.

In the Netherlands a movement is underway to assure the improvement of both access to knowledge and contracting research projects for these so-called disadvantaged groups. Science Shops and Institutes for Socially Oriented Research are operating within or in close relation to universities, where scientists and students, mostly on a voluntary basis, conduct research projects to meet the needs of social groups (Jacobs, 1983). By a pluriform programming of policy research within the knowledge household, accessible to all political actors involved, as well as for the different types of knowledge and information they need and produce, the knowledge household becomes a public one. The public knowledge household can be seen as an essential framework for a dialogue within a pragmatic approach to public policy making on both technological and physical problems, which affect a broad spectrum of society.

FINAL REMARKS

The debates on both technological development and planning are showing the central role of the three characteristics of decision making in relation to knowledge utilization, mentioned in the introduction: planning and decision making as a continuous process, the involvement of many actors, and different types of knowlege playing an important role. These characteristics require a different conception of policy research. As such we have discussed the management of the knowledge household developed in relation to Dutch planning practice. The public knowledge household is seen as a knowledge and information framework for a pragmatic approach to public policy making based on a dialogue between the three communities: political government agencies, research institutions and society in its pluriformity. This is not only desirable but even necessary for long-term decisions to be made on both urban and regional development and technological development, with a great impact on society as a whole. It is seen as a foundation for a society in which democracy is one of the crucial values as well as a condition which enables social implementation of planning and public decision making on these subjects.

Finally, as concluded by the Dutch labor-psychologist Vreeman (van Empel, 1984): "confrontation of different opinions and types of knowledge often results in creativity and innovation" central issues of this conference!

FOOTNOTES:

1. In a study on the patterns of research utilization in private and public sectors, Zaltman and Desphande point out the multitude of contextual factors which define the use and useability of research outcomes (Zaltman and Desphane 1983).

2. In Dutch "household" is a term used both in the same sense as in English and in a metaphorical way. In both cases it is used for the sum total of internal affairs, both structures and processes, e.g., for the affairs of a state or a company. In German it has come to signify budget.

REFERENCES

Caplan, N. 1975. The use of social science knowledge in policy decisions at the national level. Paris: O.E.C.D.

Caplan, N. 1982. Social research and public policy at the national level. In Report S.V.O.-Workshop: Social Science Research and Public Policy-Making: A Reappraisal. Windsor: Nelson.

Castells, M. 1978. City, class and power. London: The MacMillan Press, Ltd.

Cuilenburg, J.J. van. 1982. Zuinig met, zuinig op informatie, een beschouwing over informatisering en de "informatiesmenleving," Amsterdam: Oratie V.U.

Desphande, R., and Zaltman, G. 1983. Patterns of research use in private and public sectors. Knowledge 561-576.

Empel, F. 1984. Arbeidpsycholoog vreeman: het wordt een nieuw proletariaat. Haagse Post 30 Juni:8-11.

Edwards, A. 1974. Nieuwe bruggen tussen planningstheorie en planningspraktijk, een beschouwing over planning in "maatschappelijk handelingsperspektief. Beleidsanalyse 1:27-32.

Gunsteren, H.R. van. 1976. The quest for control: a critique of the rational-central-rule approach in public affairs. London: Wiley.

Habermas, J. 1958. Technik und wissenschaft als "ideologie" Frankfurt am Main: Suhrkamp.

Habermas, J. 1981. Theorie des Kommunikativen Handelns. Frankfurt: Surhkamp.

Hall, P. 1981. Great planning disasters. Penguin Press.

Hoed, P. den, Salet, W.G.M. and Sluijs, H. van der. Planning als onderneming. WRR Voorstudie 34 (Staatsuitgeverij) 's-Gravenhage.

Houten, D.J. van. 1984. To plan or not to plan, het planningsdebat heropend. Beleidsanalyse, 1:4-9.

Jacobs, H. 1983. Planologisch onderzoek in het spanningsveld tussen belangenbehartiging en wetenschappelijkheid (University of Technology Eindhoven, BUR-ROPS Werkdocument 2.

Jacobs, H. and C.W.W. van Lohuizen. 1983. How to program social policy research: a view from planning. Paper for the meeting of the ISA Research Committee on Sociotechnics, Antwerp Belgium, September 6-7, 1983.

Jacobs, J. 1974. The death and life of great American cities: the failure of town planning. New York: Pelican Press.

Jansen, J.L.A. 1981. Democratie is te belangrijk om aan technocraten over te laten. TNO-Project 12:454-458.

Knott, J. and Wildavsky, A. 1980. If dissemination is the solution, what is the problem? Knowledge 537-579.

Kruithof, C.L. 1975. Onwetendheid en sociaal gedrag; een wijsgerig-sociologische studie over beperkte informatie. Deventer, v. Loghum Slaterus.

Läpple, D. and Maan, E. 1984. Onderzoek, beleid en volkshuisvesting; op zoek naar een vruchtbare samenwerking tussen onderzoek en beleid (BUGA) 's-Gravenhage.

Lindblom, Ch.E. and Cohen, D.K. 1979. Usable knowledge: social science and social problem solving. New Haven, CT: Yale University Press.

Lohuizen, C.W.W. van. 1977. Op zoek naar onderzoek voorbij de ivoren toren. Oratie: THE.

Lohuizen, C.W.W. van. 1980. Van onderzoek naar plan en werkelijkeid; notities over vertaalbaarheid. Stedebouw en Volkshuisvesting 499-509.

Lohuizen, C.W.W. van. 1982. Planning studies and planning-policy. Paper IFHP Congress, Oslo.

Lohuizen, C.W.W. van. 1983a. Utilization-focused research and planning. Paper International Conference Systeemgroep Nederland: Problems and Actors and Actions, Amsterdam, April.

Lohuizen, C.W.W. van. 1983b. Van praktijprobleem naar wetenschapsprobleem. Stedebouw en Volkshuisvesting 64:575-583.

Lohuizen, C.W.W. van. 1983c. Spanning in onderzoek en beleid: Cahiers Beleidsonderzoek 2. Amsterdam: SISWO.

Lohuizen, C.W.W. van. 1984. The knowledge household and policy making. Framework paper for the Workshop on Utilization focused Research and Planning to be held at the University of Technology Eindhoven, December 1984.

Nelkin, D. 1974. Scientific knowledge, public policy and democracy, a review essay. Knowledge 1:106-112.

Nelkin, D. 1978. Public participation in technological decision making, a review of administrative reforms. In the U.S. Proceedings of the Conference Science, Society and Education, Vrije Universiteit Amsterdam 67-81.

Rich, R.F. 1979. The pursuit of knowledge. Knowledge 1:6-30.

Ringeling, A.B. 1981. De relatie tussen onderzoek en beleid, Markerwaard van onderzoek naar beleid. SPO CAHIER 2 (SPO-NIROW) 's-Gravenhage.

Ringeling, A.B. 1983. De instrumenten van het beleid, oratie Erasmus University Rotterdam. Samsom, Alphen a/d Rijn.

Rip, A. 1981. Toch een ministerie voor wetenschap en technologie instellen, TNO Project 12:459-464.

Smits, R.E.H.M. and Geurts, J.L.A. 1982. Technology policy and the utilization of knowledge. S.T.B.-TNO, Apeldoorn.

Tuininga, E.J. 1981. Sterkere greep van de samenleving nodig op ontwikkeling van technologie. T.N.O.-Project 12:439-445.

Weiss, C.H. 1977. Using social research in public policy making.

Weiss, C.H. 1980. Knowledge creep and decision accretion. Knowledge 1(3):381-404.

Weiss, C.H. 1982. Policy research in the context of diffuse decision making. Report S.V.O.-Workshop: Social Science Research and Public Policy Making: A Reappraisal. Windsor (Nelson).

Weiss, C.H. 1983b. Ideology, interests, and information: the basis of policy positions. In eds. D. Callahan and B. Jannings, pp. 213-245, Ethics, the social sciences, and policy analysis. Plenum.

Yin, R.K. and Gwaltney, M.K. 1981. Knowledge utilization as a networking process. Knowledge 4:555-580.

MANAGING HIGH TECHNOLOGY
An Interdisciplinary Perspective
B.W. Mar, W.T. Newell, and B.O. Saxberg (Editors)

THE LINK BETWEEN TECHNOLOGICAL LIFE-CYCLE AND THE FLOW OF INFORMATION IN AN R&D DEPARTMENT

Arnoud De Meyer

European Institute of Business Administration
Boulevard de Constance
77305 Fountainebleau Cedex, France

INTRODUCTION

When one focuses on the management of research and development of engineering in a small or medium-sized company, one is soon confronted with the inherent interdisciplinarity of their efforts. While in larger companies one can imagine technological efforts and research limited to a well-defined discipline, the development of new products or new processes in a smaller organization will require the cooperation of technologists with different backgrounds and disciplines, and the involvement of parties such as suppliers, customers, or contract researchers, who do not belong to the inner circle of the company. So the first assertion on which this paper is built is the statement that any kind of development effort in a small or medium-sized enterprise must be interdisciplinary. A central issue to this interdisciplinarity is the question of with whom this cooperation should be organized.

The second assertion has to do with the kind of interdisciplinary cooperation. The simplest level of interdisciplinary effort is information exchange. Before people actually start to plan together, to invest together, to work together, and to evaluate past performance together, they will organize some kind of information exchange.

Once one accepts these two assertions, a central issue becomes the question of with whom should the company organize the interdisciplinary information exchange, which channels of information flow should be opened, stimulated or discouraged? With whom should one cooperate to get access to the best sources? Phrased that way part of the problem of interdisciplinary effort can be reduced to the problem of finding the people who can provide the company with the best information. And this is, in innovation research, a rather well-known problem.

The flow of information, in particular the type of sources of innovative ideas and information used during the problem solving process, and the type of channels through which technological information reaches the technologist or industrial scientist, has indeed been a focus of research interest since the very first studies concerning the management of research and development.

SOURCES AND CHANNELS OF INFORMATION IN THE R&D PROCESS

The importance of good communication and optimal information flows has been stressed throughout the whole history of the study of management of the innovation process. Already in one of the first analyses, Carter and Williams (1957) came to the conclusion that of the 24 factors they were able to identify as characteristics of a technologically progressive company, the six most important factors (such as the high quality of incoming information, willingness to share knowledge) had something to do with the information flow. Other landmark studies such as the Myers and Marquis study (1969), the Langrish et al. (1972) study of Queen's Award winning innovations or the second SAPPHO study (Rothwell et al., 1974) reaches similar conclusions. The SAPPHO researchers summarized it as follows: "Not only must management ensure that information flows smoothly throughout its own organization, but also that the firm is aware of development outside its own environs, particularly in the specific area concerned with innovation, including the behavior of the competitors."

If one accepts this conclusion, it becomes of the utmost importance to the management of innovation to study ways of improving the information flow, and the studies concerning this topic are numerous. In a recent survey of publications about communication in an R&D department, we have classified (De Meyer, 1983) the results of empirical studies concerning the information flow in an R&D environment into 4 categories: sources of information, channels through which information is acquired, the role of individuals in the information flow, and the influence of organization and architecture upon communication.

However, given the importance of communication, the results of the empirical research provides the innovation manager with hardly any rules of thumb to steer the information flow. As an example, in Table 1 the results of some studies about sources of information have been summarized (after some transformation to obtain comparable data). The least that can be concluded from these studies is that a clear answer to the question "which are the most important sources" is not available. The number of externally generated ideas range from 11 percent to 65 percent of the total number of ideas. For those cases where sufficient data are available, the personal experience as a source of information ranges from 8 percent to 66 percent of the total number of information messages.

A main conclusion of these studies is, consequently, that there is no such thing as an optimal portfolio of sources or channels through which information reaches the R&D department. The differences in results suggest that there might be some contingency factors which influence the optimal portfolio of sources and channels. The studies of Olsen (1975), Van Hippel (1982) and VanderWerf (1982) suggest that the source of innovation might be dependent on the subindustry. Olsen concludes for example that in the textile industry the suppliers play an important role in initiating the innovation process. To Van Hippel the bulk of the ideas will come from those outside parties who may benefit most from the innovation. Some authors make a distinction between the initiation of the innovation project and the problem solving phase (Myers and Marquis, 1969; Uterback, 1971), between product and process innovation or between small and large companies (Rothwell and Robertson, 1973).

Though each of these factors clearly has some influence, none of them provide a real answer to the question of how the information

Table 1
Sources of Information During the Innovation Process
(in % of the total number of sources)

	Personal	Internal in the Company	External to the Company
Baker e.a. (1967)	64	25	11
Myers and Marquis (1969)	66	12	22
Utterback (1972)	7	47	46
Langrish e.a. (1972)	-	35	65
Gibbons & Johnson (1974)	36	30	34
Allen (1977)	8	42	50
Souder & Chakrabarti (1978)	-	84	16
Nystron (1979)	-	62	38

flow in an R&D department should be organized. The model presented in the following sections tries to contribute to the explanation of some of these contradictions by relating the structure of the information flow to the characteristics of the technological life-cycle. In this paper it is indeed argued that all contingency factors have one common underlying factor, namely the maturity of the technology and the impact of this technology maturity on organization, task and information flows.

THE TECHNOLOGICAL LIFE-CYCLE

To explain the differences in results concerning the flow of technological innovation, it is our opinion that one should not rely primarily upon an external contingency factor, but rather that one should explore the characteristics of the technology itself. Recently a model has been developed which describes the evolution of productive unit and which considers the characteristics of the product technology as well as the process technology (Utterback and Abernathy, 1975; Abernathy, Clark and Kantrow, 1983). This productive unit is defined as "an integral production process that is located under a common management to produce a particular product line." The essential idea of a productive unit is, as the classical formulation runs, that it "tends to evolve and change over time in a consistent manner" (Abernathy, 1978, p. 48).

The evolution over time of such a productive unit is characterized by a shift in the emphasis on the kind of innovation activities, the degree of flexibility, the vertical integration of the production process, and the degree of standardization of the products.

Briefly, one can distinguish three phases (see Abernathy, 1978, for a more detailed description):

1. A <u>product phase or fluid state</u> characterized by:

   a product-performance-maximizing attitude and competition of the basic of product characteristics

   a high frequency of product innovations, mostly based on expectations about developing and expanding user-needs, rather than detailed market research

   generally a large variation in the number of competitors, because the entry fee (investments in know-how and capital goods) can be relatively low at this stage of the technological life-cycle

   an entrepreneurial approach by management and an organic organization

   a noncoordinated, flexible, nonstandardized production process, built around general purpose

2. A <u>transitional phase</u> characterized by:

   the gradual emergence of one or more dominant product designs

   emphasis on product-variation rather than radical product innovations

   a segmented production-process in which automated sections exist besides manual operations

   a gradual increase in task specialization and formal procedures

3. A <u>process phase or specific state</u> in which:

   competition is based mainly on price, which consequently leads to an emphasis on cost reductions and the streamlining of the production process

   process innovation is strongly emphasized

   products become more and more standardized

   efficiency, scaling up and vertical integration are emphasized as strategic objectives

4. Finally one can add a fourth stage in which the productive unit reaches a state of maturity, in which neither product nor process innovation is of primary strategic importance.

Though the question of what the mechanisms are, due to which productive units evolve along this technological life-cycle, or which difficulties have to be overcome in order to go from the third to the second or first stage is an interesting one, here we will limit the use of these four stages to classes, or groups in which productive units can be categorized based on their innovation behavior. One of the most important consequences of the technological life-cycle concept is that to whatever class a productive unit belongs, the nature of its product and process innovation and the characteristics of its manufacturing process and organizational flexibility should be tuned to each other. Our central hypothesis is that the characteristics of the information flow are also influenced by the stage of the technological life-cycle in which the productive unit can be classified.

A MODEL FOR THE FLOW OF TECHNOLOGICAL INFORMATION

The core of the model presented here, implies that the most important sources of technological information and the dominant element of the information network are related to the class or group to which a productive unit (P.U.) belongs, with respect to its innovation behavior. This model is described schematically in Table 2.

Table 2
Relation Between the Information Flow in an Industrial Innovation Department and the Group to Which a P.U. Belongs, Based on Its Innovation Behavior

	PRODUCT PHASE	TRANSITION PHASE	PROCESS PHASE	MATURE PHASE
dominant source of information	scientific and technological research users/ customers	users competition suppliers	suppliers production department commercial department	other markets scientific research

Group 1: The Product Stage

In P.U.'s of the first group, ideas emerge from the fusion of the results of recent scientific and technological research and the rising market demand or the first crystallization of the user needs. The most important sources are located among the users on the one hand, and the sources of recent scientific and technological results on the other hand. The last ones are situated in (1) universities and science-oriented research institutes, (2) some competitors who act as technological foreruners, (3) in some industries the component suppliers who put research into practice at the front of technological knowledge (e.g. the producers of electronic components), or (4) companies who are active on markets which differ completely. On the market side, given the vague definition of the market needs, the detection of user requirements with traditional market research techniques appears to be less important. That detection of the gradual emergence and definition of these needs through a close relationship with the users appears to be more appropriate.

Internally the network of oral contacts reflects the orientation towards the technology. The number of contacts within the company with other departments is lower than for P.U.s of other groups. The contacts are rather oriented towards their own innovation department. Written sources, in particular those which reflect the recent scientific and technological advances, e.g. scientific journals, professional journals and reports, rather than business journals or textbooks, will form a major part of the information portfolio of the technologist.

Group 2: The Transition Stage

However, as soon as the dominant design emerges and the transition stage starts, the attention of the technologist will gradually shift from the front of technological and scientific progress towards competition and suppliers.

The gradually emerging dominant design results in a diminishing need for radical product innovations, for which one would have to consult sources of technology and science. The competitors, on the contrary, are the key figures in the definition of the dominant design. They supply an abundance of information about the parameters of the dominant design, and are consequently a source of ideas for incremental product innovation.

The start of the first cost-reduction programs and process innovation, the design of the first islands of automation, the need to supplement the know-how about the product with company-owned knowledge about the production process forces technologists of these P.U.s to look for information with the suppliers and specialized service companies.

This is a group where contacts with universities decrease strongly for two reasons. On the one hand, one does not need them any more to pursue radical product innovations, on the other hand highly specialized research institutes will not as yet be consulted in order to resolve typical technical problems which can arise when one tries to exploit a production process optimally (e.g., the behavior of materials in extreme conditions of heat, velocity or the avoidance of the influences of consequences of inertia forces, etc.).

The typical process problems which one can encounter at this stage can be solved with information and ideas from suppliers and engineering companies.

The market remains an important, if not the single most important source of information. Customers of P.U.s in this group, are offered a product whose basic characteristics are defined, which satisfies the needs, but to which quite a lot of incremental product variations and improvements are conceivable. Once the basic design has become clear, it is far easier for them to formulate their wishes and requests in a clear way, and these requirements constitute a major source of information to the technologist, while refining his product design.

This second group, which is somewhere in the middle of a flexible, organic organization, characteristic of the first group, and the well-structured, mechanistic organization of the third group, surrenders some of its slack and flexibility, for an increase in productivity and a more opaque structure. This is reflected in the streamlining of the channels through which information flows into the organization.

Group 3: The Process Stage

Group 3 is, contrary to the first two groups, characterized by a well defined organizational structure, an emphasis on process innovation and incremental product innovations, and a lower perceived environmental uncertainty.

The required technical information mainly concerns the implementation of process improvement, and is sought for internally, with suppliers and with specialized research institutes and laboratories performing research on the optimization of production processes. Given the fact that process know-how is pre-eminently company-specific information, the internal source, in particular the production department, becomes more important than in the first two groups.

The customer requests remain important in order to realize incremental product improvements. These needs are easy to formulate and can be discovered through traditional market research techniques. To the technologists it is not that important any more to design a product together with the users but rather to understand their requests as accurately as possible. These data can easily be supplied by the commercial department, a market research department, or a technical support department. The informal contacts with the customers can consequently be replaced by formal channels (e.g., internal written reports, or memos originating in the commercial department).

The wide gap between the technological needs of this group and the related scientific supply results in the fact that science-oriented journals become unimportant. Supplier brochures, journals which reflect the improvements of production process and journals of professional organizations become more important in the portfolio of the literature relevant to the technologist. Complementary to the formalization of channels and the growing tendency to consult internal sources, internal reports become more important to the technologist in the P.U.s of this group.

Group 4: The Mature Phase

P.U.s from the fourth group do not consider innovation as a priority at all. The proper technologies, regarding the product as well as the process, have become more stabilized and R&D is limited to a kind of maintenance of the technologies in use. It is however of the utmost importance for these P.U.s to recognize and evaluate the potential of new technologies which could render their core-technologies obsolete. To realize this, there should exist technological antennae able to receive signals continuously about technological evolutions, and one should develop the ability to internalize these potentially threatening technologies.

These antennae should be tuned to a variety of broadcasting sources located in scientific and technological research. This has something to do with the above described statement that in the origins of a new technology, the symbiosis between academic and industrial research is intense. Moreover these antennae will have to be oriented towards unknown markets for the P.U.

The existing channels, even more than in group 3, are aimed at one's own company. They are consequently, completely unsuitable for internalizing information about technologies and markets. To get this kind of information inside the P.U. one has to open destructive channels, e.g. the hiring of new employees for key positions in the company, or the creation of research teams with universities, which can be located at the boundary or even outside the P.U.

Conclusion

Reviewing the model, one can see that the evolution from phase 1 to 4 is characterized by:

- a shift from external scientific research sources to internal technological sources
- an evolution of personal (oral) and externally oriented channels to impersonal and internally oriented channels
- a growing tendency to formalize the information channels
- an increase in the importance of the suppliers as a source of technological information
- a shift from scientific journals towards business journals and process orientated journals.

METHODOLOGY AND EMPIRICAL RESULTS

The model described has been empirically tested through means of a questionnaire. In 22 companies or company units which could be clearly identified with a productive unit, a questionnaire was distributed among technologists with the aim of discovering the internal and external information network, as well as the use of a number of information sources. A total of 202 technologists responded to the questionnaire.

The productive units were classified into four groups by a panel of experts on the basis of a number of characteristics such as: their emphasis on process rather than product innovation, the degree of flexibility of the production process and the degree of customization of the product, the importance of cost and/or product characteristics as a competitive weapon, the perceived uncertainty in the environment, the time span between two consecutive introductions of new products as well as the evolution of the time span, and the appearance of new technologies. Information about these topics was obtained through questionnaires filled out by a senior executive of the companies considered. Though the sample of the companies has never been stratified, the categories result in seven "phase 1", five "phase 2", six "phase 3", and four "phase 4" productive units.

RESULTS OF THE QUESTIONNAIRE

In the first part of the questionnaire the technologists were asked to indicate the importance of fourteen potential sources. In Table 3 the relative number of respondents who considered the source concerned to be important to their jobs is shown both for the total group and for each group separately. For only five sources (customer, suppliers, universities, research institutes and the commercial department) the hypothesis that the distribution of the importance attributed to these sources was independent of the categorization of the respondents in groups can be rejected at a 5 percent confidence level.

Table 3
Relative Number of Respondents Considering the Source Concerned to be Important

Source	Total (n=202)	Group 1 (n=85)	Group 2 (n=60)	Group 3 (n=40)	Group 4 (n=17)	Significance Level
Customers	50.0%	45.9%	56.7%	53.9%	42.2%	.04
Suppliers	41.6	41.1	33.3	48.7	58.8	.05
Universities	14.9	11.8	10.0	23.0	29.4	.05
Research Institutes	9.4	4.7	6.7	22.5	11.8	.01
Libraries	15.3	11.8	16.7	22.5	11.8	n.s.
Fairs	20.3	18.8	21.7	25.0	11.8	n.s.
Technical Colleges	4.0	0.0	8.3	5.0	5.9	n.s.
Computerized Databases	3.5	1.2	1.7	7.5	11.8	n.s.
Commercial Department	28.2	20.0	26.7	42.5	41.2	.02
Production Department	45.5	42.4	36.7	45.0	58.8	n.s.
Competition	40.6	37.6	36.7	45.0	58.8	n.s.
Technical Literature	50.5	52.9	48.3	55.0	35.3	.10
Former Colleagues	11.9	9.4	18.3	12.5	0.0	n.s.
Conferences	12.9	9.4	8.3	15.0	17.6	n.s.

The more detailed results show:

1. Respondents from the second group consult universities considerably less than their colleagues from the other groups. Moreover the respondents from the fourth group are the most interested in universities. Further examination of this result revealed that in each of the phase 4 companies considered, some projects were initiated with universities through which it is hoped to acquire the new technologies needed to overcome the technological maturity.

2. Respondents from groups 3 and 4 in particular consider the commercial department to be important. This is what one would expect from the model: given the fact that their productive units are better structured, that tasks are more clearly specified, that the flexiblity of the organization has decreased in favor of efficiency, then the contacts with customers will be pre-eminently along procedural lines, in other words, through a well-defined commercial department.

3. The customers clearly belong to the "stars" among the sources. They are particularly important in groups 1 and 3.

4. Suppliers are, as one would expect, more important to respondents from groups 3 and 4. This does not imply that they are not important to other groups. It is striking that, even in group 1, more than 40 percent of the respondents consider this source to be important. Further examination of the results revealed however, that this high proportion is due, in particular, to some respondents who have recently introduced microelectronics into their products, and who in a first period had to rely heavily on component suppliers. If one eliminates these responses, the relative number of respondents from the first group who consider suppliers to be an important source drops to 32 percent.

5. With respect to the research institutes, the results are influenced by the respondents of one particular company, active in the agrobusiness, which relies heavily on the agricultural research institutes. This reliance on research institutes in the agricultural field is not however a typical Belgian phenomenon, but can be found in other countries as well.

6. The result contradicting the model is the one concerning the production department. One would expect this department to be a less important source to respondents from the first group, than to respondents from the third group. The results of the questionnaire do not allow any distinction to be made between the groups.

The Use of Literature

The questionnaire allowed a calculated estimation for the number of times per year written sources were consulted (Table 4a). Though

Table 4a
Frequency of Consultation
of Written Sources (per year)

Group	Frequency
1	293
2	184
3	251
4	234

$F = 2.97;\ p < 0.05$

Table 4.b
Use of Literature: Number of Persons
Who Consult the Type of Literature Considered at Least Once a Week
(in % of the total number per group)

	Total Sample	Group 1	Group 2	Group 3	Group 4	Significance Level
Textbooks	34.2	45.9	21.7	32.5	23.5	n.s.
Scientific Journals	27.2	32.9	10.0	35.0	41.2	0.01
Professional Journals	45.5	56.5	30.0	45.0	47.0	0.05
Supplier Brochures	56.4	63.5	50.0	52.5	52.9	n.s.
Journals of Professional Organizations	7.4	3.5	5.0	10.0	29.4	0.05
Journals of Branch Organizations	4.5	2.4	3.3	7.5	11.8	n.s.
Free Journals	22.3	32.9	16.7	15.0	5.9	0.05

the absolute number has not, as such, primary importance, it gives an idea of the relative importance of written sources to the respondents.

If one distinguishes between eight types of written sources (text books scientific journals, popular journals, supplier brochures, journals of professional and research organizations and free journals) the difference in reading behavior is mainly due to four types. Respondents from the first group read more scientific and professional journals (in particular the first ones), which reflect recent scientific and technological developments. Respondents from the fourth group read journals of professional organization more often than the others (which usually reflects more process-oriented developments). Free technical journals (which are more often promotional brochures) are frequently consulted by respondents from the first group (Table 4b).

CONCLUSIONS

In this paper we have presented a model which tries to explain some of the inconsistencies which exist in the empirical studies about the flow of technological information in an R&D department. We suggest that the underlying factor which explains the inconsistencies is the technology, in particular its characteristics as described by the technological life-cycle.

Though the empirical data support, to a large extent, the proposed model, we are aware of the limitations of the evidence presented, which is based on a sample of 22 companies, mainly producing assembled industrial products. The main result of the study remains however the fact that a relation between the nature of the sources and the way in which they are consulted by the technologist on the one hand, and the group to which a productive unit belongs on the basis of its innovative activities on the other hand cannot be denied. Literature, nonhierarchical gatekeepers, suppliers, and professional organizations are more important to group 3.

As tentative conclusions, with respect to the management of an R&D department we suggest first that, if a productive unit is forced by the environment, or chooses as a result of its own strategic choice, to move from one phase to another, that it has to adapt its system of technology flow and the composition of its extended interdisciplinary teams. The increasing formalization of the information gathering and the communication network, and the shift of attention from users to suppliers both seem to be natural evolutions, in the case where the productive segment evolves from a product-innovative and entrepreneurial unit to a process-innovative and efficiency-oriented unit. As a second conclusion on the managerial level it is important to realize that each productive unit could have an optimal portfolio of sources of information and transfer channels. This does not imply that for each unit of the same group, the information portfolio will be identical. Factors such as industry, size etc., also enter into the game as secondary factors, but a deliberate analysis of the optimal portfiolo, given the characteristics of the technology considered, is a useful exercise.

REFERENCES

Abernathy, W.J. 1972. The productivity dilemma, roadblock to innovation in the automobile industry Baltimore: Johns Hopkins University Press.

Abernathy, W.J. 1983. Industrial renaissance, producing a competitive future for America. New York: Basic Books, Inc.

Allen, T.J. 1970. Communication networks in R&D management. R&D Management 1:14-21.

Allen, T.J. 1977. Managing the flow of information. Cambridge, MA: MIT Press.

Allen T.J., Tushman, H.L., and Lee, D.M.S. 1979. Technology transfer as a function of position in the spectrum from research through development to technical service. Academy of Management Journal 22:697-708.

Baker, M.R., Siegman, J., and Rubenstein A.H. 1967. The effects of perceived needs and means on the generation of ideas for industrial research and development projects. IEEE Transactions on Engineering Management 14:156-163.

Benzecri, J.P. 1973. L'analyse des données. Paris: Dunod.

Carter, C.F., and Williams, B.R. 1957. Industry and technical progress. London: Oxford University Press.

De Meyer A. 1983a. Management of technology in traditional industries: a pilot study in ten Belgian companies. R&D Management 13:15-22.

De Meyer A. 1983b. Management van informatiestromen in een industriele onderzoeksafdeling. Doctoral dissertation, State University of Ghent (Belgium).

Ettlie, J.E. 1976. The timing and sources of information for the adoption and implementation of production innovation. IEEE Transactions on Engineering Management 23:62-68.

Gibbons, M., and Johnston, R. 1974. The roles of science in technological innovation. Research Policy 11:220-242.

Langrish, J., Gibbons M., Evans W.G., and Jevons F.R. 1972. Wealth from knowledge. London: Macmillan.

Myers, S., and Marquis D.G. 1969. Successful industrial innovations: a study of social factors underlying innovation in selected firms. Report for the National Science Foundation, NSF 69-17.

Nystrom, H. 1979. Creativity and innovation. New York: Wiley.

Olsen, P. 1975. Equipment supplier - producer relationships and process innovations in the textile industry. Doctoral dissertation, Harvard Business School, Boston.

Rothwell, R., Freeman, C., Harlsely, A., Jarvis, V.T.P., Robertson, A.B., and Townsend, J. 1974. SAPPHO updated - project SAPPHO phase II. Research Policy 2:204-225.

Rothwell, R., Robertson, A.B. 1973. The role of communication in technical information. Research Policy 2:204-225.

Shotwell, T.K. 1971. Information flow in an industrial research laboratory - a case study. IEEE Transactions on Engineering Management 18:124-131.

Souder, W.E., Chakrabarti, A.K. 1978. The R&D - marketing interface: results from an empirical study of innovation projects. IEEE Transactions on Engineering Management 25:88-93.

MANAGING HIGH TECHNOLOGY
An Interdisciplinary Perspective
B.W. Mar, W.T. Newell, and B.O. Saxberg (Editors)

TECHNOLOGICAL ASSETS MANAGEMENT

David Bnaya

Interdisciplinary Center for Technological
Analysis and Forecasting
Tel-Aviv University, Israel

INTRODUCTION

Technology has a growing impact on business that transcends the traditional issues of corporate R&D management and the development of new products. The past decade reveals managers' growing awareness of the need to incorporate technological issues in several levels of decision making:

> Strategic decision making has to take into account technology as one of the major factors in the environment of business. Thus scenarios and technological forecasting are becoming part of the environmental awareness data package prepared for strategic planning (Klein and Linneman, 1983). On the other hand, there is an increasing perception by managers of the need to place technological decisions in the context of the overall corporate strategy. Strategy formulation seems the appropriate process for reconciling technology as an autonomous force that presents opportunities, threats, constraints to the stragegic plan with technology as part of and the driving force behind that same strategy. The concept of corporate strategy formulation provides for the integration of a company's technological policy with its general business orientation. While some research was done in this field (Rosenbloom, 1978) there is still no proven prescription for merging technology with strategy. How to do this remains an art!
>
> Manufacturing technology emerged as an important issue in competitiveness. The last neglect of emerging computerized manufacturing by a significant part of managers led to gradual erosion of the competitiveness of many industries all over the world. This attitude is changing now, with the emergence of several methods of integrated, computer assisted manufacturing.
>
> R&D management and the management of the company's technology base are seldom coordinated. There exists a considerable body of research (Clarke, 1981) concerning the various aspects of R&D management. The management of the technology base is a less than frequent research topic. Top management starts to realize that there is a need to better coordinate the R&D action with the building of the technology base. A strong technological infrastructure is a competitive asset and it is a neces-

sary weapon for the implementation of a strategic response to technological threats.

Our awareness of the need to manage the technological assets of a corporation has developed gradually during the years while doing research in technological innovation and in related fields for some high-tech firms. In most firms technology management, if at all considered, is often regarded as a synonym for R&D management, the main concern of which are the selection and the management of research, development and engineering projects. In this context new technology is developed for the sake of the existing development portfolio. Major development and engineering projects are most of the time successful in acquiring the necessary technology. The areas that R&D management seldom cater to are:

projection and process technology
utilization of existing technological assets
internal technology transfer and technology diffusion

It seems that there is a need to manage the technological assets of a firm in the same sense that other assets (capital, equipment, human resources, etc.) are managed. Nevertheless in developing a method for technological assets managements one has to consider that these are different from other assets in some aspects, for example:

Technology is an intangible asset that may remain in the firm's possession even if sold or leased.

Process and product innovation may turn and rend existing technology obsolete overnight and therefore reduce its value as an asset.

There is a duality of technology (process and product).

Decades of research in technological innovation (Kelly et al., 1978) resulted in an understanding of some of the phenomena involved with some management tools, mainly concerned with managing R&D. None of them attempts the management of technology as a corporate asset. Technology assets management (hereafter TAM) is suggested as a methodology to cope with this problem. It is believed that some parts of the work, besides the term technological assets management presented are novel, we refer mainly to:

managing all technology a firm has as a corporate asset, subjects to assets management techniques

establishing the need for a corporate wide technology information system to serve as a management aid for ths stake holders in TAM

The objectives of this paper are to define technological assets management (TAM) and describe a method for its implementation.

PRESENTING TAM

TAM is a management concept and methodology that ties together corporate goals and technological policy with the day-to-day management of technology in one integrated system, the objectives of which are:

to carry out corporate technology policy

to encourage and enhance the company-wide utilization of its existing technological assets

to maintain and enlarge the corporate technology base by the adaption of new technology (acquired from sources outside the corporation or developed in house)

to enable a short response time to contingencies of technological nature by the maintenance of an adequate technology management and information system

The technological assets of the firm include all know-how the firm has. The easily identifiable ones are protected by patents, licences, registered designs, copyright, trademarks and trade secrets. But on top of these TAM regards all know-how and technology, the firm owns as a manageable asset to be used, sold, lent, leased, or discarded.

TAM is an iterative process within which management strives to reconcile the demands for technology and the demands for corporate resources in the framework of a comprehensive business strategy. At this stage four aspects of TAM to be presented henceforth are identified:

Technology inventory - ways of stock taking
Technology utilization
Addition of new technology to the corporate technology base
Trade in technology

TECHNOLOGY INVENTORY

Making an inventory of technology is a natural first step in assets management. Management in a well run company knows where the company's assets are. Normally there is a strict inventory control of materials, work in progress and finished goods. Means of production, although less severely controlled, nevertheless normally management has quite a good control of capital equipment and other tools of production. This is not the case with technology because there is neither a clear definition of the technologies to be included in the inventory nor an easy way to locate them.

Technology resides with the people in the company who develop, maintain and do something with it. It was observed (Dror and Bnaya, 1984) that there exist knowledge centers (defined as a group of experts in a specific field with some designated equipment and software) which are the real carriers of the accumulated knowledge. Although this observation is limited to a certain matrix organization it seems plausible to assume that similar knowledge and know-how carrier entities exist in most industrial-technological companies. Our conclusion is that in reality an inventory of technology should be a list of people and their residence knowledge centers (or the like, e.g. centers of excellence, know-how centers, service centers, etc.).

Defining the technologies to be included in the inventory is not straightforward. The level of relevant technology is company dependent. There are obvious technological differences between a paper or a steel mill and an aerospace or a computer company. The main rule seems to be that the choice of technology has to be relevant to the actual and projected activities of the company (Rosenbloom and Kantrow, 1982; McGlauchlin, 1968). Therefore, in an inventory, there may co-exist technologies in various levels. By levels of technology we mean an arbitrary measure of complexity, for example:

- one step in a process (e.g. vapor deposition, crystal growing, etc.)
- a simple part (a resistor, nuts, bolts)
- a complex part (a chip, a printed card, an actuator)
- sub-assemblies (an amplifier, engine)
- products - car, TV set, fighter aircraft
- systems - computer + peripherals, dairy products distribution

Some approaches to making an inventory of technologies in a company are:

- scanning the organization tree and locating the technology carriers in the organization

- locating the technologies applied in the production process

- scanning the technologies in the products the company makes and identifying those technologies the company actually 'owns' (to be distinguished from technologies embodied in parts acquired from suppliers)

We found the last mentioned technique to be very time consuming. Nevertheless it may be worth the effort because it starts with and focuses on the main issue, the technologies in the company's products.

A directory of knowledge centers, or technologies, seems to be an appropriate form of an inventory of technologies. In order to be useful the directory contains information regarding at least the following points:

- description of the technology
- leading experts
- group size and major achievements
- exact technology capabilities (what it can do and how long it will take?)
- points of contact

TECHNOLOGY UTILIZATION

Technology utilization is a primary objective of TAM. To put to work the company's existing technologies is one of the major management problems of the decade. How to bring all parts of a company to use the existing in-house know-how is the other face of the same problem. A year ago (Business Week, 1982) TRW launched a company wide effort to enhance the company wide use of its advanced technology base in its defense and electronic division. The objectives of this move are the company's Energy & Automotive divisions which could profit handsomely from an infusion of aerospace technology. TRW is not the only company trying to put to work in-house technology company wide, many tried it with different rates of success.

How and when to measure a technology's contribution to the company's bottom line is still an unsolved problem. How one acounts for the millions invested in a not yet used technology (for example, IBM's two decades of lately abandoned R&D efforts in Josephson's junction computer technology) or for the billion dollar contract the company landed on account of its mastery of some new technology (e.g. laser-gyros for Hughes)?

To enhance the use[1] of in-house technology by project managers, production engineers, etc., there seems to be two (complementary) measures open to management. Firstly to encourage[2] voluntary action by technology developers and users. Second to take the initiative by systematically looking for opportunities to use inhouse technology.

The under utilization of a technology is perceived as a problem of TAM if there exists an opportunity to use it. The potential need for a technology may be recognized by making a technology audit (Figure 1) for each product and process. In a technology audit there is a list of all function of a product (or steps of process) and the technology used for it. The possible alternatives to each technology are listed too. A screening process of the technology inventory vis a vis the technology audits, may discover new opportunities for technology utilization.

In order to have some measure of technology utilization we propose to sum up once a year, for each technology:

its contribution to the company's main activities
licenses, patents, etc.
fees (and awards) received for technology sale and technology transfer.

BUILDING OF THE TECHNOLOGY BASE

From the technology manager's point of view a new technology becomes an asset only when it is mastered by a knowledge center to the extent that it contributes to the firm as part of product, process or some service. The definition includes any technology nesessary for the ongoing R&D activity itself. But technology that is in the stage of development by the firm is an asset yet to be. The point of the addition of a new technology is to satisfy an existing or anticipated requirement for it. The requirements for new technologies usually arise when the ongoing activity (projects, production, etc.) has problems that cannot be overcome. Anticipating technology need in order to develop it in advance may be a contribution and a major objective for TAM.

There are principal ways for adding new technologies to the technology base:

Figure 1

TECHNOLOGY AUDIT

Product / Process: TACTICAL RADAR - MANPACK

Function	Tech. in Use	Alternative Technology					
		T_1	T_2	T_3	T_4	T_5	T_6
Antenna	Reflector	Phased Array					
Receiver	Non-coherent	Pseudo-Coherent	Coherent				
Data Measurement	Beam Splitting	Monopulse					
⋮							
⋮							
Data Transmission	Microwave	Cable	Fiber Optics	Laser Beam			
⋮							
Packaging	Sheet Metal	Plastic	Composite	Plywood			

Aquire new technology either in the form of patents, licenses or know-how, or buying the whole company that developed it. This last option may be the best in case of radically new technologies to the firm, as in this case besides the technology one acquires also the necessary marketing skills.

Develop new technology as an in-house R&D task or contracting it out.

TRADING TECHNOLOGY

Buying and selling technology and know-how is part of a burgeoning trade in intellectual property which includes the licensing, leasing, and outright sale of patents, designs, copyright, trademarks, etc. Most technology transfer is a commercial activity done by willing partners, the technology donor and the technology receiver, each for its own reasons. Contrary to popular belief the monetary value of technology transfer among western industrial companies by far exceeds the volume of technology transfer either from North to South or West to East or both of them. Trading technology by a company we consider as an intrinsic part of Technological Assets Management. To make the most of technology ownership, to earn fees and royalties, on a technological asset, is a main concern with the technology manager. Buying technology is an efficient and time saving way to add new technologies to the company's arsenal. Therefore in most companies one may observe two kinds of technology trade, one is the routine transactions which are part of the regular TAM activity occuring several times a year, the other is the exceptional strategic move. Following is a short treatment of some aspects of trade in technology:

earning a return on existing technology
strategic reasons for selling technology
technology acquisition as a strategic measure
pricing technology

Most of the return will be in the form of cash income from the sale of patents and licenses, from royalties and fees for technical assistance. Some of the returns may be in the form of technology swap and cross licensing which result in the addition of new technology to the technology inventory of the firm. Return on technology is mainly net income. This is different from income generated by the sale of the product, in this case net income is only a small percentage sales volume. This difference has to be accounted for when one compares income from sales of technology to income from the sales of products and services.

Reasons of business strategy may bring a company to ease its hold on its proprietory technology and be ready to sell:

The distribution of licenses may be an effective measure for making a market when a company has succeeded in developing a novel technology for which no market exists.

Commencing a novel market by preemptive licensing is another possiblity (the distribution of licenses in genetic engineering SRI may be a case in point).

Penetration of restricted and protected markets by licenses, joint ventures and wholly owned subsidiaries may be the most frequent reason for technology sale.

Cross licensing is mainly a defensive measure that is quickly becoming a habit with defense contractors on account of mandatory second sourcing.

Acquisition as a strategic measure to take when a new technology presents a threat to the company's existing business is recommended for consideration in a study (Cooper and Schnendel, 1976; Business Week, 1983) of strategic response to the threat by 22 big firms in traditional industries. While 15 firms (out of the 22 firms studied) made major efforts to establish positions in the new technology, only two of them (United Aircraft in jet engines and Parker in ball-point pens) enjoyed long-term success as independent companies participating in the new technology. Major commitments made to the new technologies by the other firms often resulted in substantial early success but they were unable to continue it. The reasons for this inability are not obvious. The lack of long-term success may be an indication of the difficulty of an existing organization to carry out the business strategy formulated by them for the new technology. Acquisition of the new technology together with the company that developed it, may be a way to acquire not only the technical capabilities but also an organization attuned to competition in the new field.

The buying of mature technology is part of a technological catch-up strategy to be used by firms that enjoy a strong marketing position in a protected market and are willing to build a competitive position in a technology new to them. This strategy was widely adopted by Japanese firms (1960-1980) and is still used by firms that enjoy the same advantages. A quantitative model for evaluating this strategy was developed (Raz et al., 1982).

Pricing technology is a problem as the market for it is far from a perfect one. Usually the price is fixed by negotiation. There are some countries where limits to fees and royalties are set by law (Unido, 1977). There exists some models for pricing (Schwartz et al., 1983) but apparently much more research is to be done.

SOME ORGANIZATIONAL ASPECTS

TAM is a continuous management process (Fig. 2) which strives to fill the gap and to act in the void between corporate strategy and the day-to-day activities concerning technology. Therefore, the technology manager has to act on the levels of strategy, planning and operation.

These activities have different time constraints. The planning horizon for technology depends on the products and technology involved (i.e., the life cycle of personal computers is 6 months to 5 years while that of utilities stretches from 30 to 50 years). Operational management of technological assets is s continuous process performed daily while strategic planning, at its best, is a corporate ritual performed annually. The reconciliation of the demands imposed by the various time constants seems to be a potential problem area.

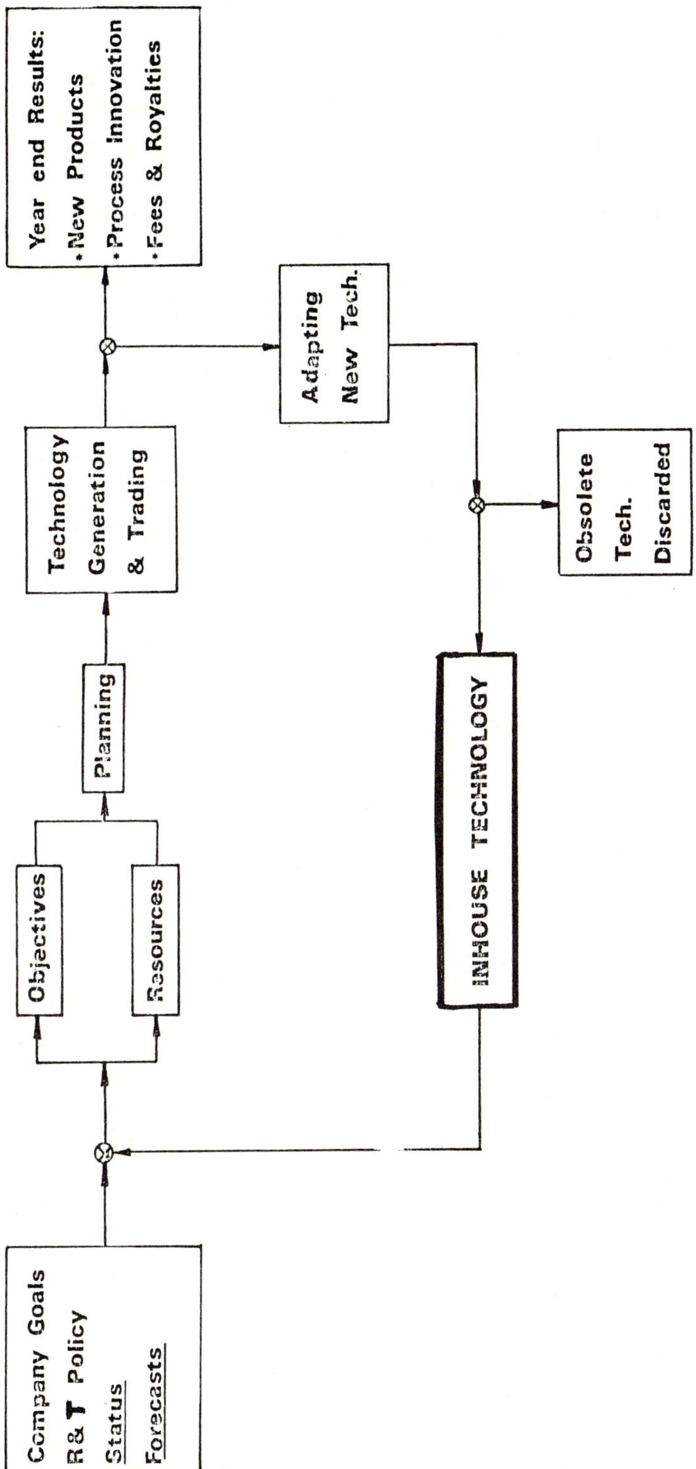

Figure 2
THE TAM PROCESS

A technology information system to support TAM seems to be a necessary management aid. Four groups of stakeholders in the TAM cycle were identified; the corporate planner, the technology manager, the technology developer, and the technology user. Each stakeholder has specific functions and objectives they require both structured and unstructured information. The system must provide access to well-defined, detailed data such as the description of existing technologies and to broad, aggregated data such as that needed to corporate policy planning. A subesquent paper (Erez and Bnaya, 1984) describes such a system in greater detail.

DISCUSSION

During research done in the last years in conjunction with some medium and small size high-tech firms, it was observed that although these firms were good at managing their assets (fiscal, manpower, production, etc.), nevertheless they were less than excelling at technology management. While technology policy, or R&D policy and its objectives were quite clear at least at the level of management, the very meaning of technology management was obscure. Under technology management some understand selection of R&D projects, others management of a considerable development project or even managing the R&D function in the firm including productivity motivation researchers. A computerized search in the literature and a manual indicated that there may be some interest in the issue of technology management at the company level.

Whether technology is manageable at all or is an autonomous force, this is a question of philosophical importance. This question was excellently treated elsewhere (Winer, 1976) and, one cannot do justice to it in a short paper.

The methodology of this work is descriptive only in the sense that some of the ingredients of TAM are based on actual cases (i.e. technology audits, technology transfer, knowledge centers, etc.). Beyond that stage the work is prescriptive by implying that the TAM idea, when applied to a corporation or a high-tech firm, will result in a sufficient technology base, a capability for adequate response time to technological threats and opportunities and with better utilization of in-house technology.

The term technological assets is defined as all technology having any useful quality for the firm and which is mastered by it.

The underlying assumption to the proposed TAM methodology is: "In the above defined sense technological assets are manageable." This assumption may be contested. Even cases of (hopefully) successful use of TAM methodology will not prove that this assumption is uncontestable. On this account, one has to regard the proposed methodology only as modest attempt to rational problem solving at the corporate or company level.

Additional research areas have been identified:

Asset management aspects of technology

value of technological asset
measurement of technology utilization
criteria for discarding 'old' technology.

Strategic aspects of TAM

methods for integration of tam into the formal long-range planning process
ways to discover technology gaps
developing criteria for the strategic value of technological assets

Criteria for the evaluation of the performance of TAM and the technology manager

information flow and needs for managing technology
technology users' information needs
reliable and low cost technology forcecasts
analyzing market information for technological content

FOOTNOTES

1. We are aware of the inherent problems of technology diffusion (Cooper and Schnendel, 1976; Business Week, 1983) with due respect to theory the proposed measures may be of practical value to management.

2. We are aware of the fact that some companies are introducing some kind of incentive system to encourge technology transfer. Mainly these are 'licensing for a fee' kind of agreements involving funds transfer from technology receiver to technology developer.

   While this measure seems to enhance the marketing zeal of knowledge centers, it may be counterproductive by acting as an inhibitor to technology acceptance. It seems more effective to reward both parties of a successful technology transfer by management.

REFERENCES

Business Week. 1983. Union Carbide, Battling slow grow by licensing technology. September 19.

Business Week. 1982. TWR leads a revolution in managing technology, November 15.

Clarke, T.E. 1981. R&D management bibliography, Vancouver, B.C.: Stargate Ltd.

Cooper, A.C. and Schnendel, D. 1976. Strategic responses to technological threats. Business Horizons, February:61.

Dror, I. and Bnaya, D. 1984. Knowledge centers - a technology & engineering hybrid. R&D management 14(2):81-91.

Erez, B. and Bnaya, D. TAMIS - a technology information system, WP-53, ICTAF. Tel Aviv University.

Kelly, P. et al., eds. 1978. Technological innovation: a critical review of current knowledge. Atlanta: Georgia Tech.

Klein, H.E. and Linneman, R.E. 1983. Environmental assessment in formal planning processes. Presented at ISF 83, The 3rd International Symposium on Forecasting.

MCGlauchlin, L.D. 1982. Long-range technical planning, Harvard Business Review July-August 54.

Raz, B., Steinberg, G. and Ruina, A. 1982. A quantitative model of technology transfer and technology catch up. MIT/Center for International Studies.

Rosenbloom, R.S., and Kantrow, A.M. 1982. The nurturing of corporate research, Harvard Business Review January-February:120.

Rosenbloom, R.S. 1978. Technological innovation in firms and industries: an assessment of the state of the Art. In Kelley, P. et al., eds., Technological innovation: a critical review of current knowledge. Atlanta: Georgia Tech.

Schwartz, L., Horesh, R., and Raz, B. 1983. Trade in technology - management decision and pricing. Technology Forecasting and Social Change 23:173-184.

UNIDO. 1977. National approaches to the acquisition of technology, United Nations, N.Y.

Winer, L. 1976. Autonomous technology. Cambridge, MA: MIT Press.

Section IV :

THE KEY TO HIGH TECHNOLOGY MANAGEMENT

MANAGING HIGH TECHNOLOGY
An Interdisciplinary Perspective
B.W. Mar, W.T. Newell, and B.O. Saxberg (Editors)

PROJECT MANAGEMENT AND MATRIX ORGANIZATIONS: THE KEY TO HIGH TECHNOLOGY MANAGEMENT?

Brian W. Mar
William T. Newell
Borje O. Saxberg

High technology organizations highlight two basic problems of designing organizations: first, how to take advantage of specialization of human and physical resources; and second, how to achieve integration of the specialties toward a coordinated effort to achieve organizational goals. These problems are not peculiar to high technology organizations, but they are accentuated by the accelerated change and added complexity associated with high technology.

Papers presented at this conference have suggested that past research on interdisciplinary processes has only revealed a small part of the complex problem of integrating the skills and specialties needed for successful high technology efforts. The case studies suggest at least four major interfaces (interorganizational, intraorganizational, intergroup and interpersonal) where interdisciplinary differences may be a factor, but interdisciplinary difference will not always be the major issue. Prescriptions for the management of these interfaces have been examined in the preceding section, but a more comprehensive approach may be needed to solve the management problems associated with high technology. This final section analyzes some of the literature proposing more comprehensive alternatives.

SYSTEMS APPROACH

The challenges in managing high technology efforts are change, complexity, and interdependency (as has been discussed in earlier sections) involving complex systems, and meeting these challenges requires a systems approach to management. In his classic work on the systems approach, Churchman (1968, p. 29) has outlined five basic considerations to be kept in mind when thinking about the meaning of systems: (1) the objectives and performance measures of the total system; (2) the environments and constraints on the system; (3) the system's resources; (4) the activities, goals, and performance measures of the individual components of the system; and (5) the management of the system. Cleland and King (1983, p. 17), in discussing the value of the systems concept to management, contribute it to two basic elements of the managerial job: (1) the need to achieve overall effectiveness of the organization; and (2) the conflicting organizational objectives with which they are faced.

The practice of the system approach usually requires the delegation of the management role to a system engineering group. This group analyzes the entire system, defines functions and requirements for each specialty group, prescribes the interfaces between the products of these groups, and establishes the tests, controls, and delivery of the final system. System engineering is a product of the military/aerospace industry and the knowledge of the practice of this management style is not widely shared (Lockheed, 1983). Education of individuals in system engineering is required if this alternative is to be effective.

SPECIALIZATION

Conflicting needs to emphasize and advance the various specialties have led to traditional management approaches of alternating between organizing on the basis of functional specialization and organizing on the basis of projects, products, geographic regions, or services to be delivered. Centralized functional specialization has advantages of better control and utilization of functional resources, but the need for centralized coordination, and increased complexity and change have led to a need to take advantage of specialization different from functions. This has been reflected in movements toward decentralization, another type of specialization in which project, product, and division managers are given autonomous control over their segments of the organization.

When one basis of designing organizations is chosen, the benefits of the other are given up. High technology organizations require skills of technical specialists who can best be employed, utilized, and developed in a functional organization. However, these same types of organizations have tasks to perform that require projects to be completed on time, within budget cost limits, while achieving high quality standards, and effectively using specialist technical resources. These goals often can best be achieved by a project type organization. But a project organization may lead to duplication and underutilization of technical specialties; and long-run technical development is difficult.

Coordination of Interdependencies

Whatever the basis used, there is a risk with any type of specialization of a consequent loss of coordination of the interdependencies that exist between organization units. In recognition of this, contemporary organization theory encourages individual line managers to improve their lateral relationships, coordination, and integration between departments and other subsystems of the organization without altering organization structure. These attempts may be seen in managerial approaches that focus on changing attitudes and perceptions, and foster improved communication. Also, recent writers in organization behavior have broadened the earlier focus on improving boss-subordinate and interpersonal intragroup relationships to give attention to problems of lateral intergroup relationships and problems of integration and coordination between organization units. Much of the organization development (OD) movement involves sophisticated methods of intervention by the personnel staff or outside consultants to help managers to improve intergroup communication and cooperation.

MATRIX ORGANIZATION

Faced in the 1960s with a situation in which both technical performance of the specialties and coordination were accentuated, the aerospace companies developed the matrix organization design and matrix management. Matrix management attempts to facilitate intergroup relations and interfaces through the use of organization structure. While there appears to be no commonly agreed precise definition of the term, the name matrix usually applies to structures which involve multiple reporting relationships. The most common form involves dual authority, a two-boss situation.

With functional and project (or product) structures representing the opposing spectrum of specialized organization designs, Galbraith (1973) described a continuum of liaison devices to foster lateral relationships, including task forces, project teams, and the matrix organization. Increased complexity and uncertainty increase the frequency of exceptions to be dealt with and consequently the information an organization has to process. At some point the organization's capacity to process information becomes overloaded. Galbraith identified four possible responses, two of which reduce the need for information processing: creation of slack resources and creation of self-contained tasks. The other two increase the capacity to process information: investment in vertical information systems, and creation of lateral relations, including matrix organization. He spelled out in detail how a matrix organization constitutes a response to complexity by dealing with communication and decision needs by establishing a type of general manager or a system engineering group in an integrating role with a dual reporting relationship. This manager is part of both the specialist technical resource department and the project department.

Cleland and King (1983, p. 277) saw the matrix form as an extension of project management in which it is a superimposition of project management on the hierarchical organization structure. In discussing the future of matrix management Sayles (1976) argued that many organizations have already adopted a matrix organization in some form without using the name. He described five kinds of matrix organization that may be found under different names. In his review article Knight (1976) cited numerous organization units and companies that that had adopted matrix structures. They were found most commonly in organizations that work on projects.

Pathologies of Matrix Structures

Implementations of matrix organization have been diverse, but often of a tentative nature, and there has been a neglect or resistance to matrix structures in many organizations. Sayles (1976) observed that this was due to a misconception that matrix structures were useful only in projects with a limited life and in R&D environments, and the fact that they are not consistent with traditionally accepted management principles such as having only one boss.

Our field research and several authors (Cleland and King, 1983; Knight, 1976; and Sayles, 1976) have shown that not many managers have a clear concept of matrix organization or an understanding of its proper implementation. It seems generally agreed that matrix organization is difficult, complex and often frustrating as compared to simpler traditional functional and product organization forms.

No organization design is without its problems, and a lack of anticipation of the difficulties of the matrix design can lead an organization into difficulties and subsequent abandonment of a matrix effort. In an attempt to forestall some of the difficulties of the matrix, Davis and Lawrence (1977) examined the diagnoses, prevention, and prescribed treatment of nine pathologies:

1. Tendencies toward anarchy, which arises from an informal matrix when people are confused about relationships between functional and project managers
2. Power struggles that can get out of hand and destroy the balance of power required to maintain the matrix design
3. Severe groupitis, or the mistaken belief that matrix management is synonymous with group decision making
4. Collapse during an economic decline, when the matrix may become the scapegoat for poor management
5. Excessive overhead associated with duplicate management costs and overstaffing
6. A tendency for the matrix form to sink to lower levels in the organization, which may come about because the matrix structure was not located at the proper level originally
7. Uncontrolled layering of matrices within matrices
8. Navel gazing, in which managers become excessively preoccupied with internal relations at the expense of paying attention to clients
9. Decision strangulation because of discussions and delays in making decisions which may arise from improper delegation, constant referral of decisions upward, or managers making unilateral decisions

Matrix Structure: A Preferred Alternative for High Technology

The various problems and limitations of matrix structures notwithstanding, the characteristics of high technology efforts make matrix structures a preferred alternative. High technology efforts were identified in Section 1 by (1) new technologies, (2) rapid change, (3) pressures from limited resources, (4) scarcity of human technical specialist resources, and (5) turbulent and uncertain environment.

The purpose of matrix structure is to provide an organization design by which the efforts of different disciplines can be applied to a common problem. This is precisely the need of high technology organizations--to apply an interdisciplinary approach to complex projects. Because of the rapid changes encountered by high technology organizations, they can take advantage of one of the main advantages of the matrix structure, which is its ability to facilitate rapid management response to changing environments.

The conditions which favor a matrix structure have been identified by several writers, including Davis and Lawrence (1977), Galbraith (1973), and Sayles (1976). Davis and Lawrence described three conditions which favor a matrix structure. First, there is outside pressure for a dual focus, such as both task and project. This may arise when organizations undertake increasingly complex tasks, and both focuses become essential to survival. It may also arise in the case of clients who may require both a focus on their unique project or product requirements, and on specialized complex technical issues. This was the situation in U.S. aerospace companies in the 1960s when the U.S. government required a project management structure as a condition of R&D contracts. A second condition is a requirement for high information processing capacity. Uncertainty, increased task complexity, and interdependence among organization units increase the amount of information that must be processed to the point where information processing capacity of conventional hierarchical organizations may be exceeded. As discussed above, a matrix structure is one way in which to increase the organization's information processing capacity. A third condition is pressure for shared human, capital, and physical resources. In a matrix organization technical specialists, equipment and facilities can be shared between several projects rather than having to be duplicated for each project with the attendant risk of underutilization. This sharing of resources also facilitates maintenance of high technical standards, because specialists retain affiliation with members of their own discipline, thus retaining one of the advantages of a functional organization.

Because all three of these conditions exist in many high technology organizations, matrix organizations may be the answer to management problems. A matrix organization can also provide flexibility, the ability to anticipate and respond quickly to changes in task or environment, which is required in high technology organizations. Managers, engineers, scientists, and other employees of high technology organizations must learn to cope with the many interfaces and interdependencies in all directions--vertical, horizontal, and diagonal--that arise within and between organizations. There are often too many of these interdependencies to be accommodated by traditional organization hierarchies; however, matrix structures can facilitate the process.

From an initial situation where the matrix organization represents organizational flexibility in meeting change, gradual deterioration in the conditions of success sets in. The increase in management personnel causes a rise in overhead costs. Continuous interaction leads to the establishment of numerous committees easily subject to the accusation of wasting time and manpower. The multiple reporting relationships introduce a confusion in authority relationships and the apparent lack of structure may result in diffuse delegation of tasks and thus loss of responsibility for project or task completion. To preserve the advantages of the matrix organization for high technology organizations requires that a continuous educational process maintains support for the matrix concept. Davis and Lawrence (1977) pointed out that the matrix organization necessitates recognition of a matrix structure, a matrix culture, matrix behavior, and matrix systems. The matrix culture is predicated on the acceptance of constructive conflict as an aspect of organizational health. Matrix behavior is predicated on the substitution of cooperation for competition in the relationship between individuals and groups and organizational units. To achieve the matrix requires management which is sensitive and knowledgeable in managing organizational interfaces.

REFERENCES

Churchman, C.W. 1968. The systems approach. New York: Delacort Press.

Cleland, D.I. and King, W.R. 1983. Systems analysis and project management, 2nd ed. New York: McGraw-Hill Book Company.

Davis, S.M. and Lawrence, P.R. 1977. Matrix. Reading, MA: Addison-Wesley Publishing Company.

Davis, S.M. and Lawrence, P.R. 1978. Problems of matrix organizations. Harvard Business Review 56(3):131-142.

Galbraith, J. 1971. Matrix organization designs. Business Horizons 14(1):29-40.

Galbriath, J., ed. 1971. Matrix organizations: organization design for high technology. Cambridge, MA: The M.I.T. Press.

Galbraith, J. 1973. Designing complex organizations. Reading, Mass.: Addison-Wesley Publishing Company.

Knight, K. 1976. Matrix organization: a review. Journal of Management Studies 13(2):111-130.

Lockheed Missile and Space Co., Inc. 1983. System engineering management guide. Defense System Management College, Fort Belvoir, Virginia, Contract No. MDA-903-82-C-0339.

Mintzberg, H. 1979. The structuring of organizations. Englewood Cliffs, NJ: Prentice Hall.

Sayles, L.R. 1976. Matrix management: the structure with a future. Organizational Dynamics (Autumn):2-17.

Section V :

INDEX OF AUTHORS AND CITATIONS

INDEX TO AUTHORS AND CITATIONS

Section VI :

SUBJECT INDEX

SUBJECT INDEX

Section VII :

APPENDICES

MANAGING HIGH TECHNOLOGY
An Interdisciplinary Perspective
B.W. Mar, W.T. Newell, and B.O. Saxberg (Editors)

GENERATIVE GRAPHICS: A MEANS FOR SETTING IN MOTION THE INTERDISCIPLINARY PROBLEM SOLVING PROCESS

Joseph Brunon

Washington State Hospital Association
190 Queen Anne Avenue North
Seattle, Washington 98109

Following Sessions I through IV at the Third International Conference on Interdisciplinary Research, Joseph Brunon acted as a facilitator for the Research Agenda Workshop, one of two workshops held to determine Interdisciplinary Research Needs. The Generative Graphics (GG) group problem-solving technique, which Brunon devised, was used. This technique permits participants' diverse, mental images of individual experiences to be externalized by a facilitator, in real time, as graphic configurations on a large sheet of paper which all participants can see throughout the course of the problem-solving session and use both to build up a storehouse of information, and also spot the interrelationships among information clusters as these become evident.

Brunon maintains that a problem solving group needs to externalize the mental images of its participants, which are based on highly abstract symbol systems characteristic of the participants' disciplines. Only after externalizing those images and actively working with them will the participants be able to communicate verbally on a mutually comprehensible level of abstraction for effective interdisciplinary problem-solving. The GG technique, along with its various derivatives, enables this necessary image-working to take place.

As a result of his experiences at the Third Conference, Brunon's convictions about image-working were reinforced and sharpened. Moreover, he was prompted to consider (and begin a paper on) the whole question of formats for conferences which address the development of generalizable approaches to interdisciplinary problem-solving, observing that the results of such conferences are usually quite disappointing. The causes of breakdown in interdisciplinary, group problem solving are interpreted by Brunon as follows:

On a spectrum of abstractions, stretching from concrete, direct experiences of a particular phenomenon being considered, all the way to the totally abstract symbol system of words and numbers being used to describe the given phenomenon, there is a wide range of cues. At conferences, presenters who deliver papers pertaining to their own discipline are able to do so at a relatively high level of abstraction using a symbol system of words, numbers, and cryptic graphic configurations to get their points across. This abstract symbol system encompasses well assimilated, mental images of direct experiences and highly developed skills. It is not surprising, therefore, that content specialists from a particular discipline, accustomed to interacting with like-trained colleagues, will naturally tend to communicate at the highest level of abstraction characteristic of that discipline. After all, they have spent life-long careers refining their individual skills in manipulating the symbol systems characteristic of that discipline, and it is expeditious for them to resort to the highest level of abstraction.

Similarly, content specialists from other disciplines have spent life-long careers refining individual skills in manipulating symbol systems, characteristic of their discipline, which are equally abstract, but different. To a content specialist from the same discipline, the abstract symbol system being delivered by a presenter provides instant cues for modifying the mental images of phenomena, fully internalized by that recipient in relation to the discipline. For that content specialist, the process of listening to the presenter and observing whatever visual materials are used is really a simple search for readily available cues which bear a relationship to well established mental images of a multitude of aspects of the discipline. Consequently, someone from the same discipline may offer support for the presentation, challenge parts or all of it, or suggest specific modifications to the ideas put forth. No such instant cues, however, are available to someone from another discipline. For that individual to participant in the interdisciplinary problem solving process, the band of cues on the spectrum of abstractions needs to be much broader, reaching far into the realm of concrete images to clarify the basic concepts of the other discipline. Consequently, for effective interdisciplinary communication to take place, the abstraction level on the spectrum must be at a point where the symbol system elements that relate to the subject matter under consideration are capable of being manipulated with skill by all the interdisciplinary content specialists facing a given problem-solving task.

The GG technique comprises both visuospatial/holistic and verbal/linear symbol systems interacting in real time. It provides problem solvers with a wide band of cues on the spectrum of cognitive abstractions. It can be expected, therefore, to have a high probability of engaging the mental images of individuals from a wide variety of disciplines, enabling them to externalize, compare, contrast, modify, and otherwise work with these images. It is the wide band of cues provided by GG that appears to be the essence of its effectiveness in the interdisciplinary, group problem solving context. In his paper, entitled, "Generative Graphics: A Means for Setting in Motion the Interdisciplinary Problem Solving Process," Brunon proposes that GG can serve not only to raise group awareness and understanding of issues. It also can follow through--providing a basis for the design of an entire system of graphic/verbal configurations (elaborated by way of computer technology) for enhancing both the qualitative and quantitative aspects of interdisciplinary decision making. Working drafts of Brunon's paper are scheduled for completion by early 1985, with an expected presentation to the Fourth International Conference on Interdisciplinary Research.

THIRD INTERNATIONAL CONFERENCE WORKSHOP REPORTS

Two workshops were held at the conference, one to define a future research agenda and the other to suggest concepts for engineering research centers.

RESEARCH AGENDA WORKSHOP

This conference was the third in a series of international conferences held to discuss interdiscipilnary research. The participants in this conference were selected to represent the views of four organizations:

INTERSTUDY - An informal organization consisting of research administrators and researchers studying interdisciplinary research groups

COLRAD - College of Research and Development of the Institution of management Sciences (TIMS), a large formal organization with a strong industrial membership

SOCIETY FOR THE SOCIAL STUDIES OF SCIENCE (4S) - An organization of humanists and social scientists that study the process of science from the disciplinary perspective of each member. Interdisciplinary groups of scientists and engineers are one subject of study.

ASSOCIATION FOR INTEGRATIVE STUDIES - An organization of educators interested in the teaching of interdisciplinary courses. Recent interest is developing in interdisiciplinary research.

The participants at the conference reported on their observations of interdisciplinary research (IDR), but there was inadequate time to debate the models or classifications of IDR. The issues explored were:

1. Are interdisciplinary problems unique to the interface between individuals from different disciplines, or do interfaces between individuals from different cultures, different organizations, or different groups have the same characteristics?

2. Are interdisciplinary problems being studied by members of the organizations represented at this conference of interest to those managing such processes, and

3. Will improved communications alone facilitate improved performance by interdisciplinary groups?

These issues have not been resolved and research is needed to clarify these issues if further research on interdisciplinary processes is to be justified. This conference spent less time than previous conferences discussing the definition of the term interdisciplinary and suggesting alternative nomenclature for the term.

Many participants felt that interdisciplinary processes are needed when existing disciplines cannot respond in a timely fashion, such as, when there are shortages of human resources with the proper education, experience, or skills, or when the time for solutions of a problem is less than the time required for a discipline to evolve the necessary mix of education. Interface processes (IP) are also effective if the problem requires only a short-run solution. The problems encountered at the interfaces between disciplines are assumed to be sufficiently unique that they should be the subject of special studies.

The workshop participants identified fifteen issues that may be possible research topics. These can be classified into three major groups: TAXONOMY, PROCESS, AND PRESCRIPTIONS. More than one half of the issues identified concerned taxonomy of interdisciplinary processes. These can be summarized as:

1. The need to improve the understanding of the process involved in the development and evaluation of disciplines. Also, a methodology is needed to predict the potential for cooperation between any two disciplines.

2. The need to establish a language that can be used in describing the characteristics of the interface between disciplines or groups

3. The need for a classification scheme to identify various types of interfaces and interdisciplinary interactions

The majority of the remaining issues can be classified as process issues related to the understanding of the evolution of disciplines and interdisciplinary groups. If interdisciplinary groups are the precursors of new disciplines, then understanding of disciplinary processes is critical. If interdisciplinary groups are temporary, then their dynamics become critical. Research to define the stages of development and fate of interdisciplinary groups is needed. The process issues are summarized below:

1. Measures of the outputs of interdisciplinary or interface processes are needed to support the development of models of these processes.

2. An evaluation of existing models of interface processes such as conflict resolution, information management, system engineering, etc., is needed to identify needed research in interface and interdisciplinary processes.

3. A much larger data base of empirical case studies that can be used to validate existing and newly developed models of interdisciplinary or interface processes must be developed.

NSF needs to build a data base of IP case studies to test these assertions and resolve IP issues. One alternative is to fund third parties to develop these case studies. Another is to require that each NSF project that has an IP component maintain a journal of the IP process that can provide an automatic case study for third party analysis. One project would be needed, in this case, to define the protocol and format for the journal record.

The final issue is the need to develop prescriptions for more effective applications of interface or interdisciplinary processes. Once interdisciplinary groups can be classified and the dynamics of each type can be understood, prescriptions for effective manaement of each type of group can be formulated and validated. Can existing management concepts such as matrix organizations be applied to IP? Are major IP problems those of conflict resolution? The need for empirical evidence to support the assertions concerning IP is great.

It is common knowledge that most interface processes require a substantial period of "wallowing" or learning to be effective. The goal of this research agenda was to identify the existing knowledge that can reduce this period of learning, to develop methods to make this knowledge available to those practicing interface processes, and to help practitioners recognize their needs for these data. Where insufficient knowledge exists, research issues have been identified. The fragmentation of groups generating the needed knowledge of interface processes was demonstrated by the limited knowledge of the conference participants. Each participant had information that was unknown to the others and the existing mechanisms did not readily permit information exchange. Thus one final issue identified was the need to improve the exchange of information concerning interdisciplinary and interface processes.

ENGINEERING RESEARCH CENTER WORKSHOP RESULTS

Using the NSF Program Announcement for Engineering Research Centers as a point of departure, this workshop examined the evaluation criteria proposed in the Announcement and formulated concerns and suggested designs that would satisfy these criteria. Only criteria that dealt with the process of implementation were discussed; criteria related to subject matter or equipment necessary to support a subject area were excluded.

Industrial concerns can be summarized as the need to insure excellence in the selected subject area, to define clear and realistic goals, and to facilitate industrial/university cooperation. Industrial support will depend on the ability of the center to meet its stated goals in a timely manner, while good cooperation will help define center goals that satisfy the needs of the center staff, as well as the industrial collaborators.

University concerns can be summarized as the impact of a center on the fabric and goals of universities to search for new knowledge in areas that the faculty select. The center requires that a university conform to goals and processes specified by NSF. Large state universities that have faculties dominated by humanities and arts will have different views on these centers than universities dominated by professional schools or engineering faculties.

The ideal environment and conditions needed for a center to be successful are:

1. University-wide support for the mission and goals of the center

2. A position in the university hierarchy that maximizes the support of the university administration in the operation and resource allocation to the center

3. A center director that has the prestige, the experience, and the drive to initiate the center and insure that it will succeed

4. A critical mass of faculty members to meet the educational and reseach goals of the center, willing to devote large blocks of effort to the center and to the director of the center

5. A center with well-defined goals and tasks that are realistic and can be achieved within the available resources

6. A center with defined plans and policy for:

   a. management succession
   b. long-term support
   c. detailed work/product plans
   d. conditions that would terminate the center
   e. conditions that would alter the major goals of the center

7. An advisory group to guide the plans and performance of the center

Given the concern expressed by university representatives, the concept of an impact statement for a proposed center was suggested. Since each host university may represent a different environment for a center a generic design that would satisfy all types of environments may not be possible. The ideal configuration of a center may be (1) a center leader with Nobel laureate stature, and an entrepreneurial style to assemble strong industrial cooperation and support; (2) a faculty that is committed to working full time towards the goals of the center; and (3) a student body that can be made to recognize that courses or research in the center is an essential part of their education.

Since universities are traditionally a strong department, strong individual faculty member type organizations, a center that appears like a department will fit most easily into the university structure. Interaction with industry is a nontraditional relationship; therefore, center positions that permit industrial participants to be labeled as students or faculty members will be easiest to create. The greater the deviation from the normal organization configurations found in a university, the greater will be the time and resources required to establish and maintain the center. Thus a center that can be created as a new department with a new faculty will be more stable than a cross-disciplinary unit created from the existing departments. The department can be internally cross-disciplinary but it should not attempt to integrate existing faculty members.

AUTHORS/ADDRESSES

Paul P. Appasamy
Program in Technology Assessment
Room 4520 East Engineering Bldg.
Ann Arbor, MI 48109

Donald Baldwin
Director Grants and Contract Services
University of Washington
Seattle, WA 98195

Philip Birnbaum
School of Business
Indiana University
Bloomington, IN 47405

David Bnaya
Head of Technology & Systems
Tel Aviv University
ISRAEL

Joseph Brunon
Wash. State Hospital Assoc.
419 Occidental Ave.
Seattle, WA 98104

Alok Chakrabarti
Chairman, COLRAD
College of Business & Administration
Drexel University
Philadelphia, PA 19104

Nicolas Danila
Professor and Consultant
Institute of Public Management
Centre de Recherche et d'Etudes Doctorales
46, rue Barrault
75634 Paris Cedex 13
France

Gyorgy Darvas
Inst. for Research Organization
Hungarian Academy of Sciences
G-1051 Budapest, Munnish Ferenc utca 18
Hungary

Arnoud De Meyer
Institut Europeen d'Administration
Boulevard de Constance
77305 Fontainebleau Cedex
FRANCE

Israel Dror
Chief Engineer
State of Israel
Ministry of Defence
Rafael
P.O.B. 2250, HAIFA

Prof. Aant Elzigna
Science Council of Canada
100 Metcalfe Street
Ottawa, Ontario
KlP 5Ml CANADA

Renee C. Friedman
Institute for Policy Research & Evaluation
N-253 Borrows Bldg.
University Park, PA 16802

Prof. Robert C. Friedman
Director, Institute for Policy Research and Evaluation
Pennyslvania State University
University Park, PA 16802

Eliezer Geisler
Department of Industrial Engineering
Northwestern University
Evanston, IL 50201

Torsten J. Gerpott
Hochschele der Bundeswehr Hamburg
Post fach 70 08 22
D-2000 Hamburg 70
GERMANY

Harvey Gold
Dept. of Statistics
Box 8203
North Carolina State University
Raleigh, NC 27695

Mrs. Shirley Gold
Natural Resource & Community Div.
P.O. Box 27687
Raleigh, NC 27611

Richard Goulet
Project Manager, OIR
National Science Foundation
Washington, D.C.

Sidney W. Hess
VP Atlas AeroSpace Division
ICI America's Inc.
Valley Forge Corp. Park
Valley Forge, PA

Hans Jacobs
Vakgroep
Urbanistiek en Ruimtelijke Organisatie
H.G. 10.25 - postvak 20
Postbus 513
5600 MB Eindhoven

Dr. Julie Klein
111 Linden Court
Ypsilanti, MI 48197

Brian Mar
Dept. of Civil Engineering, FX-10
University of Washington
Seattle, WA 98195

Raymond Miller
Social Science Program
San Francisco State University
San Francisco, CA 94132

Nicholas Mullins
Dept. of Sociology
Virginia Tech University
Blacksburg, VA 24061

Dr. William Newell
Management and Organization, DJ-10
University of Washington
Seattle, WA 98195

Peter Norden
IBM
c/o 20 Amherst Place
Harstdale, NY 10530

Roy L. Payne
M.R.C. Social & Applied Psychology Unit
University of Sheffield
Sheffield, S10 2TN, England

Alan Pearson
Manchester Business School
University of Manchester
Booth Street West
Manchester M15 6PB, UK

Alan Porter
School of Industrial & Systems Engr.
Georgia Institute of Technology
Atlanta, GA 30332

Julianne Prager
Executive Director
Corporate Technical Planning & Coordination Department/3M
220-4E-01 3M Center
St. Paul, Minnesota 55144

Emery Reeves
Space Craft Engineering
TRW
1 Space Park
Redondo Beach, CA 90278

Arie Rip
Chemistry and Society Program
University of Leiden
P.O. Box 9502
2300 RA Leiden
NETHERLANDS

Ivan Robertson
University of Manchester
P.O. Box 88
Manchester, M601QD
ENGLAND

Frederick A. Rossini
Office of Interdisciplinary Programs
Georgia Institute of Technology
Atlanta, Georgia 30332

Philip Rosson
School of Business Administration
Dalhousie University
6152 Coburg Road
Halifax, Nova Scotia
Canada B3H IZ5

Martha Russell
Executive Director
Center for Microelectronics and Information Services
University of Minnesota
Minneapolis, MN 55455

Dr. Borje Saxberg
Management and Organization, DJ-10
University of Washington
Seattle, WA 98195

C. Richard Schueller
Director Batelle, Seattle
4000 N.E. 41st St.
Seattle, WA 98195

Albert Teich
Office of Public Sector Programs
American Association for the
Advancement of Science
1776 Massachusetts Ave., N.W.
Washington, D.C. 20036

Charles W.N. Thompson
Northwestern University
Department of Industrial Engr.
and Mgmt. Sciences
Evanston, IL 60201

Graham Toft
Inst. for Interdisciplinary Engr.
Studies
Purdue Univeraity
Lafayette, IN 47907

Dr. B.G. Tweedy
CIBA-GEIGY Corporation
Greensboro, NC 27419

Frank Vince
Control Data Corporation
8100 54th Ave. S. /HQWIZY
Minneapolis, MN 55440

Observers:

Stanley Abramson
1190 Collgate Drive
Monroeville, PA 15146

Dick Barth
Faculty of Commerce
University of British Columbia
Vancouver, BC V6T-1YB

John Barnes
U.S. Dept. of Agriculture
CSRS, Room 113 West Auditors
Washington, D.C. 20251

John Hallaq
College of Business & Economics
University of Idaho
Moscow, ID 83843

Michael Gieske
Winona State University
Winona, MN 55987

Linda Jones
Kananaskis Centre for Environmental
Research
University of Calgary
Calgary, Alberta, Canada T2N 1N4

William MacDonald
13651-108th Ave.
Edmonton, Alberta T5M 2C7
CANADA

A.W. Schoennauer